Agricultural Research, Livelihoods, and Poverty

Other Books Published in Cooperation with the International Food Policy Research Institute

Land Tenure and Natural Resource Management: A Comparative Study of Agrarian Communities in Asia and Africa
Edited by Keijiro Otsuka and Frank Place

Seeds of Contention: World Hunger and the Global Controversy over GM Crops
By Per Pinstrup-Andersen and Ebbe Schøler

Innovation in Natural Resource Management: The Role of Property Rights and Collective Action in Developing Countries
Edited by Ruth Meinzen-Dick, Anna Knox, Frank Place, and Brent Swallow

Reforming Agricultural Markets in Africa
By Mylène Kherallah, Christopher Delgado, Eleni Gabre-Madhin, Nicholas Minot, and Michael Johnson

The Triangle of Microfinance: Financial Sustainability, Outreach, and Impact
Edited by Manfred Zeller and Richard L. Meyer

Ending Hunger in Our Lifetime: Food Security and Globalization
By C. Ford Runge, Benjamin Senauer, Philip G. Pardey, and Mark W. Rosegrant

Household Decisions, Gender, and Development: A Synthesis of Recent Research
Edited by Agnes R. Quisumbing

WTO, Agriculture, and Developing Countries: Lessons from the Indian Experience
By Ashok Gulati and Anwarul Hoda

Land and Schooling: Transferring Wealth across Generations
By Agnes R. Quisumbing, Jonna P. Estudillo, and Keijiro Otsuka

What's Economics Worth? Valuing Policy Research
Edited by Philip Pardey and Vincent H. Smith

Information and Communication Technologies for Development and Poverty Reduction: The Potential of Telecommunications
Edited by Maximo Torero and Joachim von Braun

Agricultural Research, Livelihoods, and Poverty

Studies of Economic and Social Impacts in Six Countries

EDITED BY MICHELLE ADATO AND RUTH MEINZEN-DICK

Published for the International Food Policy Research Institute

The Johns Hopkins University Press
Baltimore

Printed in the United States of America on acid-free paper
9 8 7 6 5 4 3 2 1

The Johns Hopkins University Press
2715 North Charles Street
Baltimore, Maryland 21218-4363
www.press.jhu.edu

International Food Policy Research Institute
2033 K Street, NW
Washington, D.C. 20006
(202) 862-5600
www.ifpri.org

LIBRARY OF CONGRESS CATALOGING-IN-PUBLICATION DATA

Agricultural research, livelihoods, and poverty : studies of economic and social impacts in six countries / edited by Michelle Adato and Ruth Meinzen-Dick.
p. cm.—(International Food Policy Research Institute)
Includes bibliographical references and index.
ISBN-13: 978-0-8018-8596-9 (hardcover : alk. paper)
ISBN-10: 0-8018-8596-5 (hardcover : alk. paper)
ISBN-13: 978-0-8018-8721-6 (paperback : alk. paper)
ISBN-10: 0-8018-8721-6 (paperback : alk. paper)
1. Agriculture—Research—Economic aspects—Developing countries—Case studies. 2. Agriculture—Research—Social aspects—Developing countries—Case studies. 3. Poverty—Developing countries—Case studies. I. Adato, Michelle. II. Meinzen-Dick, Ruth Suseela. III. Series: International Food Policy Research Institute (Series).
S540.E25.A37 2007
338.1′6—dc22 2006025982

A catalog record for this book is available from the British Library.

Contents

Figures

Tables

Boxes

Foreword

The impact of agricultural research on food production and economic growth in developing countries is well established, but there is still considerable debate about the extent to which these productivity increases have been translated into reductions in poverty. This volume provides evidence from a range of case studies on the impact of different types of agricultural research and technologies on the livelihoods of poor people. The studies go beyond conventional analyses of the direct effects on poor producers, to include broader societal impacts on consumers, including the urban poor in India and China. The studies also look beyond economic measures of poverty to consider the distribution of gains by wealth or poverty status and by gender.

The significance of this volume lies not only in the assessment of poverty impacts, but also in the identification of the pathways through which these impacts occur. Greater attention to these impact pathways early on in developing agricultural research programs can help to ensure that technologies meet the needs of poor women and men, that there are appropriate means of disseminating the technologies to them, and that lack of assets will not prevent adoption by poor producers. Increasingly, this calls for partnerships between national and international agricultural research centers, such as the Consultative Group on International Agricultural Research (CGIAR), nongovernmental organizations, communities, and farmers' groups, in developing innovation systems that meet the needs of the poor. In its role of developing international public goods and transnational learning for agricultural innovation, the CGIAR can contribute by providing both research on agricultural technologies and an improved understanding of how this research can improve the lives of the poor in developing countries.

Joachim von Braun
Director General, IFPRI

Preface

In recent years the international community has been placing increased emphasis on poverty alleviation in its development assistance programs. This has influenced the strategies and priorities of the Consultative Group for International Agricultural Research (CGIAR). Accordingly, the CGIAR has become more focused on both developing improved methods for assessing agricultural research impacts and using these to document and assess the nature, magnitude, and distribution of such impacts on the poor.

There have been mixed views on the extent to which the CGIAR has had an impact on poverty over the past several decades. However, anecdotal evidence indicates that impact pathways from some of the research done by the CGIAR and its partners ultimately have led to reduced poverty in some regions. Macro- or regional-level studies, such as those conducted recently by Evenson and Gollin (2003) and colleagues, indicated that poverty impacts were achieved mainly through reductions in the price of staple foods and nutritional benefits for children. These impacts were traced back to crop improvement research undertaken by the CGIAR and its partners. Some isolated quantitative assessments performed by the International Agricultural Research Center, discussed in Chapters 1 and 2 of this book, also show some of the impacts on the traditional income and consumption dimensions of poverty. But no rigorous body of work existed in the late 1990s that showed systematically whether, how, and to what extent CGIAR research had led to poverty reduction in its broadest sense. Interest in developing such evidence grew among investors in the CGIAR system in the 1990s, and in 1997 the CGIAR's independent Impact Assessment and Evaluation Group (IAEG), chaired by Jim Peacock, accepted the challenge of attempting to show in a rigorous fashion the extent of such impacts.

The IAEG considered several alternative individuals and organizations to lead such a study, under IAEG's guidance. It finally assigned the coordination and leadership of the study to a group of internationally recognized researchers at the International Food Policy Research Institute (IFPRI). The study leaders developed an assessment implementation plan that was accepted and approved by the IAEG. Work on phase I of the initiative started in 1998. After a thorough and exhaustive review of the literature on the impacts of agricultural research on poverty and methods involved in such assessments by Kerr and Kolavali (1999), the Standing Panel on Impact Assessment (SPIA), IAEG's successor, and the study leaders agreed to embark on a more empirically driven phase II analysis. This second phase consisted of a combination of macro- or country-level studies and micro-level case studies of specific research in which adoption of resultant innovations had occurred. These studies were to provide the basis for undertaking in-depth assessments of impacts and improving the understanding of the processes by which research leads to poverty reduction or prevention. Later, as individual studies were initiated, the scope was broadened from the traditional quantitative economic approach, focused mainly on income and consumption effects, to include qualitative approaches involving social scientists other than economists. The latter typically emphasized the sustainable livelihoods framework, which goes beyond the traditional economic dimensions of welfare and poverty to look at such aspects as vulnerability, power, and access to institutions.

As mentioned, this poverty impact assessment initiative commenced in 1998. It was completed in 2006 under the auspices of the Science Council's SPIA. No one expected the project to last for eight years. However, various delays, mainly related to uncertainty and temporary shortfalls in funding, slowed the assessment and synthesis work.[1] Nevertheless, earlier versions of individual components of the overall exercise were published at various stages along the way. These publications are referenced throughout this book. In this book, the approaches and results of this major impact assessment initiative are finally brought together and synthesized, ably edited and guided by Michelle Adato and Ruth Meinzen-Dick.

One of the objectives of this important initiative is to assess to what extent various types of benefits generated through location-specific innovations arising from CGIAR research have reached poor producers, laborers, and consumers in different locations and situations. Another objective is to learn more about why and how—that is, along what pathways—poverty was reduced because of research carried out by the CGIAR and its partners. Information on the ways in which the new technologies generate impacts is considered important in making current and future research efforts more effective in reaching the poor, which is now the main goal of the CGIAR.

As indicated in the introductory chapter, the book includes a combination of micro-level empirical case studies at the household and community levels in adopting regions as well as macro-level studies that include econometric analy-

1. When IAEG was an independent entity, evolving under the able guidance of Peter Matlon at the United Nations Development Programme, it also was an independent fundraiser, and the uncertainties and delays in funding were reflected in the timetable of the poverty impact assessment initiative. Furthermore, when IAEG became SPIA under the Science Council (then the Technical Advisory Committee), its independent fundraising function was removed. Budgeting delays for the new SPIA further delayed the project.

sis of the overall impacts of agricultural research on poverty at the sectoral, regional, and national levels.

Together the review of past assessments of poverty impacts combined with the results of the various empirical assessments reported in this book confirms that, in a number of cases and at different scales, the agricultural research of the CGIAR and its partners has had significant impacts on reducing poverty. The results, however, are partial, based on selected countries and specific research activities. They do not say anything, other than by implication, about the overall or aggregate impacts of the CGIAR on poverty reduction and prevention. The results are also mixed, indicating that the effects have been highly variable, depending on the type of research and the regions in which such research has led to actual changes in farmers' fields. Thus, in some of the cases, although the research has perhaps had positive impacts in terms of some dimensions of the sustainable livelihoods framework (for example, vulnerability and knowledge expansion for the poor to help improve their future welfare), in fact there has so far been little if any impact in terms of incomes or consumption benefits. Such limited effects are in contrast to other cases, for which both types of impacts have been substantial.

As discussed to some extent in Chapter 2 and in the final chapter, many methodological issues remain in applying poverty impact assessment approaches to agricultural research. The book provides insights on both qualitative and quantitative approaches, which, as pointed out by the authors, are complements and not substitutes for one another. The authors provide a useful overview of the advantages and disadvantages of quantitative and qualitative approaches in distilling direct and indirect poverty impacts. Because of the nature of the initiative, most of the qualitative studies were undertaken subsequent to the quantitative economic studies. Although this procedure was unavoidable, it is clear that one of the primary lessons learned is that these studies should be planned and undertaken in tandem by multidisciplinary teams of economists and biological and other social scientists, using a variety of methods to reap maximum synergies and insights on the scope for research to reduce or prevent poverty. They should also be done both ex ante and ex post and linked with monitoring and evaluation to maximize the benefits; here household panel data sets over extended periods are especially valuable both for quantitative and qualitative approaches. Of course, to do this sort of research, sizable investments in impact assessment are going to be required, perhaps many times the current levels and over longer periods, especially as the studies clearly show that the degrees of freedom to better target the poor are very circumscribed by location and culture specificities. Hence one needs to be able to cover sufficient recommendation domains using the qualitative and quantitative approaches described in this book, so that one can scale up and better define relatively homogeneous strategic research domains to guide research programs. Of necessity, these programs must be able to demonstrate high spillover potential in the context of the pursuit of an international public goods agenda. This is a salutary message for the donors to the CGIAR from the studies contained in this book. It also raises the issue of the extent to which qualitative approaches may be more relevant to national agricultural research systems than to CGIAR centers and whether resources might be more appropriately allocated to the former, with centers as mentors, collaborators, and synthesizers.

The challenges in poverty impact assessment are particularly critical when broadening the enquiry to approaches and impact dimensions that go beyond the traditional economic measures and indicators of poverty. However, SPIA believes that these challenges are worth addressing, because bringing into our assessments such dimensions as vulnerability, risk, assets, social status, gender, and mediating institutions will help to enrich the insights that can be drawn from poverty impact analysis. Such inclusiveness will also promote better understanding of the processes by which agricultural research translates into impacts on poverty in all its dimensions. This knowledge in turn can help us to improve the ways in which we choose, design, and implement agricultural research aimed at reducing poverty and preventing those who are not already in poverty from entering it. But it must be emphasized that a systematic enquiry is absolutely essential to address these issues. This requires commitment and resources. Random selection and analysis of a few case studies can, at best, offer only glimpses into the complex pathways from research to innovation to diverse effects on target groups of people. More interaction among impact assessment specialists and practitioners in the field is needed in developing common methods to ensure that results can be aggregated and compared, so that specific types of research and their likely impact pathways in specific regions can be meaningfully discussed.

As assessed in the book, the indirect poverty impacts of agricultural research on the urban poor in China and India are significant. The approach used did not involve qualitative approaches as in the micro-level case studies, but relied solely on macro-level empirical economic analysis. Of course qualitative measures of poverty are less useful in macro-level studies because they cannot be meaningfully aggregated across households or communities to the national level or compared over time. These assessments represent welcome evidence of indirect effects, but there remains a need for more studies that compare and contrast the relative direct and indirect impacts of agricultural research on poverty. These studies should include further disaggregation of the poor beyond the economists' traditional penchant for considering only consumers and producers. Poor net buyers and poor net sellers of the commodities affected would be one useful disaggregation that could help illuminate the distributional outcomes among the rural and urban poor, and the extent to which they represent positive-, negative-, or zero-sum games. As poverty becomes more urbanized, it is conceivable that the CGIAR's future poverty impacts may be largely determined by the extent of the indirect impacts on the urban poor who are net buyers of staples. This shift in attention may be facilitated by a greater focus on labor-intensive commercial farms with large marketable surpluses than on subsistence-oriented smallholders. To the extent that many smallholders are also net buyers, they would also stand to benefit from such a focus.

A weakness of these quantitative economic approaches to assessing poverty impacts is their inability to capture the importance of vulnerability and empowerment dimensions and gender effects. However, among their strengths is the ability of the econometric approaches used to facilitate the attribution of impacts to research in general and to CGIAR research in particular.

The book highlights the diversity and dynamics of dissemination pathways and the challenge of ensuring they are inclusive of the poor, especially women. Both formal and informal dissemination pathways have strengths and weaknesses—apparently there are no panaceas. However, the studies did show that informal social networks were most consistently important to the poor across a range of culture, environments, and innovations. But even here some favoritism crept in to exclude the disadvantaged on occasion. Hence there is a need for program managers to be proactive to better ensure inclusiveness.

Although livelihoods analysis was a useful starting point for the qualitative studies, in most cases authors found the need to draw on other tools as well. In particular the livelihoods approach does not consider cultures, politics, power relations, or experience, all of which were factors influencing poverty outcomes, to a greater or lesser extent, depending on the context. Also its overreliance on capital concepts was found to be limiting.

In none of the micro-level cases was it found that the technology options that were assessed precluded larger, more affluent farmers from using them. Nor were productivity gains the necessary pathways out of poverty (if broadly defined to include aspects related to vulnerability and lack of empowerment) in all the micro-level case studies, although they clearly were in the macro-level case studies in China and India. The poverty outcomes seemed to be largely determined by the fact that in the micro-level studies involved in the sample, productivity gains, especially of the staples, were not always major contributors to household incomes. This serves as a reminder of the relevance of livelihoods analysis at this level. Indeed, evidently the downward pressure on staple prices from macro-productivity gains meant that the indirect benefits to poor net buyers were inversely related to the welfare of poor net sellers. The authors raise the important strategic question from this observation of whether the opportunity for direct poverty impacts of staple grains research on farm households is diminishing. They also conclude that diversification out of agriculture is associated with larger income gains in most of the cases studied. These results provide support for the recent inclusion of high-value commodities and postharvest options in the research priorities of the CGIAR.

The micro-level case studies illustrate the value of inclusive strategic partnerships in an innovation systems framework that enhances research impacts on the poor. These studies often involved cross-sectoral linkages and participatory research. The micro-level case studies in the book also emphasize the importance of the vulnerabilities of the poor in shaping their perceptions about and adoption of innovations. The inference drawn is that scientists should minimize the risks inherent in new innovations to make them more adoptable. Unless such precautions are taken, the authors fear there may be large opportunity costs in terms of foregone impacts on the poor. To the extent that risk-productivity trade-offs are intrinsic, minimizing risk might itself come at a high opportunity cost. We believe the jury should still be deliberating this question. More lessons for the future are just beginning to be drawn and considered in terms of implications for best practices. This book—especially the final chapter—provides a roadmap to guide us on this journey.

SPIA and the Science Council thank all those who participated in this initiative. Particular thanks go to Jim Peacock, Peter Matlon, and the early members of the IAEG, who conceived and shaped this project; and to Peter Hazell and Lawrence Haddad, who did the initial conceptualizing and study design that led to the case studies reported in this book. Michelle Adato and Ruth Meinzen-Dick deserve special thanks for taking the lead in the second phase of the initiative and moving it on to completion, eventually editing this book and writing several of its chapters. And, of course, we thank the authors of the chapters and of all the earlier papers that resulted as part of the initiative. We also thank the project's advisory team—Anthony Bebbington, Jere Behrman, and Robert Chambers—and the four anonymous reviewers of the book, who helped to shape the final product. Our gratitude is expressed to Per Pinstrup-Andersen, former director general of IFPRI, and Joachim von Braun, the current director general, for their continuing guidance and support over the life of the initiative. Finally, special thanks go to the funders of this endeavor, taking special note of CGIAR members the U.K. Department for International Development (DfID), the World Bank (CGIAR Finance Committee), The Netherlands, Denmark, Australia, and the International Fund for Agricultural Development (IFAD).

Hans Gregersen
Former chair
CGIAR Science Council,
Standing Panel on Impact Assessment

Jim Ryan
Current chair
CGIAR Science Council,
Standing Panel on Impact Assessment

References

Evenson, R.E., and D. Gollin. 2003. *Crop variety improvement and its effect on productivity: The impact of international agricultural research.* Oxon, U.K.: CAB International.

Kerr, J., and S. Kolavalli. 1999. Impact of agricultural research on poverty alleviation: Conceptual framework with illustrations from the literature. Environment and Production Technology Division Discussion Paper 56. Washington, D.C.: International Food Policy Research Institute.

Acknowledgments

Many people helped to expand this project beyond a narrow impact assessment, to provide a broader perspective on the relationship between agricultural research and poverty.

We thank Lawrence Haddad and Peter Hazell for bringing us in to lead the second-phase study, recognizing the value of a sociological perspective for achieving its goals. They provided sound advice throughout the process and reviewed each case study. We were lucky to have Anthony Bebbington, Jere Behrman, and Robert Chambers as members of our external advisory committee, reviewing the overall study and each case study and providing their valuable insights and critiques from the earliest design stage to the final reports. Ruben Echeverria, Hans Gregerson, and Tim Kelly (and others mentioned in the preface) were part of the group that had the vision to undertake this project and which provided various forms of support along the way. They, along with four anonymous reviewers, provided many comments that improved the quality of the manuscript.

We are especially indebted to all the case study researchers and authors, for the exemplary quality they brought to the research and for their innovation and commitment in responding to the challenges of designing and conducting studies using a common conceptual framework and of working across disciplinary lines. Their many research assistants and other colleagues also made important contributions and are acknowledged in the individual case study chapters.

Many people at IFPRI provided assistance at several stages of the process. Natalia Smith provided research assistance during the book's production; Chelo Abrenilla, Patty Arce, Lynette Aspillera, Patricia Folkes, Ginette Mignot, Nelly Tioco, and Jay Willis provided administrative and other technical support; John Whitehead, Mary Jane Banks, and Uday Mohan patiently shepherded the book through various stages of production.

We appreciate the financial support of the following institutions: the U.K. Department for International Development (DfID), the Standing Panel on Impact Assessment of the CGIAR (SPIA), the Australian Centre for International Agricultural Research (ACIAR), the Danish International Development Agency

(DANIDA), the Government of the Netherlands, the International Fund for Agricultural Development (IFAD), and the Food and Agricultural Organization of the United Nations (FAO). DfID in particular had a significant influence on the study, insisting on the integration of social and economic analyses. The Rockefeller Foundation provided support for a workshop on institutional learning and change in the CGIAR.

We appreciate all these contributions. The views expressed, however, are those of the chapter authors.

Agricultural Research, Livelihoods, and Poverty

1 Introduction: Evolving Concerns in the Study of Impact

MICHELLE ADATO AND RUTH MEINZEN-DICK

Questions about how agricultural research has affected poverty have long been the subject of scholarly debate but are increasingly being asked in new ways by policymakers, funders of agricultural research, and agricultural researchers themselves. Until recently, poverty reduction was a secondary goal of agricultural research. The primary focus was on increasing the yields of important food staples, a successful strategy for substantially increasing food supplies and reducing food prices. Poverty reduction has largely been assumed to flow from this primary productivity goal. Despite this contribution to production, however, poverty still "abounds even in countries that have national surpluses" (Hazell and Haddad 2001, 1). For many developing countries, simply growing more food is no longer a pressing national objective. Food availability—at the national level—has been achieved through some combination of production and trade. According to Pinstrup-Andersen and Javier (2001, v), "the challenge of agricultural research now lies in developing strategies that more explicitly address the needs of the poor." This volume addresses this challenge by studying the impact of agricultural research on poverty and pointing to means by which the contribution to poverty reduction can be increased.

The debate over the impact of agricultural research on poverty has endured for more than four decades, acrimonious at first, then more tempered, with increasing nuance as the amount of empirical evidence increased and changes occurred in the nature of technologies and the focus and methods of agricultural research.[1] Buttel and Raynolds (1989) and Bebbington and Thiele (1993) summarized the arguments and underlying studies from Asia and Latin America, claiming positive and negative impacts over the decades.[2] Critiques of these

1. The term "technologies" describes the output of agricultural research and is broadly defined to include varieties of crops, livestock, and fish; improvement of germplasm; and management practices for agroforestry and natural resources, among others.

2. Studies making one or more arguments regarding social differentiation, wealth concentration, or ecological degradation include Johnston and Cownie (1969), Wharton (1969), Frankel (1971), Byres (1972), Cleaver (1972), Griffin (1974, 1975), Ophuls (1977), Perelman (1977),

technologies pointed to processes of social differentiation and concentration of wealth, focusing on the following concerns: the development of high-yielding varieties (HYVs) for better-quality lands and more favored agroecological zones; the requirements for expensive inputs (fertilizers, pesticides, irrigation) that either were unaffordable or else involved too much risk for small farmers; evictions of sharecroppers and tenants as agriculture became more profitable; consolidation of larger landholdings as large farmers were able to adopt more quickly than small farmers; incentives for increased mechanization of planting and harvesting as farm size increased; increased land prices; ecological problems from intensified monocropping, expanded irrigation, loss of genetic diversity; and exclusion of poorer farmers from access to credit, technical support, and other services that could have helped them to take advantage of the new technologies.

Counterarguments supported by different studies have argued that HYVs have been adopted by small farmers and tenants at least as rapidly as by large farmers, and that where smallholder adoption came later, it has caught up; yields of small farmers have been comparable to those of large farmers; mechanization, where it has occurred, is more the result of distortions in the price of capital, creating an incentive to mechanize; HYVs have increased labor input per hectare, thus increasing labor demand; and productivity increases have lowered food prices, which benefits the poor, who spend a higher proportion of their income on food than the non-poor.[3] Writing on the Latin American experience, Bebbington and Thiele (1993) conclude that positive and negative findings are not necessarily contradictions, but rather point to the variation across the *campesino* sector. They cite studies showing that some *campesinos* have used agrochemicals and new varieties as a means of surviving the impoverishing effects of wider processes of commercialization and land subdivision (Lehmann 1984; Rigg 1989; Field 1990; Bebbington 1992), and studies that look at reasons for differences within the *campesino* sector, such as the nature, presence, and style of technical assistance; quality of educational services; availability of credit; access to dynamic and growing markets; and forms of local organization and market integration that help small farmers take advantage of modern varieties (Figueroa and Bolliger 1985; Cotlear 1989; Echenique and Rolando 1989; Rigg 1989).

Later studies pointed to the maturation of agricultural research and technologies over time, arguing that early socioeconomic dislocations are no longer occurring, and that international and national agricultural research centers have

---

Mooney (1979), Pearse (1980), Oasa and Jennings (1982), Echenique and Rolando (1989), Thiesenhusen (1989), and Agarwal (1997). Studies making one or more arguments countering these claims and/or pointing to the benefits of agricultural technologies to the poor and changes over time include Ruttan (1977), Barker and Hayami (1978), Hayami (1981), Pinstrup-Andersen (1982), Lipton and Longhurst (1985), Horton (1987), Rigg (1989), Barsky (1990), and CIAT (1991).

3. Buttel and Raynolds (1989) observe that these largely refer to the case of rice in Asia.

increasingly emphasized sustainability and ecological concerns, pest management practices that reduce the need for pesticides, farming systems research, genetic resource conservation, technologies adapted to less favored lands, varieties requiring fewer purchased inputs, and farmer participatory research. Other studies point out that farmers on their own have experimented with, adapted, and benefited from HYVs. Finally, many economic studies and literature reviews by the Consultative Group on International Agricultural Research (CGIAR) and other researchers find that when increased productivity is combined with increased agricultural employment, lower food prices, and increased off-farm employment, agricultural research has contributed to significant reductions in rural poverty (Pinstrup-Andersen and Hazell 1985; Lipton and Longhurst 1989; Walker and Ryan 1990; Hazell and Ramasamy 1991; David and Otsuka 1994; Tribe 1994; Kerr and Kolavalli 1999; Rosegrant and Hazell 1999; Hazell and Haddad 2001).[4]

However, the paths of causality are complex and highly contingent. The benefits do not necessarily materialize for poor people, and effects can be negative for particular groups among the poor, such as women. The strongest argument made is that the relationship between agricultural technologies and poverty is contextual. Four reviews of the literature conclude that whether technology benefits poor people depends not as much on the characteristics of technology per se, but rather on contextual and socioeconomic conditions (Buttel and Raynolds 1989; Bebbington and Thiele 1993; Kerr and Kolavalli 1999; Hazell and Haddad 2001). Furthermore, the changing global environment and institutional context for agricultural research mean that approaches that were beneficial under certain sets of conditions and in particular regions in the past may not be beneficial in other regions (for example, in Africa, with high concentrations of poverty and a distinctive set of agroecological, institutional, and socioeconomic conditions) in the present.

The context within which agricultural research is undertaken is changing rapidly. Under market liberalization, where markets function as intended, improvements in agricultural productivity in any one country will not generate large indirect impacts on poverty through food price reductions. In addition, in many countries, agriculture has shrunk significantly in its economic importance relative to other sectors, and both the poor and non-poor farmers are diversifying their income sources so that farm income and agricultural wage earnings often account for minority shares of total household income (Tacoli 2002). The effects of agricultural production on employment and poverty may thus not be as significant as they once were, though they often generate related activities, such as provision of inputs, processing of outputs, or maintenance of capital goods.

Another contextual factor is that some types of agricultural research are becoming more privatized with the advent of biotechnology and stronger as-

4. See also some of the studies mentioned in note 2.

sertion of intellectual property rights over genetic resources. But research on many crops and livestock that poor people in developing countries grow and eat is not attractive to the private sector. Nor is most research on natural resource management practices focused on improving productivity on small farms and in less-favored regions. Finally, the social and political context of farming is changing in many countries, with increasing expectations of local people and the international community that governance and public spending will be more responsive to local needs and that local participation in decision-making will be increased. Worldwide, public investments in agricultural research totaled nearly US$21.7 billion in 1995 (Pardey and Beintema 2001).[5] The changing context means that publicly funded agricultural research must pay more attention than ever to poverty reduction.

To have an effect on poverty, agricultural researchers must be cognizant of how agriculture and new agricultural technologies fit into livelihood and income strategies of different types of farmers—with different social and economic statuses—engaged in a wide range of livelihood strategies and operating under different conditions of vulnerability and within different institutional and political environments. Furthermore, there must be greater attention to the effects of agricultural research on different dimensions of welfare—for example, vulnerability, power, access to institutions—that cannot easily be measured using such standard poverty indicators as income and consumption.

An increased focus on poverty also means that impact assessment—both ex ante and ex post—needs a new approach. This requires combining strong evaluation designs that generate good data, research methods that integrate economic and social analysis, and sufficient capacity to undertake the assessments. An institutional willingness to learn from and apply the results of such assessments elsewhere is also important.

Toward this evolution, CGIAR's Standing Panel on Impact Assessment (SPIA) asked the International Food Policy Research Institute (IFPRI) to develop and coordinate a multicountry project on poverty impact assessment. The first phase of the research synthesized the literature on the linkages between agricultural research and poverty. In this review, Kerr and Kolavalli (1999) concluded that even where agricultural research generated sizeable yield and productivity gains (as with the green revolution), gains to poor farmers are not assured. As noted earlier, this review found strong evidence that whether poor farmers benefit depends on underlying socioeconomic conditions. Enabling conditions include an equitable distribution of land and income, secure ownership and tenancy rights, efficient input and output markets that serve all farmers,

5. Of US$17 billion in public sector spending in 1990, $8.5 billion was spent by developed countries and $8.8 billion by developing countries. CGIAR centers spent an additional $286 million. Private-sector spending is also substantial. U.S. firms alone spent $2.8 billion on agricultural research in 1990 (Pardey and Alston 1995).

research and extension systems that are geared toward both small and large farms, and scale-neutral technologies. The review also found that agroecological conditions affect the distribution of benefits (for example, many of the early benefits of the green revolution went to farmers with irrigation). The evidence indicates that improved varieties have raised employment, though less so recently than in the 1960s. Furthermore, changes in wages are difficult to track because of other factors that influence wages: nonagricultural-sector wages, economic policies, and increased numbers of jobseekers and migrants. Another key finding was that growth in agricultural productivity can stimulate growth in the nonfarm rural sector, which can contribute to poverty reduction. However, poverty reduction through growth takes considerable time and depends on the presence of many other conditions unrelated to agriculture.

Kerr and Kolavalli (1999) also identified several areas in which impact assessments are weak. First, most studies measure only the direct effects on farmers who adopt the technology, not the indirect effects on others. Many studies also fail to distinguish between the effects of agricultural research and numerous confounding factors. This problem is greatest with regard to measuring the distribution of income across different types of farms or between farm and labor income. This is the topic upon which much of the literature on the negative distributional consequences focuses. Other research argues that, despite greater food availability at lower prices, if increased production and lower food prices come at the cost of lower wages and income for poor farmers, the poor cannot escape their poverty. Finally, it is difficult to compare studies, as they use different methods, ask different questions, and define problems differently.

To address the shortcomings identified in this first-phase review, a second phase of the project initiated seven new empirical case studies of the impact of agricultural research on poverty. Five of these studies share a common set of questions, conceptual framework, and methods, designed to assess poverty impacts at household and community levels within adopting regions. Two additional studies take a more macro-level perspective and conduct econometric analysis of the overall effects of agricultural research on poverty at regional and national levels.

The purpose of the second phase of the project was to clarify the relationship between agricultural research and poverty, as well as to provide guidance toward ways to target and implement research so that it is more beneficial to the poor. Three main objectives emerged (IFPRI 2000):

1. To refine and test best-practice methods for quantitatively assessing the impact of agricultural research on the poor.
2. To develop and test methods of social and economic analyses aimed at understanding the context in which new technologies are released and adopted, and how agricultural research affects broadly defined concepts of poverty and a wide range of social and economic outcomes.

3. To strengthen the capacity of agricultural research institutions to undertake integrated economic and social assessments of the effects of agricultural research on poverty.

Beyond Income and Consumption: New Methods for Understanding Poverty

To date, the vast majority of poverty impact assessments of agricultural technology have relied on quantitative economic approaches (see Kerr and Kolavalli 1999; Pingali 2001). Quantitative analyses rely on either secondary data from official sources or surveys, usually at the household level. These approaches have the advantage of consistency, so that comparisons can be made across households or sites, as well as over time (for example, official time-series data on the number of poor falling below established income thresholds at the state level, or panel data sets of household surveys). These studies are able to track changes in poverty over several decades for rural and urban people. Through econometric techniques and economy-wide models, they are also able to link changes in poverty to underlying driver variables, including agricultural research. The analyses in the China and India case studies in this volume (Chapters 8 and 9), for example, capture a full range of direct and indirect economic pathways through which agricultural research can affect the poor, including agricultural productivity growth, changes in rural wages and employment, rural and urban labor markets, food price effects, and intersectoral growth linkages.

There are, however, important limitations to studies that rely solely on quantitative economic approaches. Because the effects of technology will always be mediated by social and cultural processes, the disciplinary perspectives of sociology and anthropology—usually though not exclusively employing qualitative research methods—are indispensable to understanding why people adopt or do not, and why we find impacts or not, and what types of impacts on whom. Furthermore, recent work on poverty has highlighted its multidimensionality (Chambers 1994, 1997; Ashley and Carney 1999; Hulme 2003), aspects of which are not amenable to the quantification or abstraction from social processes that typifies most economic analyses. People's inability to influence decisions that affect their future, vulnerability to natural disasters and economic trends, and threats to physical safety and dignity are dimensions of poverty that are perpetuated by a lack of assets, information, connections, political voice, and time. These conditions are in turn often rooted in social processes that exclude certain groups of people from such sources of power. Techniques from other social sciences, such as sociology, are needed to analyze these aspects of poverty, and why people can or cannot benefit from new technology.

Five of the case studies in this book take this multidimensional approach to understanding poverty—both in terms of outcomes and the processes generating those outcomes—and combine standard measures of poverty with aspects identified by informants in the studies. Poverty is viewed in these five case studies as dynamic, with an emphasis on vulnerability—that is, the threats to livelihoods from shocks or trends that people face or fear, and with which they may be unable to cope, thus throwing them into, or deeper into, poverty. Furthermore, in addition to measures of the extent and severity of poverty, attention is given to social differentiation among the poor by class, ethnicity, gender, and other locally specific differences. Where panel data were used, the duration of poverty and its dynamics (movement in and out of poverty) are also considered. Until now this more complex approach to understanding impacts has been largely absent in CGIAR's impact assessment work.

Seven pilot case studies were selected to develop new approaches for assessing poverty impacts at different scales and to tease out the linkages between agricultural research and poverty. There were several selection criteria. Taken together, the set of case studies were designed to

1. Provide significant representation of important types of recent agricultural research and of geographic areas;
2. Demonstrate the different channels through which agricultural research can affect the poor, including intrahousehold effects, on-farm production effects, labor market effects, indirect growth, and nonfarm and food price effects;
3. Improve understanding of the conditioning economic and social factors that determine whether agricultural research benefits the poor, and provide guidelines on appropriate policies that may be needed to complement technological change to enhance favorable impacts on the poor;
4. Use rigorous methods, particularly with respect to establishing causality via proper counterfactuals while controlling for important confounding factors; and
5. Use a range of data and methodological approaches that are sensitive to a broad perspective regarding the livelihoods of the poor.

A further practical consideration in selecting case studies was that, because of time and budget constraints, priority was given to case studies for which CGIAR centers could build on ongoing or recently completed empirical work. Typically, this meant looking for strong extant data sets collected for other research purposes that could be adapted and expanded for this project.

Seven case studies cannot be fully representative of agricultural research, nor can they cover all the major variations that are likely to affect the impact of agricultural research on poverty.[6] The overall project aimed to learn about

---

6. The project was initially conceived to include a larger set of case studies chosen on a more representative basis. Had this larger project been funded, it would have been possible to attempt an overall assessment of the performance of CGIAR research projects with respect to poverty.

poverty impacts from a wide range of experiences and contexts, while developing a new approach to poverty impact assessment, not to provide an overall assessment of agricultural research's impact on poverty.[7] The case studies can also be seen as examples for developing impact assessment methods that can be applied in other studies. Still, the case studies cover a wide range of characteristics (see Table 1.1). They cover several different types of agricultural research (from crop breeding to natural resource management), research approaches taken (from conventional top down to participatory), methods of technology dissemination (from public sector extension to nongovernmental organizations [NGOs] and the private sector), the dimensions of poverty considered (from quantitative to qualitative), and levels of impact assessment (from intrahousehold to national scales, and from direct to indirect benefits).

To be considered for this study, the technologies had to have been adopted over a broad enough area to potentially show some impact. However, the general assessment by centers that developed the technologies was that these case studies did not represent the most "successful" cases, and the wide variation in results seems to bear this out. Another important concern, given the involvement in the evaluation work of some of the CGIAR centers that developed the technologies under evaluation, was to ensure that the studies were not undertaken in ways that biased their outcomes. Several factors worked to reduce the potential bias:

1. All but one project selected involved some degree of completed or ongoing research that could not be easily altered to enhance their poverty reduction impact;
2. Some of the studies dealt with broader research aggregates than individual technologies, making it difficult to anticipate the size and direction of the net impacts on poverty (for example, IFPRI's India and China studies); and
3. Each research team involved independent researchers from universities, which helped to assure independence in data collection and reporting of results.

As the results report both positive and negative findings, it appears that a reasonable level of independence was achieved.

Five of the seven case studies use household- and community-level data within an integrated social and economic analysis—structured around a liveli-

7. Although the focus was on the impact of the CGIAR, the case studies were not limited to CGIAR research. One study (on Bangladesh vegetables) was developed by an international center not part of the CGIAR; one (on Zimbabwe maize) dealt with private sector maize breeding; and one (on urban poverty in China and India) dealt with aggregate national research.

hoods framework—whereas two employ econometric analysis of secondary data at district or higher levels of aggregation. The former provide more detail on the mechanisms by which agricultural research affects the poor, particularly in terms of the direct effects within adopting regions, whereas the latter better quantify the magnitude of impacts on poverty, particularly indirect effects, both within and beyond the confines of the adopting regions. Of the detailed case studies, only that on rice in Bangladesh had been adopted on a large enough scale to see major indirect effects, but the aggregate studies of China and India also provide evidence on indirect effects of agricultural research (though not of particular technologies).

Overview of the Case Studies

The seven case studies include two studies each for rice (Bangladesh and China/India aggregate impacts) and maize (Zimbabwe and Mexico), one on high-value products (fish and vegetables), one on natural resource management (soil fertility replenishment), and one focusing on indirect impacts of aggregate agricultural research on urban poverty (in India and China). Four were from Asia, two from Africa, and one from Latin America. A brief description of each case follows.

Modern Rice Varieties in Bangladesh

This study, reported in Chapter 3, assesses the impact on poverty of the technological changes in rice cultivation made by the International Rice Research Institute and its national partners in Bangladesh. It is the study of a "green revolution," albeit one that has occurred later than those observed in other places, such as India, China, or Southeast Asia. Rice is the most important crop in Bangladesh, being grown over almost 75 percent of the cultivated area. High population densities, small farm sizes, and frequent floods and cyclones have contributed to serious widespread poverty and vulnerability in rural Bangladesh. Modern varieties (MVs) with higher yield potential were first introduced more than 20 years ago, and 47 varieties have since been released for different agroecological conditions. The varieties were developed and released following a top-down breeding process with little farmer involvement. Formal dissemination relied on government extension services, but farmer-to-farmer dissemination has played a major role in the rapid expansion of MVs over the past 15 years. MVs now cover two-thirds of the area farmed in rice.

Because of the importance of rice and the increases in both yield and labor requirements of MVs, agricultural research has had both direct effects on adopting farmers and indirect effects on employment and prices. This case study analyzes direct on-farm benefits, indirect effects through employment and rice prices, and positive or negative impacts of mediating institutions, such as

TABLE 1.1 Case studies of the impact of agricultural research

Country	Technology	Research pattern	Dissemination pattern	Levels of analysis	Case study leader	Lead CGIAR center
Bangladesh	Modern rice varieties	Conventional CGIAR → NARs	Government extension, farmer-to-farmer	Direct and indirect impacts within rural areas; nationally representative sample of farm households	Mahabub Hossain	IRRI
Bangladesh	Polyculture fishponds Improved vegetables	Conventional	NGOs, groups	Intrahousehold and household; direct and some indirect impacts within adopting regions	Kelly Hallman	IFPRI
Kenya	Soil fertility management	Participatory	Government extension, NGOs, community groups	Intrahousehold and household; direct impacts among adopters	Frank Place	ICRAF

Zimbabwe	Modern maize varieties	Private sector	Government extension, commercial	Intrahousehold and household; direct and indirect impacts within adopting regions; food price effects	John Hoddinott	IFPRI
Mexico	Creolized maize varieties	Farmer adaptation	Government extension, farmer-to-farmer	Intrahousehold and household; direct impacts among adopters	Mauricio Bellon	CIMMYT
India	Agricultural research, productivity growth, and poverty reduction	Conventional CGIAR and NARS		Direct and indirect benefits; regional and national; rural and urban	Shenggen Fan	IFPRI
China	Agricultural research, productivity growth, and poverty reduction	Conventional CGIAR and NARS		Direct and indirect benefits; regional and national; rural and urban	Shenggen Fan	IFPRI

NOTES: CGIAR, Consultative Group on International Agricultural Research; CIMMYT, International Maize and Wheat Improvement Center; ICRAF, World Agroforestry Centre; IFPRI, International Food Policy Research Institute; IRRI, International Rice Research Institute; NARS, National Agricultural Research Systems; NGO, nongovernmental organization.

the markets for land, labor, water, and credit. Nationwide panel data existed for 1987, 1990, and 1995 that were supplemented in 2000 with a resurvey and qualitative data collection on institutional change.

Improved Vegetable and Fishpond Management in Bangladesh

This case study (Chapter 4) assesses the impact on poverty of two different technologies that were disseminated by NGOs in Bangladesh: new vegetable varieties (supported by the Asian Vegetable Research and Development Center and the Bangladesh Agricultural Research Institute) and polyculture fishpond production (supported by the Worldfish Center with the Fisheries Research Institute and the Mymensingh Aquaculture Extension Project). In the vegetable program, the NGO provided the technology with credit and training to groups of poor women for two years before the start of the panel study in 1996. The NGO promoting one of the fishpond programs organized groups of landless women to lease fishponds, beginning in 1993. The other fishpond program is a governmental project that began in 1988. The project focuses on those with private fishponds and hence has worked mainly with men from wealthier households. These differences allow for the examination of the impact of different technologies and dissemination processes on men and women from households with different bundles of assets. Gender aspects are particularly important in the highly patriarchal context of Bangladesh. The case study supplemented a rich multiround survey data set with qualitative data that examine different dissemination pathways (especially through NGOs and women's groups), vulnerability to natural and other disasters, and a wide range of poverty reduction outcomes, including empowerment of women.

Agroforestry-Based Soil Fertility Replenishment Interventions in Western Kenya

This study assesses the impacts of low-cost agroforestry-based soil fertility replenishment (SFR) systems on the livelihoods of poor farmers in western Kenya (Chapter 5). The improved fallow system involves the broadcasting of tree or shrub seed into an existing maize stand, using species that produce significant amounts of nitrogen and other nutrients and that reduce weeds. The biomass transfer system involves the use of a common shrub that farmers harvest from roadsides or from plantings on their farms, applying the leaves at planting time and later as mulch. The study collected new data to examine how SFR technology affects farmers' assets, why different groups of farmers adopt (or adopt differently), and what the effects were on a range of livelihood outcomes. The study also compared diverse technology dissemination methods being promoted by government and NGOs to evaluate their effectiveness in reaching the poor, and it examined the effects of participation in dissemination on human and social capital.

HYV Maize in Resettlement Areas of Zimbabwe

This case study examines patterns of diffusion and impact of two generations of HYV maize in selected resettlement areas of rural Zimbabwe (Chapter 6). Unlike the other case studies, which assess mainly public sector and CGIAR involvement in research and dissemination, this study involved a major private sector player, Seed Co, which initially worked in cooperation with the government of Zimbabwe. The first-generation hybrids, released in the early 1980s, provided dramatic increases in yield and were widely adopted by smallholders. Second generation hybrids released in the 1990s were developed to resist diseases important to commercial farmers. These were not as widely adopted by smallholders. Impact is assessed in terms of selected livelihood outcomes, including incomes, vulnerability, assets, and nutrition. A key characteristic of the study is that it was conducted on resettlement areas during a period of political turmoil. Because all settlers were given similar allotments of land, initial differences in assets among respondents were not as great as in other cases, though economic differentiation did occur. The study built upon a unique household survey data set that provides detailed information for the same households in 1982–83, 1987, and annually from 1992 to 2000. This allowed an examination of the dynamics of poverty, the nature of vulnerability, and the responses to drought in terms of diversification of livelihoods and investment in various assets.

Creolization of Tuxpeño-Derived Maize in Mexico

The study documents how farmers in lowland tropical Mexico cross maize varieties from the Centro Internacional de Mejoramiento de Maíz y Trigo / International Maize and Wheat Improvement Center with local landraces to create "creolized" varieties (Chapter 7). This adaptation of research system outputs enables farmers, particularly poor ones, to better meet their needs by combining the preferred characteristics of improved varieties and landraces. The widespread use of creolized varieties is important for assessing the impact of the improved varieties, because studies that only examine the direct adoption of improved varieties will underestimate their impact. This study collected new data to reveal patterns of diffusion and adaptation, assess poverty impacts, and increase understanding of how improved, creolized, and local varieties respond to the needs and livelihood strategies of different groups of farmers in two states—Oaxaca, with very high levels of poverty, and Chiapas, with somewhat more commercial farming. The study also examined local seed distribution systems to understand how the institutional context within which technology is developed, disseminated, and demanded affects adoption. The aim of the study was to understand and narrow the gap between what farmers want and what breeders offer.

Agricultural Research, Productivity Growth, and Poverty Reduction in India and China

These two studies use subnational secondary data for recent decades to measure the impact of public investments in agricultural research and development (R&D) on agricultural productivity growth and poverty in India and China (Chapters 8 and 9). The first study uses econometric models to track the different channels through which agricultural R&D in rice affects rural poverty (Chapter 8). The study traces the parentage of some key crop varieties to calculate in approximate terms the contribution of CGIAR's own research to productivity growth and poverty reduction. The second looks at the magnitude and channels through which aggregate agricultural research has helped to reduce urban poverty (Chapter 9). Together, these studies examine the direct and indirect effects on poverty on a huge scale that includes more than two billion people and a significant share of the total agricultural output in the developing world.

At this national scale of analysis, qualitative measures of poverty are less useful because they cannot be meaningfully aggregated across households and communities to the national level or compared over long periods. For these reasons, the India and China case studies relied exclusively on econometric analysis of official income-based poverty data. Strengths of the approach include the ability to track the different channels through which agricultural R&D impacts the poor in rural and urban areas, statistically control for other factors that influence the outcome, analyze the sources of change over long periods of time, and compare investments in agricultural R&D to other governmental investments. Weaknesses include an inability to capture multiple dimensions of poverty or triangulate findings against more in-depth, micro-scale evidence.

Key Themes in This Volume

For agricultural research to have an impact on poverty, it must first be adopted. The decision to adopt does not easily fit into a conventional econometric model. Although asset holdings are clearly important to technology adoption, other factors are also crucial and are more challenging to capture in a quantitative regression framework. Three main factors are cited in these case studies as affecting adoption: whether the technologies were perceived by farmers to increase or decrease their vulnerability, whether poor farmers have the requisite assets to make technology adoption worthwhile, and the nature of mediating institutions.

Clearly it is not entirely possible to separate adoption from the nature of the dissemination process. Dissemination processes have a significant impact on who is reached with technologies and how well people are able to take advantage of them. Dissemination methods have increasingly diversified over time. Although direct visits by extension to farmers still take place, dissemination now involves a wide array of methods in which farmers are trained collectively and farmers teach one another. There is no one best method of dissemination that applies to all regions or even all groups of farmers within one region. Each method observed was mediated by local histories and social dynamics. It is thus important to conduct sufficient research on local cultural and power relationships to understand how people interact and learn before determining the most appropriate means of dissemination.

Attitudes toward and trust in institutions is a key factor in facilitating or hindering dissemination. The cases in this volume cover a wide range of disseminating institutions: conventional extension services, NGOs, the private sector, community-based organizations, and informal social networks for farmer-to-farmer dissemination. The range of outcomes of these different disseminating institutions highlights the importance of understanding local conditions and forming appropriate partnerships to enable agricultural research to reach the poor.

The case studies identify a wide variety of direct impacts on adopting households. These include positive effects (such as increased production and income), less obvious positive outcomes (such as increased knowledge and power of women), and negative effects (such as decreased soil fertility and availability of wild vegetables due to agricultural intensification).

However, the micro-scale case studies have indicated that the direct impacts of agricultural research on productivity and income for the poor were limited by a number of factors. First, constraints on adoption would of course limit their direct benefits. In some cases, technologies that required high levels of assets—for example, finances, land, or labor availability in the context of HIV and AIDS—limited access of the poor. The case studies also indicate that the poor are not necessarily excluded, especially if the technologies or their delivery are designed to build on the assets that they do have.

But these studies were also conducted during an era in which declining real food prices, especially for basic grains, limits the direct income gains to poor producers from agricultural technologies. Diversification of livelihood strategies out of agriculture further constrains the direct contribution of agricultural research to incomes. Yet the technologies could still play an important role in poverty reduction through increased stability, contributing to food security, and providing a launching pad into other activities. If we recognize poverty as being more than low incomes, then contributions of new agricultural technologies in such areas as reducing vulnerability and strengthening the knowledge of poor women and men need to be taken into consideration when evaluating technology.

The effects are felt not only by households that adopt the technologies; indirect impacts on other households are also important. Most notable are reductions in food prices made possible by greater productivity, which can help the urban as well as rural poor who are net purchasers of food. The final two case studies indicate that these indirect impacts of agricultural research on poverty have been substantial in China and India.

Even with the range of methods employed in the case studies in this volume, many important effects of agricultural research are difficult to capture. It is easiest to identify the productivity and income changes for farm households (although even this requires careful attention to establishing a counterfactual). We identified many positive and negative consequences that farmers and their neighbors attributed to the technologies, which would be difficult to quantify or first require detailed qualitative research before they can be quantified. Knowing what poor people themselves value is important to ensure that the studies capture the right factors. For other indirect effects on nonadopting households, including major changes in employment and food prices that play a large role in poverty reduction, it is difficult to identify what is attributable to agricultural research, but the case studies in this volume provide some indications of how this can be done.

This book is ultimately an effort to advance the study of impact in a number of ways. It takes into account new types of technology; longer-term effects of older technologies; the evolution of systems of agricultural research based on new information and new paradigms; and improved methods for studying impact. The methods used address varying scales (from intrahousehold to macroeconomic), provide comparability across case studies of different types of technologies in different situations, rigorously control for conditioning and confounding factors, and measure poverty in a large number of economic and social dimensions. This approach reflects a progression in including advances in econometric techniques; using diverse and complementary qualitative research methods; and integrating social and economic analyses to capture evolved understandings of poverty, vulnerability, social exclusion, and the structures and processes that reproduce these conditions or provide pathways for change. It is fairly well established now that impacts depend on socioeconomic context. A second purpose of this book then is to demonstrate how ex ante and ex post studies on the impact on poverty can take into account socioeconomic context, how context affects adoption and outcomes, and how to adapt technology to be more appropriate to different contexts. The point of this book is not to draw sweeping conclusions about whether agricultural research reduces poverty—a set of seven case studies cannot do this, despite their breadth and thoroughness—but rather to show the reasons why different types of technology do or do not affect different measures of poverty, vulnerability, and well-being. We hope that this knowledge can be put to use to generate greater impacts on poverty in the future.

The book is structured in the following way. Chapter 2 describes the conceptual framework and research methods used in the seven case studies in more detail. Chapters 3–7 present the methods and findings of the case studies using a livelihoods approach and mixed research methods, from Bangladesh, Kenya,

Zimbabwe, and Mexico. Chapters 8 and 9 present the methods and findings from macro-scale econometric studies in India and China. Chapter 10 synthesizes the results from all the studies, including findings on adoption, dissemination, and poverty impacts. It then draws conclusions about research methods for studying poverty impact, about institutional learning, and about future directions for agricultural research.

References

Agarwal, B. 1997. Gender, environment, and poverty interlinks: Regional variations and temporal shifts in rural India, 1971–91. *World Development* 25 (1): 23–52.

Ashley, C., and D. Carney. 1999. *Sustainable livelihoods: Lessons from early experience.* London: Department for International Development.

Barker, R., and Y. Hayami, eds. 1978. *Economic consequences of the new rice technology.* Los Banos, Philippines: International Rice Research Institute.

Barsky, O. 1990. *Políticas agrarias en América Latina,* Santiago: CEDECO.

Bebbington, A. J. 1992. *Searching for an indigenous agricultural development: Indian organizations and NGOs in the central Andes of Ecuador.* Working Paper 45. Cambridge: Centre of Latin American Studies, University of Cambridge.

Bebbington, A., and G. Thiele, with P. Davies, M. Prager, and H. Riveros. 1993. *Nongovernmental organizations and the state in Latin America: Rethinking roles in sustainable agricultural development.* London: Routledge.

Buttel, F. H., and L. T. Raynolds. 1989. *Population growth, agrarian structure, food production, and food distribution in the Third World.* New York: Academic Press.

Byres, T. J. 1972. The dialectic of India's green revolution. *South Asian Revolution* 5: 99–116.

Chambers, R. 1994. The origins and practice of participatory rural appraisal. *World Development* 22 (7): 953–969.

———. 1997. *Whose reality counts? Putting the last first.* London: Intermediate Technology Publications.

CIAT (Centro Internacional de Agricultura Tropical). 1991. *CIAT in the 1990s and beyond: A strategic plan.* Cali, Colombia: Centro Internacional de Agricultura Tropical.

Cleaver, H. M., Jr. 1972. The contradictions of the green revolution. *American Economic Review* 62: 177–186.

Cotlear, D. 1989. *El desarrollo campesino en los Andes: Cambio tecnológico y transformación social en las comunidades de la Sierra del Perú.* Lima: Instituto de Estudios Peruanos.

David, C., and K. Otsuka, eds. 1994. *Modern rice technology and income distribution in Asia.* Boulder, Colo.: Lynne Rienner.

Echenique, J., and N. Rolando. 1989. *La pequeña agricultura: Una reserve de potencialidades y una deuda social.* Santiago: AGRARIA.

Field, L. 1990. *Adaptación socio-económica campesino indígena de la Sierra Norte: Los sistemas agrícolas en la Sierra Norte del Ecuador. Opciones para analizar su desarrollo. Informe final.* Quito: Centro Andino de Acción Popular.

Figueroa, A., and F. Bolliger. 1985. *Productividad y aprendizaje en el medio ambiente rural: Informe comparativo.* Río de Janeiro: ECIEL.

Frankel, F. R. 1971. *India's green revolution.* Princeton, N.J.: Princeton University Press.

Griffin, K. 1974. *The political economy of agrarian change.* Cambridge, Mass.: Harvard University Press.

———. 1975. *The political economy of agrarian change: An essay on the green revolution.* London: Macmillan.

Hayami, Y. 1981. Induced innovation, green revolution, and income distribution: Comment. *Economic Development and Cultural Change* 30: 169–176.

Hazell, P., and L. Haddad. 2001. *Agricultural research and poverty reduction: Food, agriculture and the environment.* 2020 Discussion Paper 34. Washington, D.C.: International Food Policy Research Institute.

Hazell, P., and C. Ramasamy. 1991. *The green revolution reconsidered: The impact of high yielding rice varieties in South India.* Baltimore: Johns Hopkins University Press.

Horton, D. 1987. *Potatoes: Production, marketing, and programs for developing countries.* Boulder, Colo., and London: Westview Press and Intermediate Technology Publications.

Hossain, M., D. Lewis, M. L. Bose, and A. Chowdhury. 2003. *Rice research, technological progress and impacts on the poor: The Bangladesh case.* Summary report. EPTD Discussion Paper 110. Washington, D.C.: International Food Policy Research Institute.

Hulme, D. 2003. Conceptualizing chronic poverty. *World Development* 31 (3):403–423.

IFPRI (International Food Policy Research Institute). 2000. Impact of agricultural research on poverty reduction: An integrated economic and social analysis. Phase II. Proposal prepared on behalf of the IAEG/SPIA. Washington, D.C.: International Food Policy Research Institute.

Johnston, B. F., and J. Cownie. 1969. The seed-fertilizer revolution and labor force absorption. *American Economic Review* 59: 569–582.

Kerr, J., and S. Kolavalli. 1999. *Impact of agricultural research on poverty alleviation: Conceptual framework with illustrations from the literature.* EPTD Discussion Paper 56. Washington, D.C.: International Food Policy Research Institute.

Lehmann, A. D. 1984. Economic development and social differentiation in the Andean peasant economy. Cambridge: Department of Applied Economics, University of Cambridge.

Lipton, M., and R. Longhurst. 1985. *Modern varieties, international agricultural research, and the poor.* Washington, D.C.: World Bank.

———. 1989. *New seeds and poor people.* Baltimore: Johns Hopkins University Press.

Mooney, P. R. 1979. *Seeds of the earth.* Ottawa: Inter Press.

Oasa, E., and B. H. Jennings. 1982. Science and authority in international agricultural research. *Bulletin of Concerned Asian Scholars* 14: 30–44.

Ophuls, W. 1977. *Ecology and the politics of scarcity.* San Francisco: Freeman.

Pardey, P. G., and J. M. Alston. 1995. *Revamping agricultural R&D.* 2020 Vision Brief 24. Washington, D.C.: International Food Policy Research Institute.

Pardey, P. G., and N. M. Beintema. 2001. *Slow magic: Agricultural R&D a century after Mendel.* Policy Report. Washington, D.C.: International Food Policy Research Institute.

Pearse, A. 1980. *Seeds of plenty, seeds of want.* London and New York: Oxford University Press.

Perelman, M. 1977. *Farming for profit in a hungry world.* Montclair, N.J.: Allanheld, Osmun.

Pingali, P. 2001. *Milestones in impact assessment. Research in the CGIAR, 1970–1999.* Research Report. Mexico, D.F., Mexico: International Maize and Wheat Improvement Center. <www.cimmyt.org/Research/Economics/map/impact_studies/milestones/pdfs/milestones.pdf> (accessed September 2005).

Pinstrup-Andersen, P. 1982. *Agricultural research and technology in economic development.* London: Longman.

Pinstrup-Andersen, P., and P. B. R. Hazell. 1985. The impact of the green revolution and prospects for the future. *Food Review International* 1 (1): 1–25.

Pinstrup-Andersen, P., and E. Javier. 2001. Foreword. In *Agricultural research and poverty reduction: Food, agriculture and the environment,* P. Hazell and L. Haddad, eds. Discussion Paper 34. Washington, D.C.: International Food Policy Research Institute.

Rigg, J. 1989. The new rice technology and agrarian change: Guilt by association? *Progress in Human Geography* 13 (2): 374–399.

Rosegrant, M., and P. B. R. Hazell. 1999. *Rural Asia transformed: The quiet revolution.* Washington, D.C.: Asian Development Bank and International Food Policy Research Institute.

Ruttan, V. W. 1977. The green revolution: Seven generalizations. *International Development Review* 19: 16–23.

Skoufias, E. 2003. Economic crises and natural disasters: Coping strategies and policy implications. *World Development* 31 (7): 1087–1102.

Tacoli, C. 2002. *Changing rural-urban interactions in Sub-Saharan Africa and their impact on livelihoods: A summary.* Working Paper 7. London: International Institute for Environment and Development.

Thiesenhusen, W., ed. 1989. *Searching for agrarian reform in Latin America.* London: Unwin Hyman.

Tribe, D. 1994. *Feeding and greening the world.* Wallingford, U.K.: CAB International.

Walker, T., and J. Ryan. 1990. *Village and household economies in India's semi-arid tropics.* Baltimore: Johns Hopkins University Press.

Wharton, C. R., Jr. 1969. The green revolution: Cornucopia or Pandora's Box? *Foreign Affairs* 47: 464–476.

2 Integrating Social and Economic Analyses to Study Impacts on Livelihoods and Poverty: Conceptual Frameworks and Research Methods

MICHELLE ADATO, RUTH MEINZEN-DICK,
PETER HAZELL, AND LAWRENCE HADDAD

Most past studies of the impact of agricultural research on poverty have analyzed only some of the potential pathways by which the poor are affected by new technologies, often leaving out some of the more important indirect impacts. Many studies have used methods that were insufficient for controlling for the numerous conditioning and confounding socioeconomic factors that affect poverty outcomes and mask the true impact of new technologies. Most have also limited themselves to narrow income or nutritional measures of poverty and have ignored its broader social aspects. There is a need for more holistic approaches and comparative case studies such as those reported in this book. We begin with a review of past impact studies and lessons learned,[1] leading to a discussion of the conceptual frameworks and methods used in the case studies in this volume.

Challenges of Assessing the Impact on Poverty

Agricultural research can have an impact on poor people in diverse and complex ways, posing many challenges for those who aspire to assess its effect on poverty. Not only are there many direct and indirect pathways to examine, where outcomes are not always beneficial for the poor, but impact assessments can be examined in a variety of ways, depending on whether one considers poverty solely in terms of income and nutrition status or from a broader perspective that includes other material and social dimensions.

Figure 2.1 portrays some of the key pathways and complexities involved in tracing the impact of agricultural research on the poor. Investments in agricultural research must first lead to appropriate technologies for farmers, but whether these are adopted and how they affect agricultural production also depends on other conditioning factors, such as the prevailing state of rural infra-

1. In this review we draw heavily on the findings of the first phase of the IFPRI poverty impact study introduced in Chapter 1, captured in the literature review by Kerr and Kolavalli (1999).

FIGURE 2.1 How agricultural R&D affects the poor

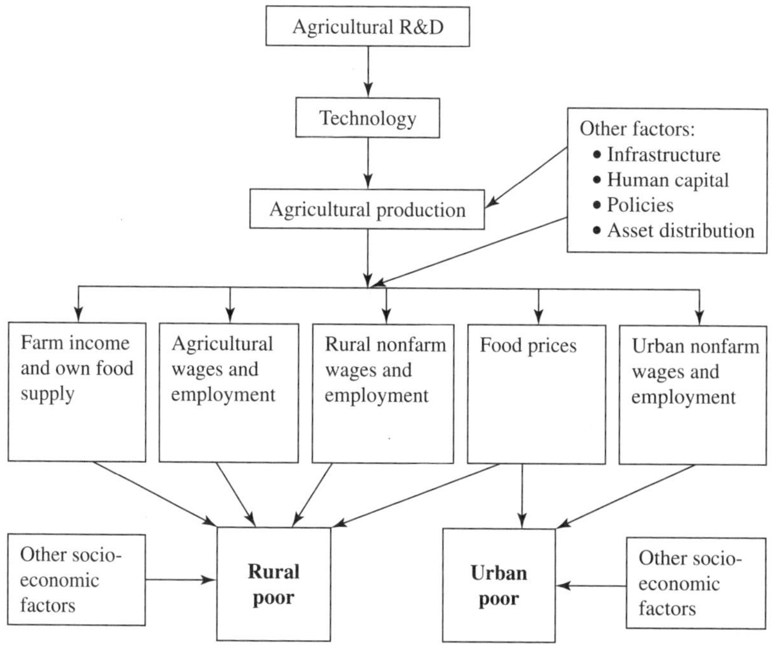

structure and institutions, human capital, agricultural policies and prices, and the distribution of rural assets (for example, land).

Technology-driven increases in agricultural production have the potential to affect the poor through several channels. The most direct impact on the rural poor is through increases in their own-farm production. This increase may involve producing more food for own consumption, or increasing the output or reducing the production costs of marketed products that increase farm income. Increases in agricultural output can also directly impact the rural poor within adopting regions by increasing agricultural employment and wages, which can be especially beneficial to small farmers and landless laborers.

Growth in agricultural output and income can also induce additional income-earning opportunities for the poor in their local nonfarm economy. These may take the form of increases in wages and nonfarm employment and new opportunities for nonfarm self-employment, which can increase livelihood options for both the rural and urban poor living in adopting regions. Many of these impacts within adopting regions can also strengthen the social standing of poor people, reducing their overall vulnerability to economic and social shocks and contributing to their asset base and social, political, and market empowerment.

Agricultural growth can also generate powerful indirect benefits that reach well beyond adopting regions. By stimulating broader nonagricultural growth in a country, new technologies can increase migration opportunities for the rural poor and income-earning opportunities for the urban poor, and can generate lower food prices for everyone.

But these benefits do not necessarily materialize for the poor. There are numerous conditioning factors that help determine who benefits. These factors work in myriad complex and often conflicting ways, and the outcomes are difficult to determine a priori. Some of these conditioning factors are economic in nature, but others involve social, political, and cultural considerations. For example, poor farmers will only obtain on-farm benefits from new technologies if they adopt them. Adoption requires the new technologies to be appropriate and profitable for local farming conditions and the farmers to have access to the necessary knowledge, assets, and inputs. Even when poor farmers do benefit from significant productivity gains, these benefits are not always shared equitably among household members and need not translate into improved nutrition and well-being for women and children.

Many yield-enhancing technologies increase total on-farm employment, but whether the former translates into higher wage earnings for the poor depends on the characteristics of the local labor market. If labor is abundant in the adopting region, then the additional employment will have little effect on wages, and there will be limited incentive for farmers to invest in labor-replacing machines. But if labor is scarce, then wages will rise more sharply and labor-displacing machines may become attractive. Even where employment earnings do increase, social and cultural barriers may prevent some kinds of poor people from benefiting (for example, there may be gender, language, and caste barriers). Similar problems may arise in the nonfarm economy, and the opportunities from new growth may not be available to poor people who live in villages with poor access to towns, who have limited education and skills, or who are disadvantaged by gender or other social identity.

Technological change can lead to an increase in the aggregate food output, but it does not always translate into lower food prices for everyone. If the national demand for the affected product is trending downward (that is, export opportunities are constrained by trade policy or by high transport costs), then the output price will fall. But these price reductions may not be very large in an open economy with low transport costs, and more countries now fall into this category than before, because of recent rounds of market liberalization policies. The food price benefits may also be enhanced if technological change leads to a reduction in production costs per unit of output, as farmers can then maintain or increase profits even at lower sales prices. But whether consumers benefit from these lower costs depends on whether the food marketing and distribution system is sufficiently competitive that cost savings at the farm gate are passed up through the marketing chain. In some cases, the cost savings are simply captured as additional profits in the marketing chain. Moreover, although cost-reducing technologies enable adopting farmers to compete at lower prices, they penalize nonadopters whose costs remain high. These may include many poor farmers in regions where the new technology is not suitable.

These examples illustrate how conditioning factors can exert a strong influence on the outcomes for the poor. Given the variety of ways in which the poor may be impacted by agricultural research and the powerful influence of many conditioning factors, it is difficult to predict whether poor people will experience a net benefit or loss from agricultural research. Moreover, most poor people have complex livelihood strategies and are often part-time farmers, nonfarm entrepreneurs, laborers, and migrants, but full-time consumers. They may simultaneously gain or lose in these different dimensions, so that the net impact can remain ambiguous. A poor farmer, for example, might be able to gain from increased on-farm production as a technology adopter, but may lose or gain from increases in agricultural wages or reductions in food prices, depending on whether he or she is a net buyer or seller of labor or food. Again, a small nonfarm business entrepreneur might gain from cheaper food, but business profits might fall or rise, depending on whether hired labor costs rise faster than sales, and how the incomes of his or her customers fare as a result of the technological change. Net effects may vary widely for different types of poor households, some gaining more than others and some not gaining at all. Understanding household livelihood strategies and the diversity of these strategies across poor households is therefore necessary for assessing the impact of technological change on the poor.

Impact Studies

Given the complexity of the factors conditioning the impact of technology on the poor, assessing impact empirically is a complex task. It is not surprising that many studies have proved inconclusive or questionable; they were simply not well designed for the task. Many studies have proved misleading because they failed to establish an adequate counterfactual situation, failed to identify the true causality of change, were not representative, were too narrow in scope and did not consider all the indirect ways in which the poor are affected, or were too short term in perspective. Some of the key analytical issues that need to be addressed in impact studies are reviewed below.

Scope of the Analysis

The direct impact of improved agricultural technologies on poor farmers has been the focus of many studies. But these are often only a small proportion of the overall effects on the rural and urban poor. The direct effects are captured by poor farmers who adopt improved technologies and who produce more output, which they can consume themselves or sell. However, there are important spillover benefits to other households or regions. These include the benefits that may arise from the generation of new employment, higher wages, and less costly food. These spillover effects have received inadequate empirical attention, despite their enormous potential impact on poor people, including landless laborers, the nonfarm rural poor, and the urban poor. Intrahousehold effects are another important dimension that has received little study. Attention must be paid to the significance of social differentiation across and within households. To illustrate, recent work undertaken by the International Food Policy Research Institute (IFPRI) and others (for example, Kumar 1994; Quisumbing 2003) shows that significant biases along gender and generational lines can arise when the distribution of work within households changes and that technologies can reduce or reinforce these biases, depending on who grows or owns the crops that are affected. Assessing the impact of improved technologies at this level requires information about individuals within households.

To capture these different effects requires a research design that operates at different scales of analysis (intrahousehold, household, village, region, and nation). Different research methods are also required that are appropriate to each scale. For example, measurement of indirect benefits arising from intersectoral growth linkages and less costly food requires economy-wide models and analysis, whereas measurement of household benefits within adopting regions must take account of the diversity and complexity of household livelihood strategies.

Integrating Economic and Social Analyses

The field of social analysis (composed of various aspects of applied anthropology, rural sociology, political science, and institutional analysis) has been key in developing new ways of understanding and measuring poverty. It has promoted the idea of poverty being a process, rather than a state, with causes and manifestations that in many cases are one and the same. Social analysis has also emphasized the different ways of understanding and defining poverty. Whereas income- or consumption-based approaches have been most common, especially in economic analyses, other approaches that consider basic human needs, basic capabilities, vulnerability, social support or exclusion, and related factors can reveal more complex aspects of the lives of poor people. Many of these facets of poverty cannot be easily measured using formal quantitative techniques—ranking or scoring is only an approximation, and other methods need to be used to capture these aspects of poverty. In particular, qualitative and participatory research methods are necessary for understanding the complex conditions of poverty and the interlinkages between them.

A livelihoods perspective provides a means of broadening the understanding of poverty and drawing together the various perspectives of social and economic analyses to do a broader poverty impact assessment. Focusing on livelihood strategies requires a broad sectoral perspective that links agriculture to the wide range of livelihood strategies undertaken by the rural poor and an assessment of direct and indirect effects of agricultural research on livelihoods. Because poverty is multidimensional in scope, not all of its dimensions are easily measured (such as power, access, and exclusion). While aspects of many key relationships can be examined with conventional quantitative techniques, many aspects are more suited to qualitative analysis. Thus integrated social and economic analyses require a combination of techniques for data collection and analysis.

Establishing an Adequate Counterfactual Situation

To assess the impact of a new technology on poverty, the researcher must be able to assess what the situation would be like if the technology had not been adopted—the counterfactual situation. Many studies fail to establish an effective counterfactual situation and often rely on a simple before-and-after analysis. This approach can be quite misleading, for many other factors may have changed along with the technology.

The best counterfactual is a comparable region or group of farmers who are identical in all respects to the adopters except that they have not had a chance to adopt the technology themselves. The "gold standard" would be a quasi-experimental design in which the sites for introducing the technology are randomly allocated within the appropriate recommendation domain, with a baseline and panel data, so that we can compare before and after in the sites and households, both with and without the technology. Such situations are extremely rare, and such a design often has prohibitive ethical implications.[2] Thus most often it is necessary to use comparator groups that differ in several other attributes as well. The danger of this approach is that there may be systematic reasons why the comparator group has not adopted (for example, the technology is less appropriate to their conditions, they have a different distribution of land) and these reasons might also have affected the impact of the technology had it been adopted. Such sample biases can be controlled through econometric techniques, but the latter requires that particular types of data be collected. Establishing appropriate counterfactuals for assessing the indirect benefits of technological change is even more difficult. Sophisticated modeling or econometric approaches offer means for detangling these various effects. Qualitative approaches provide another means of examining causality of various factors, by eliciting the interpretation and understanding of those who are directly and indirectly affected. Although neither econometric models nor local people's responses alone may be fully satisfactory for establishing causality, triangula-

2. An intervention cannot be deliberately withheld for a research experiment. In IFPRI's evaluation of Programa de Educación, Salud y Alimentación (PROGRESA) in Mexico (Skoufias 2005), such a design was possible because of the gradual roll-out of the program. However, this approach is much more difficult with agricultural research, where technology may spread more rapidly.

tion and cross-checking of these methods greatly increases the trustworthiness of interpretations.

Accounting for Other Factors

Many other factors besides improved technologies affect changes in agricultural production and their impact on the poor. At the farm level, prices; risk; access to inputs, credit, and markets; education levels; and the distribution of land are just some of the factors that affect both the rate of uptake of improved technologies and the extent to which they benefit the poor. Improved technologies may fail to benefit poor farmers not because they are inherently biased against the poor but because the distribution of land or access to inputs and markets is unequal and unfair. It is only when these factors are taken into account that it becomes possible to explain why similar technologies can have very different effects on the poor in different regions or at different times. The need to control or account for other factors is even more challenging when assessing the indirect benefits to the poor. For example, changes in rural employment opportunities and wages in the farm and nonfarm sectors are affected by macro, trade, and agricultural sector policies, as well as by prevailing prices, public investments in rural infrastructure, health and education, and public employment programs. Teasing out the specific effects of production increases due to improved technologies needs to be done within an analytical framework that accounts for all these important factors. Similar problems arise in trying to assess the indirect benefits to the poor arising from changes in food prices or from improved migration opportunities. Resolving such difficulties can only occur by looking at countries over longer times, and by comparing the experiences of different countries or regions within a country (see, for example, Datt and Ravallion 1997, 1998; Fan, Hazell, and Thorat 2000).

Allowing for Time Lags

There are often long time lags between expenditures on agricultural research and the effects of that research. In addition to the long lead times inherent in much agricultural research, there can also be important lags in adoption, such as found in a number of studies of the green revolution; large-scale farmers often adopted quickly while many smaller-scale farmers took several years to catch up (Hazell and Ramasamy 1991). There may be further lags between the adoption of improved technologies and their effects on production and poverty. Such technologies as establishment of farm trees, livestock improvement, and watershed development require long-term investments that do not yield any productivity gains for some years. Most of the indirect benefits arising from improved technologies also take time, as factor and product markets must adjust. The analytical framework of a study must be sufficiently dynamic to capture and aggregate these kinds of lagged benefits.

Attention to Risk and Vulnerability

Agricultural production is inherently risky, and yields and prices can fluctuate markedly from one season to another, particularly in the rain-fed farming systems that are home to many of the rural poor. Assessments of the aggregate impact of improved technologies need to average out these random effects, either by taking enough years in "with" and "without" analyses, or by using an analytical framework that specifically includes weather and price variables. However, because such fluctuations can be an important contributor to the vulnerability of poor households, technologies that reduce fluctuations are of particular value in alleviating poverty. By contrast, those that produce high returns in a good year but low returns in a bad year (either because of yield or price fluctuations) may be less appropriate for adoption by the poor because they increase vulnerability or do not fit with their livelihood strategies. Thus variability in production and other dimensions of technology with implications for risk and vulnerability must be considered along with average returns.

Defining the Benefits

New technologies, practices, and the institutional context can potentially affect a wide range of indicators. Process indicators assess whether the new intervention is being used and used as intended. Intermediate outcome indicators assess such outcomes of the intervention as changes in crop yields, post-harvest losses, soil fertility, and improved forest management. Welfare outcome indicators assess the well-being of adopters and nonadopters of the intervention. Welfare can be measured in a number of ways (for example, income, expenditure, food consumption, nutrition status, decisionmaking ability, social support, control of resources), at a number of different levels (community, household, individual), for different types of individuals (adopters, nonadopters, farmers, nonfarm rural or urban individuals, women, men).

The conventional focus on measuring poverty relies on large-scale surveys and quantitative data analysis. A focus on understanding poverty from poor peoples' perspectives gives greater weight to qualitative approaches. Recent expansion of interest in participatory research methods has helped stimulate the improvement in understanding poverty from poor people's perspectives and the role of power in determining livelihoods choices (Norton, Owen, and Milimo 1994; Chung et al. 1997; Nabi et al. 1999; Narayan et al. 2000), but many forms of qualitative research can explore local meanings of poverty. Both quantitative and qualitative methods can be used to develop indictors of power that reflect not only intra- but also interhousehold power relationships and those between households and institutions.

The full range of analytical issues discussed above has yet to be addressed in an integrated way in the empirical literature. Studies have been much more piecemeal in their approach, looking, for example, at particular scales of analysis (adopting households, or producers and consumers in national food markets), using narrowly defined economic measures of poverty, and often failing to control for the full range of conditioning and confounding factors that influence the outcomes. The set of case studies selected for this overall study and described in Chapter 1 represents perhaps the first serious attempt to generate a body of new knowledge and understanding about the impact of agricultural research on the poor that (1) is comprehensive in its coverage of scale (from intrahousehold to macroeconomic), (2) measures poverty in a number of economic and social dimensions, (3) provides comparability across case studies of different types of technologies in different country situations, and (4) rigorously controls for conditioning and confounding factors. This effort required considerable methodological innovation as well as a carefully coordinated set of field studies and analyses.

The remainder of this chapter describes the conceptual frameworks and methods used for two types of studies that together captured all the necessary scales of impact. In the next two sections we describe the framework and methods used to assess poverty impacts within the adopting regions at the household and intrahousehold levels. The approach draws on a livelihoods framework and uses quantitative and qualitative methods in an integrated way. It is applied in the studies for Bangladesh, Kenya, Mexico, and Zimbabwe. The subsequent section describes the framework and methods used to assess poverty impacts at higher scales, including the indirect impacts in nonadopting regions and on the urban poor. The approach used here is much more aggregated and narrowly economic and uses data and techniques that extend well beyond the household level. This approach is applied in the case studies for China and India. The two approaches are complementary, each addressing poverty impacts that the other cannot adequately capture.

Conceptual Framework for the Bangladesh, Kenya, Mexico, and Zimbabwe Studies: A Livelihoods Approach

Five of the case studies used a livelihoods approach to understanding the various ways that agricultural technologies affected people's lives and well-being, taking into account the multiple dimensions of poverty and the diverse causal pathways among agricultural research, dissemination, production, and poverty (see Adato and Meinzen-Dick 2003). To date, nearly all impact assessments in the Consultative Group on International Agricultural Research (CGIAR) centers have used conventional measures of poverty based on income and consumption data and, occasionally, nutrition indicators. However, poverty assessment means more than measuring how many people live on a purchasing power of $1.00 a day or how many households consume less than 2,000 calories per person per day. Five of the case studies use the basic assumptions of livelihoods analysis to look beyond these narrow measures to consider additional aspects of poverty and well-being, for example, access to land, water, credit, or education; vulnerability to natural disasters; political rights; physical safety; and social relationships that provide economic security and social well-being. The studies also looked at self-perceptions by local communities on who is poor, what constitutes poverty, taking into account what people themselves find most problematic and what they value. The studies assume that people, whether poor or not, are agents with assets and capabilities who act in pursuit of their own livelihood goals—not passive victims or recipients of government policies and external aid—though constrained by economic, social, and political relationships.

The concept of "livelihoods" has become increasingly popular in development studies as a way of conceptualizing the economic activities poor (and non-poor) people undertake in their totalities. The focus in the 1970s on employment and jobs has given way to the realization that, although job creation in the formal sector continues to be one important strategy for poverty reduction, the reality for poor people in the global South is that survival and prosperity depends on the simultaneous pursuit of diverse activities, by different family members, taking advantage of various opportunities and resources at different times. As Chambers (1997, 163) wrote:

> They maintain a portfolio of activities. Different members of the family seek and find different sources of food, fuel, animal fodder, cash and support in different ways in different places at different times of the year. Their living is improvised and sustained through their livelihood capabilities, through tangible assets in the form of stores and resources, and through intangible assets in the form of claims and access.

Livelihood activities may be composed of, for example, year-round or seasonal formal-sector employment, informal trading or sale of labor, home gardens and food processing, livestock production, cultivation or use of natural or common property resources, labor exchange among family or neighbors, contracted "home work," borrowing, scavenging, stealing, and begging. They may be on or off farm, include local or international migration, involve elderly household members or children, be legal or illegal. For poverty analysis and poverty reduction interventions to be effective, it is important to understand these multiple activities and related sources of vulnerability faced by the poor, the ways in which their lives are affected by structures and institutions, and the varied ways in which development interventions may strengthen or weaken these livelihood activities. In addition to recognizing these activities, using livelihoods approaches requires an attempt to understand the processes that underlie poverty, and the social, cultural, political, and institutional contexts in which poor people live. Although the individual, household, and community are the primary levels of analysis, livelihoods approaches seek out the relevant interactions at micro, intermediate, and macro levels. Hebinck and Bourdillon (2001, 2) point out the different ways in which a livelihoods framework is used in the field of development:

> For policy makers . . . "livelihood" provides a framework that focuses on poverty within the contexts of the people who are poor, and on the processes that underlie poverty. For consultants who operate in the field of development, "livelihood" represents a framework for the formulation of development projects that focus on the people being affected by the project and the variety of ways in which they might be affected. For social scientists, such as anthropologists, sociologists and economists, "livelihood" provides a framework for a holistic interpretation of the dynamics of development and the different rhythms of change. For plant breeders, soil scientists and other technologists, the livelihood framework serves the purpose of linking their specific work and capacities with what people are capable of doing, what they are looking for and how they perceive their needs. The livelihood framework thus provides a guide for research and intervention.

Agricultural Research and Sustainable Livelihoods

Livelihoods approaches have evolved from several decades of changing perspectives on poverty, how poor people construct their lives, and the importance of structural and institutional issues (Ashley and Carney 1999). Bebbington (1999) develops a livelihoods framework concerned with rural livelihoods and poverty but not necessarily linked to agriculture or natural resources.[3] This framework invokes a broad conception of resources, considering livelihoods in terms of access to five types of capital assets: produced, human, natural, social, and cultural. These assets are vehicles for material well-being, but they also reflect and are part of the meaning that people create through their livelihood strategies, meaning which continues to influence their choices and strategies. Furthermore, they give people the capabilities and power to act, in ways that influence their access to and use of resources. This framework also requires an analysis of economic, social, and political relationships that influence poverty and wealth, and addresses relationships between intrahousehold, household, regional, and macro economies, and between households and institutions and organizations (Bebbington 1999, 2022, 2028–2029).

The "sustainable livelihoods" conceptual framework (SL framework) picks up on many of these core concepts. It is a particular form of livelihoods analysis used by a number of organizations, including the Department for International Development (DfID) of the United Kingdom, the United Nations Development Program (UNDP), as well as such nongovernmental organiza-

---

3. Bebbington credits work on environmental entitlements (Leach, Mearns, and Scoones 1998, 1999), and access to resources (Africa 1989; Bryant 1992; Ribot 1998) as influencing his conceptualization.

tions (NGOs) as CARE and Oxfam (DfID 1997; Carney et al. 1999).[4] It relies on fewer analytical concepts than Bebbington's framework above but serves as a practical operational tool for understanding livelihoods and improving development interventions. It is primarily a framework for considering relationships among causes of poverty, access to resources, diverse livelihood activities, and the role of the institutional environment. It is an organizational tool, used to classify variables and processes that must be taken into account, pose their interrelationships, and ensure that these factors and relationships have been at least considered if not ultimately included in an analysis. It does not have explanatory or predictive power, nor can it tell how to derive meaning from information collected. IFPRI was asked by DfID to use the SL framework in these studies to see how useful it would be for improving impact assessment in agricultural research. Five of the case studies presented in the volume refer to this framework and use vulnerability, assets, and institutions as principle concepts for understanding adoption and evaluating impact.

The overall conceptual framework for SL is illustrated in Figure 2.2 (see also Carney 1998; DfID 2001). The starting point is the vulnerability context within which people operate. Attention is given to the assets that people can draw upon for their livelihoods. Assets interact with policies, institutions, and processes (PIP) to shape the choice of livelihood strategies. These, in turn, shape livelihood outcomes, which feed back into the future asset base. Fleshing out this framework and operationalizing it can make impact assessments quite complicated, as the studies presented in this book show. Using this framework brought in many considerations not often included in impact studies dealing with agricultural research.[5] Figure 2.2 indicates three ways in which agricultural research factors into SL: by affecting the vulnerability context, through linkages to the asset base, and as part of PIP.

Vulnerability is a widely used concept in the physical and social sciences and, increasingly, in development studies. Dercon (2001, 66) sees vulnerability as "ex-ante poverty, i.e. before one knows what the outcome of risk variables will be." Defined in relation to poverty, most definitions capture a dynamic dimension of poverty, a notion of time and change. Vulnerability analysis incorporates risks and threats, as well as people's resistance and resilience to them. Vulnerability factors encompass, for example, (1) trends in population, resources, and such economic indicators as prices, governance, or technology; (2) shocks that include changes in human or animal health, natural disasters, sudden economic changes like price or currency fluctuations, or political con-

4. Although each organization has its own variation on the framework, emphasizing different aspects, there are many common elements.

5. UNDP's approach to sustainable livelihoods makes technology and investment one of the "drivers" (along with policy and governance) that affect local adaptive strategies, assets, knowledge, and technology, which in turn have sustainable livelihoods as an outcome (Carney et al. 1999).

FIGURE 2.2 The sustainable livelihoods conceptual framework with agricultural technologies

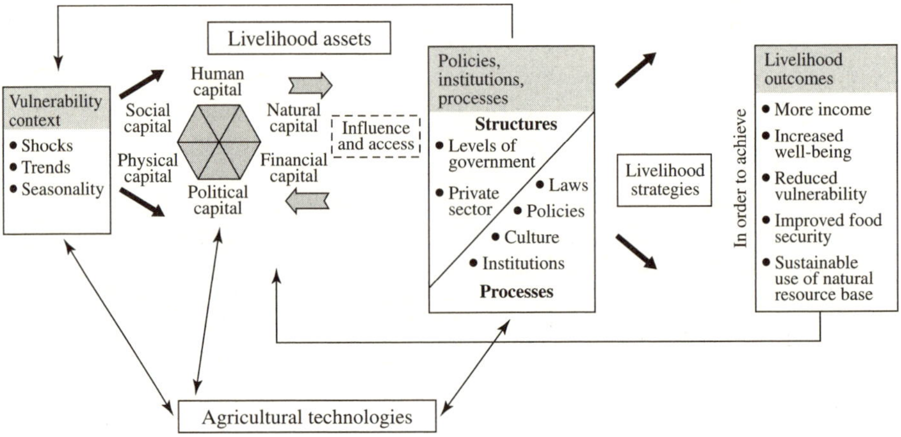

SOURCE: Adapted from DfID (2001).

flict; and (3) seasonality in prices, agricultural production, employment opportunities, resource availability, or health. Gender, class, politics, power relations, culture, ethnicity, beliefs, and other factors affect the nature and degree of vulnerability and people's resiliency. The case studies presented in this book find that it is not objective "risk" that matters, but people's subjective assessments of things that make them vulnerable. These matter because both perceived and actual vulnerability can influence people's decisions and hence their livelihood strategies. Perceptions of risk are especially important to whether people are willing to adopt agricultural technologies.

Agricultural research and technologies can reduce vulnerability, such as when irrigation reduces susceptibility to fluctuations in water supply, pest control technologies reduce vulnerability to crop or animal loss, and crop diversity is increased.[6] However, research and technologies can also increase vulnerability, such as when new varieties are more susceptible to crop failure if condi-

6. The importance of downside risk as a deterrent to adoption of new technologies is well established in the agricultural economics literature, although there is considerable variation in farmers' individual behavioral patterns (for example, Binswanger 1981; Feder, Just, and Zilberman 1985; Antle and Chrissman 1990). Perhaps the best parallel concept to vulnerability is the economists' concept of safety first. The safety-first approach assumes that households seek to minimize the probability that their income (including the value of home-produced foods) falls below some minimum (or disaster) level needed for subsistence (Roy 1952; Telser 1956). Several variants of

tions are not right, farmers have to purchase the seed every year under conditions of cash constraints, or dependence is increased on unreliable markets.

Livelihoods approaches focus on assets broadly defined. Rather than looking only at land or other classic wealth indicators, six different types of assets are considered:

1. Natural capital that includes land, water, forests, marine resources, air quality, erosion protection, and biodiversity;
2. Physical capital that includes transportation, roads, buildings, shelter, water supply and sanitation, energy, technology, and communications;
3. Financial capital that includes savings (cash and liquid assets), credit (formal and informal), and inflows (state transfers and remittances);
4. Human capital that includes education, skills, knowledge, health, nutrition, and labor power;
5. Social capital that includes any networks that increase trust, ability to work together, access to opportunities, reciprocity; informal safety nets; and membership in organizations; and
6. Political capital that includes citizenship, enfranchisement, membership in political parties, and informal political power bases—all assets that can be key in obtaining or operationalizing rights over other assets.

Agricultural technology is strongly linked to the asset base. The best-researched aspects of this relationship are the types of assets that are required to adopt new technologies. For example, much of the debate on the green revolution centered around whether large landholdings (natural capital) were required to adopt the various components of the green revolution package. Considerable policy emphasis has also been given to expanding agricultural credit and roads or transportation to permit technology adoption. Human capital, in the form of knowledge, skills, and labor power, is often required to make effective use of new technologies. It is now increasingly recognized that social capital can facilitate adoption of technologies that operate on a large spatial scale, wherein collective action is needed to coordinate the actions of individuals for common investment or adherence to rules. Many natural resource man-

---

this model have been developed and used in the applied literature, but a common outcome in adoption studies is that any new and risky technology will only be adopted if it yields an adequate increase in average income to offset its additional risk, so that its net contribution is to reduce the risk of the household falling below its subsistence income level (for example, Low 1974; Hazell and Norton 1986). Approaching vulnerability using a qualitative rather than a quantitative way limits the researcher's ability to measure trade-offs between the additional risk and the additional average return attainable from new technology. However, by avoiding the constraining influence of a specific decisionmaking model, there may be greater possibilities for learning about how farmers perceive and balance the additional risks and rewards associated with a new technology, even when they may at first appear to be acting "irrationally" from an economist's perspective.

agement practices, including integrated pest management, community nurseries, rangeland management, irrigation, forestry, and watershed management, fall into the category of technologies that are facilitated by collective action (Meinzen-Dick et al. 2002). Access to assets is, in turn, affected by the vulnerability context, for example, where AIDS and other illnesses reduce people's ability to participate in collective action or labor activities.

Even technologies that are designed to be adapted to less favored areas and poor farmers' conditions may not be suitable without their access to certain assets, for example, in the face of severe drought, where ownership of cattle may be needed as a buffer. Assets also have many gender dimensions, affecting the value of technologies to men and women; for instance, women may have less access to credit for purchasing hybrid maize but better access to social networks that allow access to open-pollinated maize.

In developing a new technology, researchers should consider the interaction and complementarities among assets and their sequencing. For example, membership in a social group may be necessary for access to rights and land, which is necessary for access to credit, which, in turn, is needed to purchase inputs to take advantage of a new technology. This understanding may lead to a different choice of intervention. There are also conflicts to be considered, for example, whether increased production might conflict with protection of the natural resource base, or whether income maximization through increased cash crop production might increase vulnerability of women through decreased production of crops used for their enterprises.

Agricultural research can shape the asset base as well. This influence is most easily seen in new equipment that becomes part of physical capital, or irrigation or soil fertility management practices that improve the natural capital of water and land. Participatory or action research processes can strengthen the human and social capital asset base when knowledge is generated and groups are formed to work together on the research.

Social assets play an important role in influencing the effects of agricultural technology, because of the ways in which social networks and social relationships facilitate and constrain technology dissemination. Preexisting social capital may be used in dissemination, or new forms of social capital may be developed for the purpose of dissemination. Social assets, perhaps more clearly seen as complex social relationships, can also interact with new technologies and produce negative consequences or otherwise complicate dissemination efforts, such as when social status determines influence in groups and access to the benefits of participation.

Formal and informal institutions and organizations, policies, laws, and customs shape livelihoods by influencing access to assets, livelihood strategies, vulnerability, terms of exchange, and other conditions. The public and private sectors, civil society, and community institutions are all relevant considerations. Agricultural research institutions are part of the institutional environment in which farmers operate. These institutions change people's livelihood options by, for example, changing the relative returns to different factors or assets and changing the distribution of the value of assets within and between households. Agricultural research also interacts with other political institutions at the global, national, regional, and local levels, and its benefits are facilitated or constrained by public policies. Land tenure arrangements, legal rights to natural capital, marketing institutions, input packages, prices, and other policies influence the ability of farmers to take advantage of technologies; they influence richer and poorer farmers, and men and women farmers differently. Taking these factors into account leads to a better understanding of the pathways of impact or obstacles in those paths.

In Figure 2.2, the arrows between agricultural technologies and vulnerability, assets, and PIP point in both directions, because each of these domains has the potential to shape technologies. Vulnerability factors may lead people to adopt new crop varieties, such as where these varieties may be perceived as having certain desirable traits that reduce risk. The vulnerability context and assets of farmers should lead agricultural researchers, including farmers themselves, to adapt technology to these contexts. Moreover, the organized efforts of farmers (social assets) as well as individual farmer's experimentation, adaptation, and innovation can lead to changes in technology that better meet the needs of resource-poor farmers (Chambers, Pacey, and Thrupp 1989). Finally, PIP can shape the decisions of agricultural researchers in technology development, for example, through commodity markets and prices, laws and policies related to land and water, education and extension, and direct financial investments in technology at the national level.

All of these influence people's livelihood strategies and the choices they make in pursuit of income, security, well-being, and other productive and reproductive goals. As discussed above, what is important about the livelihood strategies approach is that it recognizes that households and individuals may pursue multiple strategies, sequentially or simultaneously. Thus even in the context of agricultural research, we should not assume that the target individual is solely a farmer, or that people with other businesses are not involved in farming. Nor should we overlook even minor livelihood strategies, because they can be very important, especially for the poor, who often pursue many livelihood strategies either to make up enough income or to provide a measure of security. The pursuit of multiple activities can have important implications for cash and labor availability at different times of the year and is relevant to the development of specific interventions for poverty reduction.

Livelihood outcomes may include conventional indicators, such as income, food security, and sustainable use of natural resources. Outcomes can also include a strengthened asset base, reduced vulnerability, and improvements in other aspects of well-being, such as health, self-esteem, control, maintenance of cultural assets, and have a feedback effect on vulnerability status and asset base.

The SL framework draws on a number of conceptual approaches in the development literature, and in this sense it is more a synthetic framework than a new set of concepts. It should not be applied rigidly. It provides a method for thinking about the multiple and interactive influences on livelihoods, providing a "checklist" (Ashley and Carney 1999) of issues to be considered in designing research initiatives or program evaluations, to ensure that important explanatory factors are not overlooked, particularly in cases in which natural scientists, economists, and sociologists may tend to look for different things. Not everything on the list can be included in one study, so prioritization is necessary. The framework allows researchers to understand the parameters of the "big picture"; they can then narrow the scope of the study to what has the highest impact or what is most relevant to the important stakeholders (including researchers). The framework may guide researchers to consider and prioritize less visible factors and local priorities that may or may not revolve around production and consumption or even physical or financial resources, but could instead relate to education, safety, or legal rights.

Using a holistic framework requires consideration of many intervening factors at multiple levels, factors that may or may not have a major impact or relate to the technology. But the approach reflects reality. Agriculture is only one part of people's livelihoods, and agricultural research and technologies may affect only one part of a total farming system. Understanding the other factors that impinge at each point can be critical to improving the ultimate impact of agricultural research.

This approach may also identify issues that are highly salient in explaining livelihood impacts, but are either (1) too far outside the domain of the research to be a focus of study (for example, vulnerability to domestic violence in the context of a study of agricultural research); or (2) relevant but impossible to include in the study (for example, a high level of political violence that is either too sensitive or dangerous to address). In these cases, this approach can help make explicit what is not included but may still be important to understanding chains of causality or important constraints on the ability of a technological intervention to affect livelihoods. This approach involves acknowledging complexities that can be hard to manage in a study. Use of this framework implies a willingness to acknowledge that livelihoods—and the process of affecting them—are complex. It makes an effort to achieve the most comprehensive understanding of these issues possible.

The case studies in this book have shown that this approach to evaluating the impacts of agricultural research is manageable and helpful in suggesting relationships to be examined. For example, technologies that stabilize yields in the face of climatic or other fluctuations, thereby reducing vulnerability, may emerge as more valuable for improving people's livelihoods and well-being than technologies that maximize average production, but with higher fluctuations. Technologies that do not require many purchased inputs may be more effective for households with low income or poor access to transportation and market infrastructure. Those that reduce labor requirements, especially for women, may allow households to diversify into other income-earning activities or devote more time to child care, or may be more suitable for families with one or more members who are sick—an especially important consideration with the rise of HIV and AIDS. However, decreases in agricultural employment opportunities may be a negative outcome of labor-saving technologies, highlighting the need to conduct disaggregated analyses of impacts on differentiated groups.

These factors can be identified in conventional farm management and intrahousehold analyses (see Knox, Meinzen-Dick, and Hazell 2002) but are often omitted in poverty impact studies because of the difficulty in measuring them. Where there are trade-offs involved in a new technology, such as between average productivity increases and higher vulnerability, qualitative methods can improve understanding of how different categories of households and individuals assess and value those trade-offs. Qualitative methods are also useful to identify factors that might otherwise be overlooked, or to prioritize which of the many potential effects are important for poor people in the area under study.

A better understanding of the key relevant contextual conditions in a given region can lead to research into technologies better adapted to the context, needs, and priorities of target groups. The case studies that follow provide examples of these conditions, of technology traits that were desirable and undesirable, of how these conditions and technologies interrelate, and of how research can be conducted to better understand these environments and interrelationships. Knowledge databases can be built, and new research may often be necessary, undertaken as part of agricultural research or by partnerships with other research institutions, government or NGOs. The studies that follow also show how impact studies can be undertaken. They suggest that the effort and expense to understand contextual factors is worthwhile, because the alternative may be the development of technology that is not adopted by targeted groups and instead benefit nontargeted groups, leading to a waste of resources, or potentially increased inequality.

Limitations of the Livelihoods Framework

As noted above, a common livelihoods conceptual framework was used to increase comparability across the case studies. The SL framework was selected as the starting point to analyze the relationship between agricultural research and poverty. In the case studies, researchers identified various aspects of people's lives that are not captured in this framework but are important to explaining their choices and consequent livelihood outcomes. Two such aspects are culture and identity. For example, how things have been done in the past, the relationship of certain crops or practices to ancestors, their role in such rituals as festivals, and whether they are seen as appropriate for poor or rich people influence whether people adopt a new farming practice or crop variety, or how they value innovations. The consideration of cultural assets is thus important, for example, beliefs, traditions, language, identity, festivals, and sacred sites. Other aspects that can be attributed to culture may include preference for taste and texture of agricultural products, as well as status that is associated with certain varieties of crops or values that determine how certain resources should be used, or the age or gender appropriateness of cultivating certain kinds of crops or involvement in the management or marketing activities that accompany them. These cultural assets or factors may not have direct economic value but are centrally important in people's lives, choices, and well-being. They can have economic value as well, such as where assets are transformed into tourism or handicraft production. For example, in one village included in the Mexico case study, sewing was an important activity for the purpose of keeping the village supplied with dresses for its almost monthly festivals.

There is something unsettling, however, about trying to fit these aspects of culture, identity, and values into an assets or capital framework, and it may not be worth doing so. Hebinck and Bourdillon (2001, 6), two authors in the Zimbabwe study, take exception to the livelihoods framework's overreliance on the notion of capital, because of the overemphasis that it places on the material aspects of people's lives:

> One of the problems is that it is an economic metaphor that does not do justice to the nature of people's activities, which are not entirely oriented towards material gain. Although material gains are a very important aim in the notion of livelihood, "livelihood" does not span only the commoditized world and associated values. The term also incorporates the non-commoditized, non-material, and cultural part of life and sets of values that are embedded in local cultural repertoires . . . for example . . . community values determine how and under which conditions forest resources should be used, and how the rights to these resources are embedded in culturally defined relationships.

They also draw on Long's (2001) work on knowledge to critique its inclusion as part of human capital, because as such it implies that knowledge is a universal, culture-neutral resource that can be accessed as a commodity, rather than recognized as a social construct or a relationship that is redefined within a local context.

Also missing from the SL framework are the notions of power and power relationships. There is a large body of literature on these topics, for example, on intrahousehold power relationships and women's empowerment (Afshar 1998; Kabeer 1999), and the extent and nature of women's power as well as its increase or decrease can have a strong influence on livelihoods. A new agricultural technology may affect women's decisionmaking power; participation in a group involved with village-to-village technology dissemination may increase members' confidence, which involves material and nonmaterial aspects of improved well-being. Class-based power imbalances may lead to technology access—or the extension needed to benefit from the technology—being captured by richer farmers. Lack of political power may mean that poor farmers do not have access to certain marketing channels or cannot get sufficient prices for their crops. As Ashley and Carney (1999, 35) note, "the SL framework overall can convey a somewhat cleansed, neutral approach to power issues. This contrasts starkly with the fundamental role that power imbalances play in causing poverty." Institutional aspects of power can be captured through the PIP and social capital dimensions of the framework, and empowerment can be identified among livelihood outcomes. However, nothing in the framework explicitly accounts for power relationships. Also, whereas the household can be seen to be one of the institutions to be considered within PIP, and intrahousehold power relationships are part of this, individuals and class-based social groups are not institutions, and it is difficult to see where individual, class-based, or ethnic group–based power dynamics or conflict suggest themselves within the SL framework. There is also a question as to how politics fits within the framework. In one district in the Zimbabwe study, farmers have been wary of new technologies because they view the phase-out of older varieties and their replacement with new varieties as a conspiracy between Agritex officers and the private sector distributor to discredit the government. This observation argues for the interpretation of PIP to include politics as part of institutional processes, but more broadly for the incorporation of an explicit notion of politics to capture these significant dynamics, even where they are not related to formal institutions.

Finally, the SL framework does not account for the importance of historical factors. For example, a history of problems with external interventions can influence the reception of new interventions. The Zimbabwe study explores the ongoing influence of farmers' bad experience with loans taken in conjunction with the adoption of earlier varieties of improved maize. In focus groups conducted as part of the Mexico study, some villagers reported that they distrust government agents and programs, and thus the advice they give on modern maize varieties, preferring to learn from their neighbors' successes and problems. They also said they put more trust in the seeds they acquire from friends, family, and neighbors. Poor people in Bangladesh also reported that government agencies served the rich rather than the poor, and that they tended to prefer NGOs, except in one village where a bad experience with an NGO had also eroded trust in such organizations. These examples illustrate the continuing significance of historical social relationships in contemporary development initiatives. All five case studies consider people's trust of different dissemination pathways. Trust normally encompasses a historical dimension, in which trust in a dissemination institution is based on past experience, whether related or unrelated to the institution in question.

More generally, the framework diagram in Figure 2.2 suggests a snapshot approach, when, in fact, it should be seen as dynamic. Livelihood strategies, vulnerability factors, asset portfolios, and PIP are often in a state of flux, so it is important to operationalize the framework in a way that incorporates this temporal dimension. The above critiques are addressed by including these additional concepts wherever they are relevant. Thus use of the sustainable livelihoods framework does not have to be limiting; it is simply not sufficient on its own for pointing to all possible factors relevant to a study or development intervention. It must be used in conjunction with concepts, tools, and modes of analysis that have long been used in such fields as sociology, anthropology, political science, history, and economics.

SL analysis needs to draw on a set of tools that may include gender analysis, institutional appraisal, stakeholder analysis, or market analysis (Ashley and Carney 1999; DfID 2001). The framework does not explicitly address the differential conditions, assets, and strategies of socially differentiated groups. Therefore additional attention must be given to the implications of gender, ethnicity, class, or other types of social differentiation. It is likely that in a given community, livelihoods analyses will need to be conducted for different social groups, sometimes at the level of the individual rather than the household. Even key concepts often emphasized as SL principles, such as the idea that poor people themselves should be key actors in identifying and addressing livelihood priorities (Ashley and Carney 1999; DfID 1999), are not explicit in the framework. Instead, participatory poverty assessments are another tool that is recommended when using the SL framework (DfID 2001). In fact, as Ashley and Carney (1999, 36) point out, there is "no explicit mention of poverty in the SL framework" and the assumption that use of SL approaches will contribute to poverty elimination must be continually tested.

One paradoxical aspect to using the framework lies in its advocacy of participation and stakeholder analysis, while simultaneously specifying a specific set of concepts that may or may not be the choice of a particular community or set of stakeholders. This problem was experienced in the stakeholder meetings for the Kenya project, in which participation was structured around the concepts in the SL framework (for example, vulnerability and assets). This limited participation, given that some participants had difficulty in understanding what these concepts meant within the context of the framework. This reduced the time available for identifying problems and priorities in a less structured manner. Learning from this experience, in the subsequent Zimbabwe stakeholder workshop, the framework was introduced but participation was solicited in an open brainstorming session on experiences with the use of hybrid maize. The researchers brought the SL framework back into a smaller group only to organize and prioritize issues raised by stakeholders. This approach contributed to an understanding of how best to operationalize the framework, particularly when incorporating stakeholder input in planning a study.

Research Design Process

Using a common conceptual framework facilitates comparison of results across case studies. The case studies were originally developed with a variety of different research questions, conceptual frameworks, and research designs. When the sustainable livelihoods framework was adopted across the five cases, the first step was to look at how the original questions mapped into this framework. After this mapping exercise, in national workshops for each case study, researchers and other stakeholders discussed other critical questions that the SL framework raised. Because these discussions generated more questions than the case studies could address, the questions were prioritized according to their importance and linkage to the impact of the agricultural technologies under consideration. The result was a set of key questions and hypotheses. Following this step, other sources of information were identified for each of the key questions and research methods were developed to address them.

In this way, the studies demonstrated how to contextualize a common set of concerns and arrive at an assessment of the types of outcomes to which the conceptual framework points. The crosscutting themes included dissemination pathways, social differentiation, assets, and institutions. Comparability across case studies was also addressed by a set of guiding principles across the case studies, including adoption of a shared understanding of core concepts, application of a common conceptual framework, commitment to interdisciplinary perspectives, combination of qualitative and quantitative research methods, and integration of the economic and social analyses into a unified study.

Research Methods for the Livelihoods Approach

Integrating Quantitative and Qualitative Methods

The prospect of covering all questions identified as critical for assessing the livelihood impacts of agricultural research can be daunting. Especially for econometric analysis, the way in which so many factors are interrelated creates endogeneity problems that would require ever larger data sets to resolve. What is required, then, is an integrated, interdisciplinary approach that draws upon both quantitative and qualitative data collection and analysis. This section describes and assesses the methods used in five of the seven case studies.[7]

By using data from a variety of sources and methods, it is possible to cover a wide range of issues relatively efficiently. Rather than seeing this as a second-best solution, such a combined approach can actually provide a more convinc-

7. The research design covering quantitative and qualitative methods, including populations, sampling frames, and other details for each case study are found in the respective case study chapters that follow.

ing analysis than any single method, because studies have found that people respond differently to quantitative and qualitative information. Numbers are required to convince some audiences, whereas others will be more impressed by in-depth and contextual information gathered using qualitative techniques.[8] Because livelihood activities are so varied and are often intermittent or non-commoditized, surveys are likely to pick up some activities and miss others. Providing examples from their work in Zambia, Norton, Owen, and Milimo (1994, 93) argue that "Most aspects of rural livelihoods are not captured in either income or expenditure-based survey data. This is because they are neither commoditized nor evident enough to the researchers to be allocated 'imputed values.' . . . Energy (fuelwood) and herbal medicines are two examples. An element of the 'safety net' for rural people in times of stress consists of 'famine foods' that can be gathered from bush and fallow lands."

Triangulation and cross-checks on the results of different methods can improve confidence in the overall study. Use of quantitative and qualitative methods provides a richer base for analysis, in which data from each method help to interpret the other.

Establishing Counterfactuals

To fully assess the impact of agricultural research, it is important to establish a counterfactual—what would have happened in the absence of the technologies. Kerr and Kolavalli (1999) found that lack of an adequate counterfactual was a key weakness of many studies of the impact of agricultural research. This aspect was therefore given particular attention in our study. Three basic comparisons are used to establish a counterfactual. One is to conduct studies in communities where the technology has not entered. However, this approach is not reliable, as the community samples are small and many other community-level contextual factors can influence findings. A second means is to include studies of adopters and nonadopters, examining different conditions among them and controlling for any systematic differences in their characteristics that might affect their decisions and performance.[9] This second means was undertaken in the five livelihoods case studies, with both qualitative and quantitative comparisons.

A third method looks at changes over time as the technologies are introduced and adopted. Three of the case studies (two in Bangladesh plus Zimbabwe) were able to use previous surveys or existing panel data sets to do quan-

8. During one of the Bangladesh case study planning meetings, Binayak Sen summarized the complementarity of methods as "numbers give one a feeling of facts; qualitative stories give one a feeling of truth."

9. For example, if the nonadopters in the control group are less efficient farmers than the adopters even when using the same technology (perhaps they are less educated or have more limited access to credit), then assuming that all the difference in the farm productivity levels of the two groups is due to adoption of the new technology would be misleading.

titative longitudinal analysis. The qualitative methods in all five case studies take advantage of the strength of in-depth interviewing to establish plausible linkages between adoption and outcomes, including people's perceptions and experiences of changes. Ultimately, combinations of before/after and with/without, as well as insiders' and outsiders' perspectives provide the most convincing case of what changes can be attributed to the outputs of agricultural research.

Research Methods

The major data-collection methods used include surveys, focus groups, key informant interviews, in-depth household case studies, and secondary data. The case studies combine social and economic (as well as some biophysical), qualitative and quantitative, and participatory and conventional (or extractive) data. Although there is often a tendency to equate social, qualitative, and participatory data collection on the one hand, and economic, quantitative, and extractive data on the other, the studies also collected quantitative social information, qualitative economic information, and used both participatory and extractive methods for each. Table 2.1 shows the different methods used in each case study.

HOUSEHOLD SURVEYS. All the case studies include some form of survey. All except that for Mexico have panel data for the same households over a number of years, which allows for analysis of changes over time. Several of the surveys (notably those in Mexico and Zimbabwe, and the vegetable and fish study in Bangladesh) have collected data at the level of the individual household

TABLE 2.1 Methods used in the case studies

	Bangladesh rice	Bangladesh fish and vegetables	Kenya soil fertility replenishment	Mexico maize	Zimbabwe maize
Household single-round survey				X	
Household-level panel survey	X		X		
Individual-level panel survey		X			X
Focus groups	X	X	X	X	X
Participatory rural appraisal	X	X	X	X	
Household case studies			X	X	X
Key informant interviews	X	X	X	X	X
Secondary data	X	X	X	X	X

member, which allows for comparison between men and women and also helps to capture the full range of livelihood strategies within the household. Sampling to cover the range of wealth/poverty categories is critical for the surveys. Although some qualitative data are included in the surveys, researchers analyzed most survey data using econometric techniques.

FOCUS GROUPS AND PARTICIPATORY METHODS. All the case studies make use of focus groups to elicit collective experience and opinions. Separate groups are convened for men and women of different wealth/poverty categories. For example, in Bangladesh, six focus groups were held in each selected village (men and women separately for the very poor, poor, and non-poor categories of households). Preexisting survey data help in the disaggregation of wealth groupings for the focus groups, particularly in communities in which a wealth ranking exercise may be divisive or difficult to carry out (for example, because of large community size or time limitations that prevent researchers from getting sufficiently acquainted with a community to comfortably carry out such an exercise). Where possible, households selected for the surveys were included in the focus groups to improve the comparability of the information obtained by the different sources. Focus groups have the advantage of including larger numbers of participants than do other types of qualitative research, and they generate a synergy of ideas when people speak collectively. If well designed and implemented, they can be used as a primary source of data collection, or used to follow up findings from other data collection processes and inquire about puzzles or contradictions. Their disadvantage is that the views of more dominant participants may be given more weight than they should, and minority or even majority opinions from more timid or less powerful participants may not be not heard. Certain issues that are controversial may not be raised at all. These problems can be reduced with good facilitators and careful disaggregation of group participants.

The focus group meetings included a range of participatory and extractive data collection activities: seasonality mapping, identification and ranking of livelihood activities and sources of vulnerability, as well as discussions of the technologies being studied and dissemination approaches. In some of the studies (for example, those for Kenya, Mexico, and Zimbabwe) focus groups were used following a series of household case studies to further investigate issues raised (including the experiences of households not included in these studies), check whether the findings resonate or contradict, and receive feedback on the research findings. In other studies (especially in Bangladesh) focus groups were the primary means of qualitative data collection, but were followed up with in-depth interviews with individuals who participated in those groups. Analysis of focus group data was done partly by the respondents themselves, partly by the field staff's summaries, and partly by the lead researchers on each case study team.

KEY INFORMANT INTERVIEWS. Key informant interviews allowed the research team to follow up in more detail with individuals that have specialized knowledge. These individuals included researchers from CGIAR and national centers; members of NGOs and community organizations; government project staffs; extension agents; local seed distributors and shopkeepers; agricultural researchers from the private sector; community elders; chiefs; and early adopters. Semistructured interviews allowed the researchers to establish a core set of information that they planned to collect, but also to follow up on relevant topics that emerged during the course of the discussion. This information was especially important to address the policies and institutions affecting agricultural research. As with the focus groups, analysis was shared among the respondents and lead researchers.

ETHNOGRAPHY AND HOUSEHOLD CASE STUDIES. In-depth household case studies using ethnographic methods provided more detail on the complexity of household livelihood strategies, particularly in the Kenya, Mexico, and Zimbabwe cases. Researchers lived in sample villages for three to six months, spending time in the homes of a subsample of the survey households, conducting informal interviews, observing and participating in their daily activities, such as farming, extension field days, and social interactions. Long exposures of this type in communities increased trust between researchers and respondents and increased the chances of receiving candid responses and cross-checking responses with observations. Interviews were conducted with household members of different ages, genders, and roles, often privately. Participant observation provided insights that were not available from other methods and informed and refined the questions asked in other, more structured, data collection. Qualitative data were coded according to issues identified by the research questions in advance, as well as issues that emerged through the fieldwork. These data were analyzed by the research team, taking into account findings of the focus groups and surveys. The main disadvantage to this household case study method is that the number of communities and respondents was smaller than through group-based methods, depending on the research budget. However, the depth of insight compensates for this shortcoming, and the use of students from local universities provided employment and opportunities for pursuing higher education for some of our case study researchers.

SECONDARY SOURCES. Secondary data sources from government and other researchers' studies were also used to provide the basis for sampling frames, cross-check the information from the study with other regions or nationally representative samples, and even provide direct information for the study. For example, participatory poverty assessments (PPA) conducted in Bangladesh (Nabi et al. 1999) provided the basic criteria for classifying households according to poverty or wealth status. These criteria could then be applied to the households in the survey and used to select participants for the focus groups. In Mexico earlier ethnographic studies exist of communities included in our case study region (González 1993; Pérez Sánchez 1997), which provide important historical information on livelihoods, institutions, and processes. In

Kenya, previous and ongoing studies carried out in the same region as the case study provided additional information on related technologies, asset portfolios, gender issues, cultural issues in adoption, social networks, and other issues.

Integrating Methods, Data, and Disciplines

Linking the different sources of data required careful attention. Depending on the sequence of data collection, insights from the surveys were followed up in the focus group or key informant interviews and participant observation, or vice versa. In most of the cases an iterative approach to data collection was used. However, in each of the case studies the quantitative and qualitative data take a somewhat different relationship to each other and provide different types of interpretative power. In the Kenya and Mexico cases, preexisting qualitative studies informed surveys. Our case study teams then initiated new qualitative work (household case studies and focus groups) in a wider set of communities and included new issues, which were informed by the earlier work. Larger-scale household surveys were designed drawing on the findings of the new qualitative work, and household-level case studies provided a depth of understanding used to interpret findings of the surveys. In Mexico, qualitative work explored reasons for people's preferences for different maize varieties and the main risk factors they face. The importance of these risk (or vulnerability) factors and perceived advantages of maize characteristics identified in this qualitative work was tested quantitatively through a survey of a wider sample of farmers. Household studies deepened our understanding of how different maize characteristics elicit responses within the context of vulnerability, as well as such issues as people's perceptions and trust of the pathways through which seeds enter communities (whether by government channels or informal social networks) and how this influences people's choices.

The Bangladesh case on vegetable and fish technologies used an iterative process of survey data collection and qualitative data collection on intrahousehold dynamics and women's empowerment, followed by another round that collected individual-level indicators of empowerment. The focus groups built upon the analysis of the survey data, for example, looking at why—if the technologies were profitable—households had not expanded their vegetable or fish production and why household incomes of adopting households were not necessarily higher than for nonadopters. Key informant interviews then followed up on problems with the program that focus group members identified.

The Bangladesh rice case built upon a nationally representative panel survey data set, which offered a broader picture of changes over time, as well as comparisons among early adopters, late adopters, nonadopters, and disadopters (villages where modern rice varieties had been used but had to be abandoned because the village lost the necessary water control; see Hossain et al. 2002). Our study supplemented this data set with focus groups and another round of the survey in villages covering different agroecological zones.

Careful attention was given to sampling for the surveys, focus groups, and household case studies, with links between the samples. For example, efforts were made to include the same households in the surveys, in the focus groups, or in the surveys and household case studies. Thus the quantitative and qualitative data can enhance one another's interpretive power.

One of the most challenging aspects of such multimethod research involves assembling a research team with the proper mix of skills. Each of the studies had a case study leader (usually an economist) who was an international staff member of the lead CGIAR center, with extensive experience in the case study country. Although technical scientists were not formally part of the research team, they often served in informal advisory capacities and, in some cases, assisted with aspects of the study.[10] The case study leader was paired with an international social analysis team member (a sociologist or anthropologist) with extensive experience in the region who was involved in at least one other case study, which increased comparability across cases. Each case also has national economics and social analysis experts who guided the data collection and analysis and who worked with teams of less experienced researchers, engaging in training and capacity building. The field staff required strong analytic and facilitation skills to conduct the focus group and household case studies, while the key informant interviews were often conducted by the national or international social or economics experts. Finally, an external advisory committee composed of leading experts from different disciplines advised on the research design, methods, process, and analysis.

Working with interdisciplinary groups from the international to the local level provided a valuable learning process in mixed-method research and in integrating economics, sociology, and anthropology, and it can provide a model for strengthening the capacity of agricultural research institutions to address poverty in the future.[11] The challenges of developing such mixed teams lie in ensuring that members can communicate across disciplinary lines, respect one another's contributions, and find the time to integrate the findings or insights from other members into their own work. The SL approach helped in this integration because it provided a common framework into which each member could contribute but also required the contributions of others. It also helped that all members of the team were familiar with the situation of poor people in

10. In future studies, we recommend involving technical scientists formally. Although the studies did benefit from their insights, a greater degree of engagement would have likely produced additional insights, and importantly, would have increased their understanding of, and likely commitment to, this approach to impact assessment. It is particularly important following the studies to engage them in determining implications for technology development in the future—a dialog that is occurring in some of the centers.

11. The participation of international social science researchers is not necessary where local social research capacity exists, but the team should have some members with training in qualitative methods.

rural areas, and the livelihoods framework drew on concepts that they were familiar with from their experience in the field. The significant financial contribution from DfID for social analysis sent a strong signal of the increasing recognition among the international development community of the importance of interdisciplinary and mixed-method research. The additional budget that the social analysis had for both data collection and analysis meant that qualitative data collection was not seen as taking resources away from survey data collection, but was a valued addition to the project. However, largely separate budgets for the economic and social analyses can also lead to separation rather than integration of the components, requiring more vigilance for integration, including a budget and time frame that enables quantitative and qualitative researchers to work together in the data analysis and writing process.

The coordination of activities, especially among people based at different institutions, remained a challenge, which could be reduced if all members of the team were within one institution or could focus only on this project. Yet drawing from different institutions can also be an important aspect of capacity building. Ultimately, however, interdisciplinary work within each institution would be strengthened by having more disciplines represented within the research staffs.

Methodological Framework for the China and India Studies

Household-level studies can provide important insights into how agricultural research affects poverty in adopting regions, but they are not sufficient for capturing the aggregate indirect impacts that can arise through labor markets, the nonfarm economy, and food prices. These factors not only affect the rural poor in adopting regions, but can affect poor people elsewhere, including the urban poor. At these levels of analysis, conceptual frameworks and models are needed that can capture the different pathways by which technology influences aggregate economic outcomes at regional and national levels and that can control for other intervening factors affecting the outcomes (such as changes in policies, public investments, and world prices). Analytical methods are also needed that enable construction of aggregate measures of outcome (such as regional income, welfare, or poverty), and because such measures involve summing over the outcomes for large numbers of disparate households, there is an inherent need for quantitative approaches. The China and India case studies in this volume are important because they push this line of research to new levels, in terms of:

1. Developing econometric approaches that help interpret historical and cross-regional data rather than constructing more normative input-output or computable general equilibrium (CGE) type models for hypothetical simulations;
2. Assessing the impact on poverty and disaggregating it by rural and urban poor; and

3. Linking the broader impacts of research and development (R&D) investments to specific lines of plant improvement research and teasing out the share of the impact attributable to particular CGIAR centers (rice and the International Rice Research Institute in this book).

Conceptual Framework

The starting point for the China and India case studies is a conceptual framework showing how public investment in agricultural R&D is expected to influence agricultural productivity growth and poverty alleviation (Figure 2.3). Figure 2.3 shows how public investments in general, including agricultural R&D, affect rural poverty through many channels. Taking this broader view is necessary because other types of investment must be controlled for when making any attributions to agricultural R&D. The nature and level of investments is affected by political, economic, and governance impacts. The most direct (and most widely studied) channel by which public investments like agricultural research, rural education and health, and infrastructure affect poverty is through increases in agricultural production or productivity that in turn increase farmers' incomes. Indirect impacts come from higher agricultural wages and improved nonfarm employment opportunities induced by growth in agricultural productivity. Agricultural output from rural investment often yields lower food prices, again helping the poor indirectly because they are often net buyers of food. Redistribution of land caused by higher agricultural growth also has important impacts on rural poverty. In addition to their effect on productivity, public investments in rural education, health, and infrastructure directly promote rural wages, nonfarm employment, and migration, thereby reducing rural poverty. For example, improved infrastructure access may help farmers set up small rural nonfarm businesses, such as food processing and marketing enterprises; repairs shops; and transportation, trade, or restaurant services.

Investments in the rural sector not only contribute to growth, employment, and wages in rural areas but also help development of the national economy by providing labor, human and physical capital, cheaper food, and markets for urban industrial and service development. Growth in the national economy reduces poverty in both rural and urban sectors. Understanding these different effects or pathways provides useful policy insights for improving the effectiveness of government poverty reduction strategies. In particular, it provides information on how public investments in such areas as R&D can be used to strengthen weak links within and among poverty reduction channels to increase efficiency in targeting public resources on poverty reduction. Taking a broad systems approach also provides the methodological framework for controlling for other key driving factors that affect how investments in agricultural R&D affect agricultural productivity growth and poverty reduction (equivalent to establishing an appropriate counterfactual or control for measuring the impact

FIGURE 2.3 Government spending and poverty

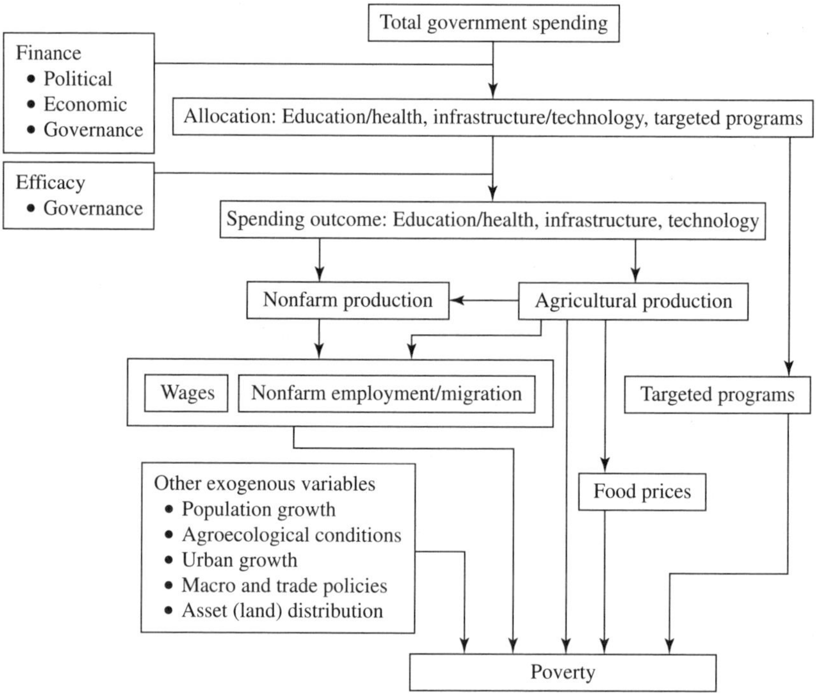

of R&D) and for controlling for the many endogeneities that can arise in an economic system such as the one described in Figure 2.3.

Based on the conceptual framework in Figure 2.3, the China and India case studies use more formal models, pooled time series and regional data, and econometric techniques to estimate the direct and indirect impacts of agricultural research on the rural and urban poor in these two countries. The studies build on earlier work by Fan, Hazell, and Thorat (2000) for India and Fan, Zhang, and Zhang (2002) for China. The studies extend that work to assess, in Chapter 8, the specific impact of the CGIAR's rice breeding research on poverty reduction in China and India, and then in Chapter 9 to assess the impact of total investments in agricultural R&D on urban poverty in those two countries.

Conclusions

There are many empirical studies of the impact of agricultural research on the poor, but most have only analyzed some of the potential pathways by which the poor may be affected by new technologies, often leaving out some of the more important indirect pathways. Many studies have used methods that were insufficient for controlling for the myriad conditioning and confounding socioeconomic factors that affect poverty outcomes. Most have also limited themselves to narrow income or nutritional measures of poverty and have ignored the broader social aspects of poverty. It has been argued in this chapter that more holistic approaches are needed; a framework for conducting such studies has been developed.

A livelihoods framework is useful for assessing poverty impacts at the household level within adopting regions, introducing many factors and relationships that are often missing from conventional reductionist approaches. This approach can provide important insights about the reality that rural households, especially the poor, face—insights that might otherwise be missed. In particular, the framework highlights the importance of different sources of vulnerability. A broad range of assets is considered, not only conventional land and financial resources. Households and even individuals are not regarded as only farmers, laborers, or business operators. Instead, a wide range of simultaneous livelihood activities and strategies is recognized. Policies, institutions, and related processes that form the environment in which livelihood strategies are pursued are considered central to the analysis. Finally, the outcomes include much more than just income levels or food security. Although there are important dimensions of people's lives that the framework does not explicitly address, these can be integrated into the framework or addressed through the inclusion of other types of analysis in the study.

Agricultural research and technologies may not play a central role when we take into account the full picture of people's livelihoods. But understanding the full picture can help develop technologies that better fit in with the complex livelihood strategies, especially of the poor.

Conducting impact studies using a livelihoods framework requires interdisciplinary teams with different skills in data collection and analysis, but with a shared commitment to the research and interest in one another's contributions. Such a framework can then provide a basis for overcoming disciplinary boundaries, help build a more complete analysis of the impact of agricultural research, and point to how technologies could further improve the livelihoods of the poor.

Household studies provide rich insights into how technology can impact on welfare outcomes at the intrahousehold and household levels within adopting regions. But they are insufficient for capturing the indirect impacts of new technologies that can arise through labor markets, the nonfarm economy, and food prices. These factors not only affect the rural poor in adopting regions, but can affect poor people everywhere, including the urban poor. At these levels of analysis, conceptual frameworks and models are needed that can capture the different pathways by which technology influences aggregate economic outcomes at regional and national levels and that can control for other intervening factors affecting the outcomes (such as changes in policies, public investments, and world prices).

Taking such an approach, the China and India studies extend on previous quantitative studies that use regional and national time series data in being able to (1) isolate the impact of plant improvement research for particular commodities on poverty reduction; (2) make attributions of the poverty reduction benefits to specific lines of CGIAR research; and (3) assess the many direct and indirect pathways by which agricultural R&D can impact on poverty including a rigorous analysis of its impact on the urban poor. At this scale of analysis, qualitative measures of poverty are less useful because they cannot be meaningfully aggregated across households and communities to the national level or compared over long periods. For these reasons, the China and India case studies relied exclusively on econometric analysis of official income-based poverty data. Strengths of the approach include an ability to track the different channels through which agricultural R&D affects the poor in rural and urban areas, control for other factors that influence the outcome, analyze the sources of change over long periods of time, and compare investments in agricultural R&D to other governmental investments. Weaknesses include an inability to capture other important dimensions of poverty or to triangulate findings against more in-depth, micro-scale evidence.

The complementary strengths of the different methods for assessing the impact of agricultural research on poverty point to the usefulness of an integrated approach. Thus, although no single method captures all of the effects, we hope this volume will point the way toward methods that can be applied to address the multifaceted relationships between agricultural research and the lives of poor people.

References

Adato, M., and R. Meinzen-Dick. 2003. Assessing the impact of agricultural research on poverty and livelihoods. *Quarterly Journal of International Agriculture* 42 (2): 149–167.

Africa. 1989. Special issue on access to resources in Africa. *Africa* 59 (1).

Afshar, A., ed. 1998. *Women and empowerment: Illustrations from the Third World.* New York: St. Martin's Press.

Antle, J. M., and C. C. Crissman. 1990. Risk, efficiency, and the adoption of modern crop varieties: Evidence from the Philippines. *Economic Development and Cultural Change* 38 (3): 517–537.

Ashley, C., and D. Carney. 1999. *Sustainable livelihoods: Lessons from early experience.* London: Department for International Development.

Bebbington, A. 1999. Capitals and capabilities: A framework for analyzing peasant viability, rural livelihoods and poverty. *World Development* 27 (12): 2021–2044.

Binswanger, H. P. 1981. Attitudes towards risk: Theoretical implications of an experiment in rural India. *Economic Journal* 91: 867–890.

Bryant, R. 1992. Political ecology. An emerging research agenda in Third World studies. *Political Geography* 11 (1): 12–36.

Carney, D., ed. 1998. *Sustainable rural livelihoods: What contribution can we make?* London: Department for International Development.

Carney, D., with M. Drinkwater, T. Rusinow, K. Neefjes, S. Wanmali, and N. Singh. 1999. Livelihoods approaches compared. A brief comparison of the livelihoods approaches of the UK Department for International Development (DfID), CARE, Oxfam, and the United Nations Development Programme (UNDP). London: Department for International Development.

Chambers, R. 1997. *Whose reality counts? Putting the last first.* London: Intermediate Technology Publications.

Chambers, R., A. Pacey, and L.-A. Thrupp. 1989. *Farmer first: Farmer innovation and agricultural research.* London: Intermediate Technology Publications.

Chung, K., L. J. Haddad, J. Ramakrishna, and F. Z. Riely. 1997. *Identifying the food insecure: The application of mixed method approaches in India.* Washington, D.C.: International Food Policy Research Institute.

Datt, G., and M. Ravallion. 1997. *Why have some Indian states performed better than others at reducing rural poverty?* FCND Discussion Paper 26. Washington, D.C.: International Food Policy Research Institute.

———. 1998. Farm productivity and rural poverty in India. *Journal of Development Studies* 34 (4): 62–85.

Dercon, S. 2001. Assessing vulnerability to poverty. Paper prepared for the DfID. <www.economics.ox.ac.uk/members/stefan.dercon/assessing%20vulnerability.pdf> (accessed September 2006).

DfID (Department for International Development). 1997. *The UK white paper on international development—And beyond.* London: Department for International Development.

———. 1999. Background Briefing. November 9. London: Department for International Development.

———. 2001. Sustainable livelihoods guidance sheets. <www.livelihoods.org/info/info_guidanceSheets.html#6> (accessed September 2005).

Fan, S., P. Hazell, and S. Thorat. 2000. Government spending, agricultural growth and poverty in rural India. *American Journal of Agricultural Economics* 82 (4): 1038–1051.

Fan, S., L. Zhang, and X. Zhang. 2002. *Growth, inequality, and poverty in rural China: The role of public investment.* Research Report 125. Washington, D.C.: International Food Policy Research Institute.

Feder, G., R. Just, and D. Zilberman. 1985. Adoption of agricultural innovations in developing countries: A survey. *Economic Development and Cultural Change* 33 (2): 255–298.

González, J. L. 1993. *Vida Cotidiana de Jicayan,* vol. 1. Oaxaca, Mexico: Celiac.

Hazell, P. B. R., and R. D. Norton. 1986. *Mathematical programming for economic analysis in agriculture.* New York: Macmillan.

Hazell, P. B. R., and C. Ramasamy. 1991. *The green revolution reconsidered: The impact of high-yielding rice varieties in South India.* Baltimore: Johns Hopkins University Press.

Hebinck, P., and M. F. C. Bourdillon. 2001. Analysis of livelihoods. In *Women, men, and work: Rural livelihoods in central-eastern Zimbabwe,* M. F. C. Bourdillon and P. Hebinck, eds. Harare: Weaver Press.

Hossain, M., M. L. Bose, A. Chowdhury, and R. Meinzen-Dick. 2002. Changes in agrarian relations and livelihoods in rural Bangladesh: Insights from repeat village studies. In *Agrarian studies: Essays on agrarian relations in less-developed countries,* V. K. Ramachangran and M. Swaminathan, eds. New Delhi: Tulika Books.

Kabeer, N. 1999. Resources, agency, achievements: Reflections on the measurement of women's empowerment. *Development and Change* 30 (3): 435–464.

Kerr, J., and S. Kolavalli. 1999. *Impact of agricultural research on poverty alleviation: Conceptual framework with illustrations from the literature.* EPTD Discussion Paper 56. Washington, D.C.: International Food Policy Research Institute.

Knox, A., R. Meinzen-Dick, and P. Hazell. 2002. Property rights, collective action, and technologies for natural resource management: A conceptual framework. In *Innovation in natural resource management: The role of property rights and collective action in developing countries,* R. S. Meinzen-Dick, A. Knox, F. Place, and B. M. Swallow, eds. Baltimore: Johns Hopkins University Press and International Food Policy Research Institute.

Kumar, S. K. 1994. *Adoption of hybrid maize in Zambia: Effects on gender roles, food consumption, and nutrition.* IFPRI Research Report 100. Washington, D.C.: International Food Policy Research Institute.

Leach, M., R. Mearns, and I. Scoones. 1998. Challenges to community-based sustainable development: Dynamics, entitlements, institutions. *IDS Bulletin* 28 (4): 4–14.

———. 1999. Environmental entitlement: Dynamics and institutions in community-based natural resource management. *World Development* 27 (2): 225–247.

Long, N. 2001. *Development sociology: Actor perspectives.* London: Routledge.

Low, A. R. C. 1974. Decision taking under uncertainty: A linear programming model of peasant farmer behavior. *Journal of Agricultural Economics* 25: 311–320.

Meinzen-Dick, R. S., A. Knox, F. Place, and B. M. Swallow. 2002. *Innovation in natural resource management: The role of property rights and collective action in developing countries.* Baltimore: Johns Hopkins University Press and International Food Policy Research Institute.

Nabi, R., D. Datta, S. Chakrabarty, M. Begum, and N. J. Chaudhury. 1999. *Consultation with the poor: Participatory poverty assessment in Bangladesh.* Bangladesh: NGO Working Group on the World Bank.

Narayan, D., with R. Patel, K. Schafft, A. Rademacher, and S. Koch-Schulte. 2000. *Can anyone hear us? Voices of the poor.* Washington, D.C.: World Bank.

Norton, A., D. Owen, and J. Milimo. 1994. *Zambia participatory poverty assessment,* vol. 5: *Participatory poverty assessment.* Southern Africa Department. Washington, D.C.: World Bank.

Pérez Sánchez, J. 1997. *Buscando el Origen de los Chatinos de Nopala.* Oaxaca, Mexico: El Autor.

Quisumbing, A. R., ed. 2003. *Household decisions, gender, and development: A synthesis of recent research.* Washington, D.C.: International Food Policy Research Institute.

Ribot, J. 1998. Theorizing access: Forest profits along Senegal's charcoal commodity chain. *Development and Change* 29 (2): 307–342.

Roy, A. D. 1952. Safety first and the holding of assets. *Econometrica* 31: 181–196.

Skoufias. E. 2005. *PROGRESA and its impacts on the welfare of rural households in Mexico.* IFPRI Research Report 139. Washington, D.C.: International Food Policy Research Institute.

Telser, L. G. 1956. Safety first and hedging. *Review of Economic Studies* 23: 1–16.

3 Rice Research, Technological Progress, and Poverty: The Bangladesh Case

MAHABUB HOSSAIN, DAVID LEWIS,
MANIK L. BOSE, AND ALAMGIR CHOWDHURY

Through large-scale adoption of modern rice varieties (MVs), Bangladesh has made notable progress in sustaining a respectable growth in rice production over the past three decades. This growth occurred despite the declining availability of arable land and the predomination of small farmers and tenants. The coverage of irrigation has expanded to more than 50 percent of cultivated land and adoption of MVs to 63 percent of the rice area. Rice production has increased from 17.6 million tons of paddy in 1975–76 to 37.6 million tons in 2000–01. Bangladesh faced a famine situation in 1974–75. Since then, it has been able to avert severe food insecurity in spite of several natural disasters, including devastating floods in 1987, 1988, and 1998.

Economic growth has accelerated since the mid-1980s and was quite impressive in the 1990s. The national income grew at 5.1 percent per year and per capita income at 3.6 percent.[1] Recent studies have shown moderate improvements in poverty rates for both rural and urban populations despite trends toward income inequality, but questions have been raised about the validity of the methodology used for assessing the changes (Muqtada 1986; Rahman and Haque 1988; Khan 1990; Osmani 1990; Hossain and Sen 1992; Rahman and Hossain 1995; Ravallion and Sen 1996; Hossain, Sen, and Rahman 2000; Khan and Sen 2001).

The literature on the adoption of modern rice varieties for Bangladesh is quite rich (Hossain 1977, 1988; Asaduzzaman 1979; Mandal 1980; Hossain et al. 1994; Magor 1996). Contrary to the general perception that small farmers and tenants would have an inherent disadvantage in adopting the input-intensive MVs (Griffin 1974; Pearse 1980; Lipton and Longhurst 1989), these studies did not find a strong association of adoption with agrarian structure. The disincentives to adopt varieties that require substantial investment in irrigation and

1. These data suggest that concerns raised in several in-depth rural studies in the 1970s and 1980s (for example, Januzzi and Peach 1980; Van Schendel 1981; Boyce 1987) that the agrarian structure would constrain the development of productive forces in Bangladesh were perhaps overstated.

chemical fertilizers under the widely prevalent sharecropping system, noted by Januzzi and Peach (1980) and Boyce (1987), were ameliorated by institutional changes required to facilitate adoption, such as renting of land under fixed-rent arrangements for MVs, rapid expansion of irrigation infrastructure with small-scale private investment on low-lift pumps and tube wells that are affordable to small farmers, and development of a market for water transactions from tube wells.

The perception that withdrawal of agricultural subsidies and privatization in the marketing of key agricultural inputs would adversely affect adoption by small and marginal farmers (Osmani and Quasem 1990) was proved unfounded by subsequent empirical studies. Private investment in irrigation spread rapidly, and small and marginal farmers got access to irrigation and chemical fertilizers through expanding and competitive markets for water and fertilizers (Hossain 1996; Abdullah and Shahabuddin 1997). The technological progress was found to have a significant positive effect on efficiency in input use, employment of hired labor, and household incomes, although it accentuated the inequality in the distribution of rural incomes (Sidhu and Baanante 1984; Alauddin and Tisdell 1986; Hossain 1988; Hossain et al. 1994).

Magor (1996) found a small fraction of rural households as vulnerable in spite of being in a land-scarce environment, and a significant group of small and marginal farm families not only had maintained their landholding but also actually increased it over the present generation through the tenancy market. Diversification of income sources and access to infrastructure were the major factors contributing to resilience against the shocks created by natural disasters (Ahmed and Hossain 1990; Magor 1996). Recent empirical studies demonstrated that the landless and marginal farm families benefited from the green revolution technology and provided a critique of the hypothesis of polarization and social conflict put forward by eminent social scientists from studies in the 1970s and 1980s (Jahangir 1979; Van Schendel 1981; Boyce 1987; Jansen 1987).

However, the studies mentioned above do not specifically address the impacts of modern rice varieties on poverty. This study was undertaken to understand the magnitude and impact of technological progress in rice cultivation on the livelihood of the rural households, particularly of the poor. Household-level quantitative and qualitative data are used to analyze the asset base of poor and non-poor households and its relationship to the adoption of improved rice varieties. We have analyzed the effect of adoption on productivity, profitability of rice farming, unit costs of production and prices, and food entitlement of low-income people. We examine how they have gained or lost from changes in the livelihood strategies and outcomes induced by the productivity growth in rice cultivation, especially as mediated by such key institutions as land, labor, credit, and water markets.

The study builds on an ongoing large-scale quantitative research project undertaken by the Bangladesh Institute of Development Studies (BIDS) and the International Rice Research Institute (IRRI) in 62 villages covering 57 districts. The study seeks to answer the following questions, mapping a set of research issues developed by IRRI-BIDS study onto a livelihoods framework (see Chapter 2):

- How do we understand the overall trend in household economy and vulnerability context, and what is their relationship with the adoption of MVs?
- What is the relationship between access to assets, technology adoption, and livelihood strategies? What are the asset constraints to adoption?
- What are the effects of the intervening organizations and institutions? How do the approaches of the public sector agencies and nongovernmental organizations (NGOs) affect livelihood strategies?
- How does adoption of MVs affect paddy yields, farm incomes, and household incomes?
- What outcomes can be detected in terms of direct and indirect effects of MV adoption on livelihoods and welfare of the poor and non-poor households?

Methods

The benchmark data for this study are drawn from a sample survey conducted in 1987–88 using a multistage random sampling method for the project Differential Impact of Modern Rice Technology in Favorable and Unfavorable Rice Growing Environments, sponsored by IRRI (David and Otsuka 1994). In the first stage, 64 unions (one union from each of the 64 districts in Bangladesh) were selected from the list of all the unions in the country, using a random number table.[2] In the second stage, data on landholdings, total population, and literacy rates were obtained for all villages in selected unions from the district reports of the 1981 population census. Two villages were chosen from each union, such that the population pressure and the literacy rate for the selected villages were close to the average for the selected unions. A census of all households in the first-choice village was undertaken to collect information on the ownership and tenure of land, adoption of modern rice varieties, and the major source of household incomes. Where the first-choice village was uncooperative, the second choice was included in the sample. Two sites were dropped at this stage because of logistical problems.

The census of the selected villages enumerated 9,874 households or an average of 159 households per village. The census was used as the sample frame for the final draw of the sample for the generation of quantitative data on the

2. A union is the smallest administrative unit in Bangladesh. Districts are divided into *thanas* (*upazilas*), and *thanas* are further divided into unions. A union is composed of several villages.

operation of the household economy. The households were classified into four landownership groups: (1) functionally landless (up to 0.2 ha of land); (2) small landowner (0.2–1.0 ha); (3) medium landowner (1.0–2.0); and (4) large landowner (over 2.0 ha). Each group was further classified into two subgroups according to whether the household was engaged in tenancy cultivation. Twenty households were then selected using the proportionate random sampling method so that each of the eight (4 × 2) strata was represented according to its weight. For a few villages, the sample size was 21 households because of a rounding error. The total sample for the 1987 survey was 1,245 households. Six households were dropped during data analysis due to incomplete data.

The selected households were interviewed in 1988 with a structured questionnaire for generating data on the demographic characteristics of all household members, the use of all parcels of land owned and operated by the household, costs and returns on the cultivation of major crops for the crop year 1987–88, purchase of inputs and the marketing of products, ownership of nonland assets, employment of working members and earnings from nonfarm activities, and the perception of changes in household economic conditions. The findings were published in Hossain et al. (1994).

All villages originally surveyed in 1988 were revisited in 2001 to generate data for the 2000–01 agriculture calendar to generate two-point random sample and household-level panel data. The sample was drawn using the classification of households by the wealth-ranking method of the participatory rural appraisal (PRA) technique. The households in the village were classified into four groups: (1) rich, (2) solvent, (3) poor, and (4) very poor. To ensure that all 1988 sample households and their offshoots were covered in the present survey, a sample of 30 households (a larger number than in the benchmark survey) was drawn from the four groups proportional to their weights, using the stratified random sampling method. All old sample households and their offshoots were covered, except those who had migrated out. New samples were drawn for the cells that were underrepresented by the old sample. The total sample size consists of 1,888 households.

The qualitative component of the research used focus group interviews, stratified by poverty ranking that is based on categories adapted from the Bangladesh Participatory Poverty Assessments (PPA) (Nabi et al. 1999) to complement longitudinal survey data collected by the quantitative study. The focus group methodology was judged to be a cost-effective means of building on an existing large-scale quantitative study while still maintaining a relatively large coverage and sample size that would be attractive to researchers more used to quantitative approaches.

The qualitative component collected and analyzed data from eight villages selected to represent different agroecological conditions (such as elevated or flood-prone land) and levels of infrastructure (such as access to tubewell irrigation and proximity to a paved road). Within the villages, separate focus groups were held representing three socioeconomic categories (non-poor, moderately poor, and very poor) divided by gender (separate male and female groups for each category), giving a total of 48 focus groups. Attendance in focus group meetings varied from seven to 10 individuals, with some members leaving before the end of the meetings. Members who participated in the focus group meetings had no systematic relationship with the sample households selected for the quantitative study. The focus group discussion questions were supplemented by selected PRA techniques, such as ranking exercises.

Rice Research and Technological Progress

Production of Improved Varieties

The major achievement of rice research in Bangladesh, as in other Asian countries, has been the development of high-yielding MVs. By 2001, the Bangladesh Rice Research Institute (BRRI) had released 41 rice varieties for different agroecological conditions, while the Bangladesh Institute of Nuclear Agriculture (BINA) and the Bangladesh Agriculture University, Mymensingh (BAU), released six. The varieties were, however, developed and released following a top-down breeding and evaluation process. Farmers' involvement in the identification of research issues and evaluation of improved germplasm has been lacking. Only in recent years have breeders used farmer participatory variety selection methodology to select advanced lines for unfavorable rice-growing environments. Many of the varieties are direct releases of advanced lines developed at IRRI, and most of the crosses made for developing the varieties contained IRRI breeding materials distributed through IRRI's International Network for Genetic Evaluation of Rice (Evenson and Gollin 1997). Almost 70 percent of the varieties released in Bangladesh have IRRI lineage (Hossain et al. 2003).

Only a few improved varieties have, however, remained popular with the farmers. In the 1970s the most popular varieties in the dry (*boro*) and the pre-monsoon (*aus*) seasons were IR8, Purbachi (released before the introduction of IRRI varieties under the Food and Agriculture Organization of the United Nations [FAO] program), and BR1 and BR3 (which were replaced in 1980s by BR8, BR14, and BR16). Since the late 1990s BRRI Dhan 28 and BRRI Dhan 29, released in 1994, have spread fast because of higher yield potential compared to the varieties released in the 1970s and 1980s. For the monsoon or wet season (*aman*), the most popular varieties in the 1970s were Paijam (Mashuri) and IR20 (IRRI Shail), which have been gradually replaced by BR11 since the early 1980s. BR11 still remains the most popular variety, although many varieties have been released since then for the *aman* season.

The 2000 household-level survey conducted for the study found the most popular varieties grown in the wet season to be BR11 (introduced in 1980),

Paijam (1960s), Swarna (an Indian variety), and BRRI Dhan 30 (1994); and in the dry season BR14 (1983), BRRI Dhan 28 (1994), and BRRI Dhan 29 (1994).

Diffusion of Improved Varieties

BRRI has used several mechanisms to transfer rice technology to farmers (BRRI 1989; Hossain et al. 2002). It has developed a network of multilocation trials with district-level extension officers of the Department of Agricultural Extension (DAE). BRRI also organizes a training course on rice production for the extension officers of DAE, who play a key role in disseminating new information and technologies.

The direct contact of farmers with agricultural extension has, however, remained weak. The household-level surveys conducted for this study found that only 12 percent of the farmers in 2000 got information on MVs from the public sector extension officials; the number was estimated at 11 percent by the 1988 survey. Furthermore, the qualitative component of the present study revealed low levels of trust and confidence in public sector services, including agricultural extension (see below). Only 3 percent of the farmers got information from the input traders or NGO workers. The data presented in this study support the argument that it has been primarily through informal farmer-to-farmer exchange and learning, rather than through official extension efforts, that the increase in MV adoption has been achieved. The focus groups reported that a few entrepreneurial farmers first try a new variety after obtaining information from extension officials or the media, while the others watch the outcome. If the experiment shows better performance of new varieties with regard to preferable traits such as yield, grain quality, duration of crop maturity, and pest resistance compared to the existing varieties, other farmers follow in adopting them. It takes about three to five years for the variety to be spread to all the suitable land in the entire village.

A major constraint on the diffusion of MVs is the production of high-quality seeds (Hossain et al. 2001). BRRI provides breeders' seed of newly released varieties to the Bangladesh Agricultural Development Corporation (BADC), which has a mandate for multiplication of foundation seeds (produced from the seed obtained from the rice breeders) and production of certified seeds through contract growers. The capacity utilization for the production of breeder seed by BRRI and of foundation seed by BADC has, however, remained limited due to (1) price control by the government and (2) lack of incentives for seed production by these public sector institutions. A few private sector companies and NGOs have recently started production and distribution of rice seeds (particularly hybrid rice seeds, which are imported mostly from China), but the size of the market is very small. In 2002, the seed supplied by the BADC accounted for only 4.2 percent of the seed requirement of MVs. The seed replacement rate has remained at a low level. Nearly 90 percent of the seed planted is obtained from the farmers' own harvest or exchanged with neighbors.

Technological Progress

Farmers started cultivating MVs in 1967 when the Bangladesh Academy of Rural Development imported IR8 seeds from IRRI and introduced them to farmers in the dry (*boro* plus *aus*) season. For the wet season (*aman*), IR20 was the first MV; it was introduced in 1970 and became known as IRRI Shail. Two other improved varieties of non-IRRI origin were introduced in the 1960s: Purbachi (Taiwan) and Paijam (known as Mashuri in India and Mansuli in Nepal, developed through an FAO rice improvement program in the 1950s).

The spread of MVs was relatively slow during the 1970s. By 1980, coverage had expanded to 16 percent of the rice area in the wet season (*aman*) and 28 percent in the dry season (*boro* plus *aus*). Diffusion in the dry season has been rapid since the mid-1980s, which coincided with changes in government policies in favor of privatization in the procurement and distribution of small-scale irrigation equipment and chemical fertilizers, liberalization of trade, and reduction in tariff for imported agricultural equipment (Hossain 1996; Abdullah and Shahabuddin 1997). Another spurt in the expansion of MVs took place in the late 1990s, with improved linkages between agricultural extension and research and collaboration between the public and the private sectors (including the NGOs) for the production of certified seeds of newly released varieties. By 2000–01, the coverage of MVs had expanded to 63 percent of the rice-cropped area, 95 percent for the irrigated dry season crop, 35 percent for the pre-monsoon drought-prone crop, and 49 percent for the rain-fed monsoon rice crop.

A dominant factor facilitating the diffusion of MVs is the private investment in small-scale irrigation equipment, such as shallow tube wells and power pumps. At the inception of modern irrigation in the late 1950s, the government placed exclusive emphasis on large-scale surface-water development projects. The projects, however, had long gestation periods, suffered from management and maintenance problems, and were unpopular with farmers because the distribution canals took up scarce land. Over time, the government shifted emphasis to small-scale projects: fielding power pumps to lift surface water from creeks and canals, and tube wells for extraction of groundwater. Since the early 1980s, the government has privatized the procurement and distribution of minor irrigation equipment, reduced import duties, and removed the restriction on the standardization of irrigation equipment (Mandal 1989; Hossain 1996). As a result, farmers have made substantial investment in shallow tube wells and power pumps that contributed to rapid expansion of irrigation facilities since the mid-1980s (Figure 3.1). The area irrigated by tube wells expanded from 53,000 ha in 1973 to 982,000 ha in 1987; it then expanded exponentially to reach 3.3 million ha by 2000. Shallow tube wells and power pumps owned by

FIGURE 3.1 Trends in area under irrigation and in modern rice varieties

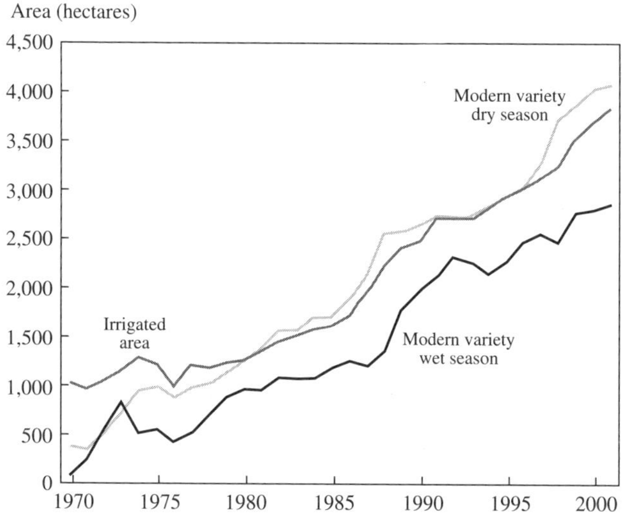

SOURCE: Bangladesh Bureau of Statistics: Monthly statistical bulletins and Bangladesh statistical yearbooks, various issues.

the farmers accounted for 71 percent of total irrigated area in 2000. The diffusion of MV *boro* rice is strongly related to the expansion of groundwater irrigation.

The average rice yield increased from 1.52 tons/ha in 1965 to 3.48 tons/ha by 2000–01, a 2.4 percent per year growth rate. Although rice-cropped area remained almost stagnant at about 10.5 million ha, the growth in yield has enabled Bangladesh to maintain a favorable food-population balance. While the population doubled during 1965–2000, rice production increased by 164 percent, from 14.3 million tons in 1965 to 37.6 million tons in 2000.

What would have happened to rice production if the modern rice varieties had not been developed and adopted? It is hard to establish the counterfactual. The yield of traditional varieties (TVs) had also increased from 1.52 tons/ha in 1965 to 2.14 tons/ha by 2000, a growth rate of 0.9 percent per year. Major factors behind the increase in yield of TVs are the increase in the use of chemical fertilizers (a spillover effect from the experience of the high returns from fertilizers on MVs), a reduction in the share of rice grown in lowest-yielding season (*aus*), and an increase in the cultivation of *boro* rice with higher yields. We assume that, if rice area had remained unchanged, total rice production could have increased at the rate at which the TV yield grew. Figure 3.2 shows the estimate of the trend in counterfactual rice production based on this assumption,

FIGURE 3.2 Long-term trend in rice production and the contribution of modern varieties

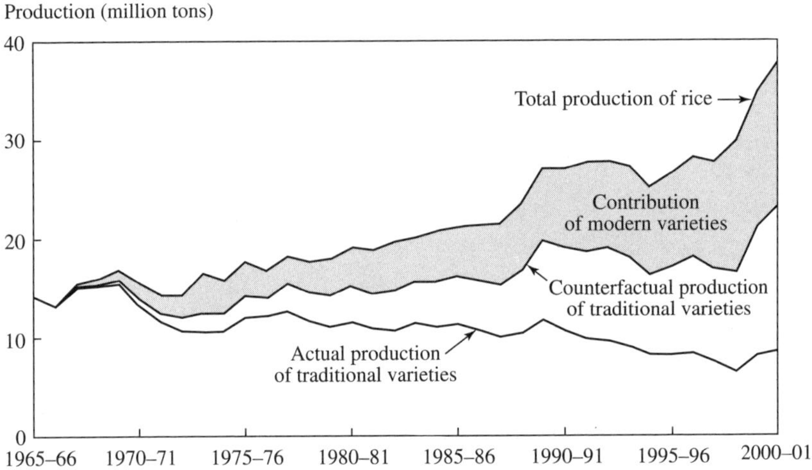

and compares it to actual production. The net contribution of MVs (actual production minus the counterfactual) had grown to 5.8 million tons by 1985 and to 13.1 million tons in 2000, which can feed about 59 million people (46 percent of the population in 2000).[3] Without this impressive increase in production, Bangladesh would have faced a growing demand-supply gap, which could have been difficult to meet with imports, given the country's precarious foreign exchange position. The market would have distributed the scarce supplies in favor of the upper income groups who could afford to pay higher prices. The increase in the real price of this dominant food staple in the diet of the low-income people would have worsened food insecurity and poverty.

Livelihood Systems

Vulnerability Context

The poor in Bangladesh face many sources of vulnerability, including trends in resource availability and depletion, seasonality in employment and health, and

3. The contribution may be an overestimate for the early period when entrepreneurial farmers might have adopted MVs on better-quality land. But it would be an underestimate for the later period. The counterfactual yield of TVs used for this estimation is biased upward because of combining all seasonal MVs together. The area reduction has been more for the relatively low-yielding TVs, such as *aus* and deepwater *aman* varieties, than for the higher yielding TVs, such as transplanted *aman* and *boro* varieties. Farmers now use chemical fertilizers in large amounts in TVs, a spillover effect from their experience in the cultivation of MVs.

shocks such as floods or human or animal diseases. The qualitative data provided information on how different categories of households cope with crises and on the effects of shocks on livelihood strategies. For example, focus groups were asked whether food security had improved as a result of changing prices and employment opportunities. They were also asked whether access to credit services from NGOs or other sources had affected their overall vulnerability to shocks. Moving away from direct references to technology adoption, there were also discussions about whether wider social changes—such as dowry (gift in cash or kind from the bride's parents) or deteriorating law-and-order conditions in the locality—had made a difference to household vulnerability. The focus groups attempted to better understand people's changing perceptions of vulnerability and how these perceptions may have influenced livelihood strategies. In general the respondents reported a positive impact of the adoption of MVs and a favorable trend in the price of rice and vulnerability reduction, a negative response on the rent seeking of officials in the delivery of credit and its impact, and a mixed response on the quality of services provided by the vastly expanding NGO activities.

Landlessness, Land Holding, and Tenancy

The household-level endowment of land is very low in Bangladesh because of extreme population pressure. Three-fourths of the population lives in rural areas. In 2001, Bangladesh supported a population of 129 million with an arable land of 8.1 million ha (BBS 2002). According to the report of the population census undertaken in 2001, the rate of population growth has declined from about 2.4 percent per year in the 1980s to 1.5 percent in the 1990s, and the rural population is still growing, despite rapid rural-to-urban migration. According to agricultural census reports (BBS 1999), the average farmholding size declined from 1.7 ha in 1960 to 0.91 ha in 1983–84 and to 0.68 ha in 1996. The latest census enumerated 17.8 million rural households in 1996, of which 5.8 million (29 percent) did not own any cultivated land, and 9.4 million (53 percent) owned less than 0.2 ha (called "functionally landless" in Bangladesh), an amount that cannot generate significant income. At the other extreme, only 0.1 percent of rural households owned parcels of more than 10 ha and 2.1 percent owned more than 3 ha.

The survey conducted for the study showed a similar pattern of landownership among households as reported by the national agricultural census. Households owning up to 0.2 ha of land, the "functionally landless," made up 47 percent of households in 1987–88; this fraction increased to 50 percent in 2000–01 (Table 3.1). The proportion of households owning more than 2.0 ha declined from 8.3 to 5.2 percent. The average size of land owned per rural household has declined from 0.61 to 0.53 ha over the 13-year period.

The distribution of operational holding estimated by our survey shows similar patterns (Table 3.2). Operational holding is defined as land owned by

TABLE 3.1 Changes in the distribution of landownership, 1987–88 and 2000–01

	1987–88		2000–01	
Area of land owned (ha)	Percentage of households	Share of land (%)	Percentage of households	Share of land (%)
≤0.20	46.5	3.9	49.9	4.7
0.21–0.40	11.9	5.6	15.0	8.2
0.41–1.00	21.9	22.8	19.5	23.4
1.01–2.00	11.4	26.0	10.4	27.1
≥2.01	8.3	41.7	5.2	36.6
Total	100.0	100.0	100.0	100.0
Average area of land owned (ha)	0.61		0.53	

SOURCE: IRRI-BIDS sample household survey.

TABLE 3.2 Changes in the distribution of landholding, 1987–88 and 2000–01

	1987–88		2000–01	
Size of holding (ha)	Percentage of farms	Share of land (%)	Percentage of farms	Share of land (%)
≤0.20	20.7	2.6	22.9	4.0
0.21–0.40	14.6	4.9	23.5	10.3
0.41–1.00	35.2	26.8	34.2	31.9
1.01–2.00	18.9	30.0	15.0	30.7
≥2.01	10.6	35.7	4.4	23.1
Total	100.0	100.0	100.0	100.0
Average area of farm (ha)		0.87		0.67
Percentage of nonfarm households		33.2		40.4

SOURCE: IRRI-BIDS sample household survey.

all members of the household minus the land rented out to other households plus the land rented in from other households for self-cultivation. The proportion of nonfarm households (those that do not cultivate any land) has increased from 33 to 40 percent. For farm households, the number of those that operate holdings of up to 0.4 ha (marginal farms) has increased from 35 to 46 percent, while the number of farms with holdings of more than 2.0 ha (relatively large farms in the Bangladesh context) has declined from 10.6 to 4.4 percent (Table 3.2). Thus marginal and small farms dominate the agrarian structure of Bangladesh, although they control a small share of land. The picture is of a trend toward pauperization rather than differentiation. The structure of land distribution and the small size of holdings imply that gains from increases in land productivity from the adoption of improved technologies would be small for a large majority of rural households (see below).

Substantial land transactions occur through the operation of the tenancy market. The information obtained from the surveys shows that tenancy cultivation is widespread and has increased over the periods covered by the two surveys. With the expansion of the rural nonfarm economy, some of the medium and large landowner households are leaving farming in favor of full-time nonfarm jobs and are having the land cultivated by tenants. The proportion of tenant farmers has increased from 44 to 54 percent, and the area under tenancy cultivation has grown from 23 to 34 percent. The majority of the tenants own some land and rent more to increase the capacity utilization of the farm establishment. It is more socially prestigious to self-employ family labor on rented holdings than to work as wage laborers on another's farm. The number of purely tenant farmers who do not own any cultivated land has grown from 14 to 23 percent and their share of land from 7 to 15 percent. Thus some landless households are getting access to land through the operation of the tenancy market. If the productivity of family labor in the tenancy cultivation is higher than the wage rate, the land-poor households would gain from the adoption of improved technologies and the expansion of the tenancy market.

Since land is extremely scarce, households look for options to increase income through more intensive use of land and the adoption of improved technologies. Investment in irrigation has been the most important means of increasing cropping intensity and land productivity. The coverage of irrigation has expanded quickly, from 24 to 60 percent of cultivated land, and the share of farm households with access to irrigation for some parcels of land has increased from 20 to 70 percent over the periods covered by the two surveys (Table 3.3). The rice area covered by MVs has increased from 33 to 70 percent. The data also show that irrigation coverage and adoption of MVs are higher on the smaller farms, but there is no systematic relationship of these two variables with the tenure status of the farm. This situation is in many ways comparable to Geertz's (1963) analysis of "agricultural involution" in Indonesian wet rice cultivation systems where, like Bangladesh, there are small, fragmented land plots. Our data point to a less pessimistic view of the structural obstacles to technological change (Januzzi and Peach 1980; Boyce 1987).

Endowment of Other Capital

The changes in the endowment of both land and non-land assets during the periods covered by the two surveys for all households, as well as for the land-poor households, are reported in Table 3.4. The land-poor households are defined as those with ownership of up to 0.4 ha. As mentioned below (see Table 3.11), this group contains all the extremely poor households (on the basis of self-perception of the respondents) and most of the moderately poor house-

TABLE 3.3 Coverage of irrigation and adoption of modern rice varieties by farm size and tenure groups

	Percentage of cultivated land irrigated		Percentage of rice area in modern varieties	
Socioeconomic group	1987–88	2000–01	1987–88	2000–01
Farm size (ha)				
≤0.4	32.5	73.0	52.7	81.2
0.41–1.0	24.9	62.2	37.8	72.6
1.0–2.0	23.0	60.6	30.7	67.8
≥2.01	22.4	48.2	27.5	62.2
Land tenure				
Owner-farmer	25.6	62.9	32.6	72.0
Owner-tenant	20.8	57.6	32.1	68.3
Tenant	27.0	58.4	43.4	70.1
All farm households	24.0	60.1	33.1	70.3

SOURCE: IRRI-BIDS sample household survey.

NOTE: This table includes only farm households: those who cultivate some land, either owned or rented.

holds. This group constitutes 58 percent of the sample households in 1987–88 and 65 percent in 2000–01.

Labor is the most abundant resource in Bangladesh. The number of members per household was very high, at 6.06 in 1987–88, but declined to 5.53 by 2000–01 due to the recent progress in fertility control. The child:woman ratio, an indicator of current fertility, declined from 84 children (up to 5 years old) per 100 women of reproductive age (16–49 years old) in 1987–88 to 58 per 100 women in 2000–01. The proportion of children (up to 15 years old) in the total population declined from 47 to 38 percent over this period.

The effect of reduced population growth has, however, not been felt on the working-aged population. The average number of earning members per household has declined from 1.81 to 1.70, but this decline was mainly due to reduction in the proportion of child labor and increased enrollment of young adults in colleges. But the number of agricultural workers has declined substantially, resulting from an increase in the number of nonagricultural workers (Table 3.4). The change in the number of earning members was, however, relatively less for the land-poor households, but there has been a similar movement of labor from agriculture to rural nonfarm activities. The fraction of workers engaged primarily in nonfarm activities has increased from 38 to 51 percent for all households, and from 42 to 54 percent for the land-poor households.

The level of education of the earning members has increased by 45 percent, but still remains low at 4.35 years of schooling on average, indicating the poor quality of human capital (Table 3.4). The amount of schooling com-

TABLE 3.4 Changes in asset base per household, 1987–88 and 2000–01

	Land-poor households[a]				All households			
	1987–88	2000–01	Change (%)	*t*-Value of difference	1987–88	2000–01	Change (%)	*t*-Value of difference
Natural capital indicators								
Own land (ha)	0.10	0.11	10	1.67	0.61	0.53	–13*	–2.16
Cultivated land (ha)[b]	0.36	0.35	–3	0.32	0.87	0.67	–23*	–6.08
Rented land[b]	0.25	0.27	8	0.53	0.20	0.23	15	1.18
Irrigated land[b]	0.09	0.22	144*	9.12	0.20	0.40	100*	8.23
Human capital indicators								
Household size (number)	5.34	5.15	–4	1.78	6.06	5.53	–8*	–4.41
Total workers	1.58	1.56	–1	0.39	1.81	1.70	–6*	–2.58
Agricultural workers	0.91	0.71	–22*	–5.33	1.13	0.83	–27*	–8.76
Nonagricultural workers	0.67	0.85	27*	4.32	0.68	0.87	28*	5.31
Education of household head (years of schooling)	2.07	2.78	34*	4.53	3.16	3.84	22*	4.55
Education of all workers (years of schooling)	2.02	3.15	56*	7.54	3.01	4.35	45*	9.66
Physical and financial capital indicators								
Total capital (US$)	176	243	38*	2.67	324	563	74*	3.33
Agricultural capital (US$)	68	87	28*	3.11	154	151	–2	0.36
Nonagricultural capital (US$)	108	156	44*	2.02	171	412	140*	3.44
Total loan (US$)	37	68	84*	2.85	58	90	55*	2.11
Institutional loan (US$)	9	39	333*	8.54	16	62	288*	3.42
Noninstitutional loan (US$)	29	23	–21	–0.83	42	28	–33*	–1.98

SOURCE: IRRI-BIDS sample survey.

NOTE: * indicates the difference is statistically significant at the 5 percent level.

[a] Land-poor households are defined as those owning up to 0.4 ha of land. All extremely poor households and most of the moderately poor households belong to this group (see Table 3.11).

[b] Numbers are for farm households only.

pleted by an average earning member was 3.01 years in 1987–88 but increased substantially to 4.35 years in 2000–01. The labor productivity could be substantially increased through improvement of human capital from this low level. The improvement in educational attainment of the labor force, although small, has facilitated occupational mobility from lower-productive agriculture to higher-productive nonagricultural activities (see below).

The endowment of physical capital (the value of non-land fixed assets, such as draft animal and agricultural and nonagricultural machinery and equipment) remained low in 2000, at US$563 for all households and US$243 for land-poor households (Table 3.4), indicating unequal access. There has been some accumulation of physical capital in agriculture, with increased investment in irrigation equipment and power tillers. In 1987–88, only 3 percent of sample households owned shallow tube wells; this number had increased to 9 percent by 2000–01. Only 1 percent owned a power tiller in 1987–88 compared to 19 percent in 2000–01. But there was an absolute reduction in the number of cattle that are used as draft power, due to the spread of mechanization in land preparation and the increased cost of maintaining cattle (a reduction in grazing land caused by increased cropping intensity). On balance, the value of agricultural capital remained almost unchanged for all households and increased marginally for land-poor households. However, rural capital accumulation has been very impressive in such nonfarm activities as transport operations, trade, and business. Ownership of rickshaws and rickshaw vans increased from 2 percent of households in 1987–88 to 5.7 percent in 2000–01. The value of non-land fixed assets increased by 74 percent, almost entirely on account of nonagricultural fixed assets.

Bangladesh has always had a substantial credit market, largely managed by professional moneylenders, rich peasants, and traders. Many analysts see the informal credit market as a source of exploitation that perpetuates semifeudal relationships (Bhaduri 1973). But in an imperfect financial market where the landless households and small farmers have difficulty gaining access to banks and credit societies, moneylenders perform a socially useful function of financial mediation. The relationships between lenders and borrowers in the informal market, and between landowners and tenants in the tenancy market, constitute an important component of social capital (Woolcock 1998; Bebbington and Perreault 1999).

The expansion in the supply of microcredit by a number of large NGOs in Bangladesh has reduced the importance of informal credit markets. Households taking credit from NGOs increased from 4 percent in 1987–88 to 20 percent in 2000–01, and the share of NGOs in total credit supply increased from 7 to 30 percent. As a result, households borrowing from informal credit markets declined from 31 to 13 percent during the period. Greater access to NGO credit was an important source of capital accumulation of land-poor households. But the credit is used for organization of small informal business and may not have contributed much to finance working capital needs for the adoption of MVs.

One of the issues discussed in the focus groups was the general perception of the kinds of assets that the participants have that they deem important for survival. The perceptions of the respondents on the relative importance of different assets of the poor and non-poor households, which we failed to capture in the quantitative survey, are shown in Box 3.1. The very poor reported good health, trust of the employer, and social networks as the most important assets they had; land was not mentioned as a significant asset because this group did not have much land. But land rented from the tenancy market figures prominently in the priority list of assets of the moderate poor, indicating that the relatively better-off among the poor would gain from technological progress in land productivity enhancement. The non-poor households reported land, house, and education as their most important assets. Women in poor households reported goats and poultry, homestead trees, and NGO membership as important assets, whereas women in non-poor households mentioned livestock and savings in jewelry as significant assets.

The data show quite clearly that people within poorer households are likely to value security-enhancing assets. The data also show that for the very poor, the human body is the most important tangible asset and the household becomes vulnerable if its working members become sick. Therefore the government's health and nutrition programs are of highest priority for this group.

BOX 3.1 Importance of different assets for livelihoods for the group

Non-poor	Moderately poor	Very poor
Owned land	Rented land	Good health
House	House	Trust of the employer
Education	Good health	Social network
Social network	Social network	Goats and poultry (women)
Political affiliation	Homestead trees	NGO membership (women)
Agricultural machinery	Goats and poultry (women)	Cottage industry skills
Livestock (women)	Education	Fishing nets
Nonagricultural machinery	Agricultural implements	Agricultural implements
Jewelry (women)	NGO membership (women)	
Cash savings in banks	Transport equipment	

Livelihood Strategies

The information obtained from the survey on occupations of rural households is reported in Table 3.5. In 2000–01, only half of the households earned a livelihood from agriculture; others were dependent on various nonfarm activities—salaried and personal services; petty trade; shop keeping; and business; and on providing labor in agroprocessing activities, transport operations, and road and house construction. The importance of nonagricultural activities as sources of livelihood increased substantially over the periods covered by the two surveys.

Very few households reported women engaged in income-earning activities. An analysis of the time budget for the four days preceding the date of the survey, however, reveals that in 2000–01 about 36 percent of the workers engaged in expenditure-saving or income-earning activities were women. The number was 40 percent in 1987–88. The number of women engaged in income-earning activities outside the homestead, which is considered unacceptable in the traditional Muslim society in Bangladesh, has declined from 7.7 percent (of all female workers) in 1987–88 to 5.7 percent in 2000–01. The decline was mostly on account of non-poor households. For very poor households, the incidence of women working outside the home was higher: 11.2 percent in 1987–88 and 10.2 percent in 2000–01. The data indicate that women's participation in income-earning activities outside the home is poverty induced.

Many households are engaged in multiple occupations. For example, a landless household may be simultaneously engaged in agricultural wage labor, tenancy cultivation, goat and poultry raising, petty trade, and transport opera-

TABLE 3.5 Distribution of workers by primary occupation, 1987 and 2000

	Percentage of households reporting it as primary occupation		Percentage of households reporting some income from the occupation	
Primary occupation	1987–88	2000–01	1987–88	2000–01
Farming	44.6	36.7	64.0	69.9
Agricultural labor	22.4	11.8	51.6	28.2
Other agriculture (livestock, forestry and fisheries)	1.2	0.9	78.0	86.2
Trade and business	8.3	12.2	31.9	32.3
Services	14.7	21.7	21.9	28.0
Nonagricultural labor (for example, transport, construction labor, handicrafts)	8.7	16.8	29.2	23.7
Total	100.0	100.0	—	—

SOURCE: IRRI-BIDS sample household survey.

tions. Even an individual worker may be engaged in two or three occupations. This practice is indicated by a much larger proportion of households reporting earning some income from the source than the proportion of workers indicating it as the principal occupation (Table 3.5). Most rural households engage in poultry and livestock raising, particularly involving women and children, but it is a marginal economic activity. Nearly 83 percent of the households reported some income from livestock and poultry raising in 2000, but few reported these as the principal or second occupation. Similarly, 28 percent of households reported some income from agricultural wage labor, but only 12 percent reported it as the principal occupation. The incidence of multiple occupations was less for households engaged in nonfarm activities and has declined over the period, indicating a trend toward specialization and relatively full-time employment in a particular occupation.

As indicated above, agriculture has been releasing labor to the expansion of rural nonfarm activities. The dependence on agriculture for livelihoods has waned substantially, with the proportion of rural workers reporting crop farming as a primary occupation declining from 45 percent in 1987–88 to 37 percent in 2000–01. The proportion reporting agricultural wage labor as the primary occupation has also dropped from 22 to 12 percent over the period. The mobility in rural occupations has been most pronounced for land-poor households whose members were initially employed as agricultural wage laborers. They have been increasingly seeking employment in rural transport operations, such as rickshaw pulling, and at the lower end of the productivity scale of service and business activities. Some of them have accessed land from the growing tenancy market and have become tenant farmers. The mobility of the labor force—the movement from agriculture to rural nonfarm activities—was facilitated by the improvement in rural roads and the increase in the level of schooling. It was also stimulated by technological progress in rice cultivation that created additional employment in trade and transport operations related to the marketing of chemical fertilizers and seeds, maintenance of irrigation equipment, and the disposal of marketable surplus of rice.

The perceptions of people on the importance of different livelihood strategies obtained from the focus group discussions are reported in Box 3.2.[4] Agricultural labor was reported by the very poor households as the most important source of livelihood, followed by nonagricultural labor, goat and poultry raising, and cottage industries. This group did not mention crop farming as a source of livelihood, as they do not have access to land. This finding indicates that the

4. People's own valuation of importance of different livelihoods sources may differ from the quantitative indicators of income. For example, a particular livelihood strategy may be very important because it is seen as stable and reliable, or fits in well with other activities (such as child care), whereas others that may be more remunerative are riskier or cannot be combined with other responsibilities.

BOX 3.2 Perceptions of the people on the importance of different livelihood strategies

Non-poor	Moderately poor	Very poor
Farming/farm supervision	Tenancy cultivation	Agricultural labor
Services	Farming own land	Nonagricultural labor
Business enterprises	Transport operation	Goat and poultry raising
Livestock raising	Agricultural labor	Cottage industry
Rental of machinery	Informal trade	Construction labor
Contractor with local government	Livestock raising	Open-water fisheries
	Cottage industry	

increase in land productivity through the adoption of technology cannot provide any direct benefit to very poor households. However, they may gain some indirect benefits from increased employment and/or higher wages. The moderately poor mentioned tenancy cultivation as the most important source of livelihood, followed by farming their own land, agricultural labor, and informal trade and business. Thus the adoption of improved technology would provide direct benefits to moderately poor households. The non-poor households mentioned services, business, livestock raising, and rental of agricultural machinery as important means of livelihood besides cultivation of land.

Structures and Processes

The qualitative data obtained from the focus group discussions helped us understand a range of intervening structures and processes that bear on the livelihood strategies of rural people. The focus is on wider issues and the process of change over time that may have relevance to farm households engaged in the adoption of MVs that the quantitative data failed to capture.

A striking finding is the generally weak relationship and the absence of trust between rural people and public sector agricultural service providers, contrary to the evidence of synergy between government involvement and private corporate efforts provided by Evans (1996). In many places, people reported DAE as the least effective among a range of governmental and nongovernmental service providers. All categories of farmers report the importance of informal farmer-to-farmer learning in the acquisition of knowledge and skills for MV cultivation. In one site covered by focus group meetings, the very poor mentioned that they learned about cultivation of MVs from the experience of working as laborers on the land of rich farmers. Some groups cited broadcasts on agricultural issues on television and radio as an important source of information.

There are similarly negative perceptions of wider public services and local government. Very few people have anything positive to say about the Union

Parishad (the lowest unit of local government): a poor male group member said, "the political leaders only come to the village at election time and give out packets of *bidi* [local cigarettes], asking for votes." They are simply remote and irrelevant to the people. The government veterinary services were also generally poor and inaccessible. Women in all categories and some men tend to be very positive about the Health and Family Planning Department. Female focus groups (both poor and non-poor) in one site were also very enthusiastic about the government's mass literacy program.

NGOs generally fare much better than public service providers in the ranking given to them, particularly by the poor group members. They are seen mainly as providers of credit and agricultural inputs. However, the very poor and some of the moderately poor are fearful of taking loans, even from NGOs, because they are worried about the pressure of having to repay the loan regularly and feel that the debts might increase their overall vulnerability. There is also a high degree of variation in the perceptions of different NGOs, which suggests that NGOs vary in the quality of services they deliver.

People perceive a decline in law and order, the quality of governance, and access to justice. One of the very poor group members said, "there is no justice; [for] those who have money and can give money, the case against them is dismissed. But we are always punished." This view is most acute among the very poor, who are particularly vulnerable. There is a hostile attitude toward the police: "when there is a conflict, they come and take money from both sides." The formal banking sector is also generally seen unfavorably. Even the non-poor groups report that it is difficult to get a bank loan without paying a rent (bribe), normally 10 percent of the loan.

Livelihood Outcomes

Table 3.6 reports the findings of the survey on household income and its composition for all households as well as for the land-poor households. The concept of income used here is comprehensive, including income received in kind and cash. A money value was imputed to production and receipts in kind at average prices for the entire sample. Household consumption of self-produced crops, livestock, forestry, and fisheries products is treated as income. For international comparison, and comparison over time in real rather than in nominal values, the income has been estimated in U.S. dollars using the exchange rate prevailing during the reference periods of the survey. The exchange rate increased by 68 percent over the period, compared to a 72-percent increase in the wholesale price index. Thus the growth rate estimated from the dollar-denominated income should approximate growth in real incomes.

The average household income increased from US$889 in 1987–88 to US$1151 in 2000–01, indicating a rate of growth of 2.1 percent per year. Per capita income has increased faster, at 2.7 percent, because of the reduction in the number of members per household—the effect of the reduction in population growth rate. The per capita rural income was estimated at US$206 in

TABLE 3.6 Growth and structure of rural incomes, 1987–88 and 2000–01

Source of income	Land-poor households			All households		
	Share of total income (%)		Growth rate	Share of total income (%)		Growth rate
	1987–88	2000–01	(% per year)	1987–88	2000–01	(% per year)
Agriculture	50.3	34.7	0.0	60.9	48.7	0.3
Rice farming	10.7	10.2	2.0	29.9	21.9	–0.4
Other crops	3.9	6.6	6.6	8.9	11.6	4.3
Noncrop agriculture	11.2	9.1	0.7	10.6	10.6	2.1
Agricultural wage	24.6	8.8	–5.4	11.5	4.5	–5.5
Nonagriculture	49.7	65.3	4.5	39.1	51.3	4.3
Trade and business	15.1	26.2	6.8	12.6	19.9	5.9
Services	15.8	16.2	2.6	14.2	12.5	1.1
Remittances	3.1	9.5	11.6	4.7	11.8	9.8
Nonagricultural labor	15.8	13.5	1.2	7.6	7.1	1.5
Total household income (US$/year)	590	803	2.4	889	1151	2.1
Household size (number of members)	5.34	5.15	–0.3	6.06	5.60	–0.6
Per capita income (US$ per year)	110	156	2.7	147	206	2.7

SOURCE: IRRI-BIDS sample household survey.

NOTES: The nominal income has been deflated by changes in the exchange rate, so the numbers are in 2000 U.S. dollars. The change in exchange over 1987–2000 is almost the same (67%) as the change in the wholesale price index (71%), so the growth rate would approximate the change in real incomes.

2000–01. The per capita income of land-poor households also increased at 2.7 percent per year over the period, from US$110 in 1987–88 to US$156 in 2000–01.

The growth in rural incomes over 1987–2000 was almost entirely on account of nonfarm activities. The share of nonagricultural activities in total household incomes has grown from 39 percent in 1987–88 to 51 percent in 2000–01. From a sample survey of 16 villages, Hossain (1988) estimated the share at 36 percent for 1982. Thus the income from rural nonfarm activities has been increasing at a faster rate than that from agriculture since the early 1980s. These findings support the general observation that the rural nonfarm economy accounts for an increasing proportion of rural employment and incomes with the development of the overall economy (Chuta and Liedholm 1979; Shand 1986; Ranis and Stewart 1993; Rosegrant and Hazell 2000; Reardon, Berdegué, and Escobar 2001).

Several aspects are noteworthy with respect to changes in the structure of household incomes over 1987–2000. First, landownership is no longer the predominant source of household income in rural Bangladesh: income originating from agriculture has declined from 61 percent to 49 percent, and from rice farming from 30 to 22 percent. Land is the dominant factor of production in these activities. Thus contribution of the increase in land productivity to improving rural livelihoods has waned over time. Second, business, services, and remittances accounted for 43 percent of rural incomes in 2000–01, a substantial increase from 31 percent in 1987–88. The most dramatic rise has been in the share of remittance income from relatives who have migrated to cities and abroad. The number of households receiving remittances increased from 8 to 19 percent over the period, and the income from remittances increased from 5 to 12 percent of household incomes. These numbers suggest that education (human capital) and the accumulation of physical capital have become important sources of livelihoods. Third, the role of the agricultural labor market in income generation is no longer of high importance. Hiring out of labor services in crop production, processing and construction activities, and generation of self-employment in manual labor–based activities (cottage industries and transport operations) accounted for only 12 percent of rural incomes in 2000–01, a sharp drop from 19 percent in 1987–88. The data show that for the land-poor households who supply the bulk of the wage labor, the income from agricultural and nonfarm labor accounted for 40 percent of the household income in 1987–88; the share declined substantially to only 23 percent in 2000–01. Informal business and cultivation of nonrice crops (mostly vegetables and fruits around the homestead) are the fastest growing source of income for non-poor households. The land-poor households were able to maintain the income from rice farming, in contrast to the decline in income from this source for the medium and large landowners, because of the increased access to land through the tenancy market.

The absolute decline in the income from rice farming—in spite of the impressive increase in rice yields and production in the 1990s—is noteworthy. An important factor is the decline in the size of landholding due to demographic pressure (see Table 3.2), so the area under rice cultivation per household has also declined. But more important is an adverse movement in the terms of trade for the rice farmers. The wholesale price index has increased by 4.3 percent per year over 1987–2000, while the paddy price increased by only 3.1 percent. The prices of major agricultural inputs also increased at a faster rate than that for paddy prices: the wage rate at 5.6 percent per year and the price of chemical fertilizers at 3.8 percent. Had the paddy price increased at par with inflation, the erosion in income from rice cultivation would have been much slower.

Determinants of Technology Adoption

A crucial factor affecting the distribution of gains from technological progress is the extent and intensity of adoption of MV rice among different groups of farmers. The literature is full of studies that analyze adoption behavior of farmers to test the hypothesis that the gains from the introduction of new technology have been unequally distributed (Griffin 1974; Pearse 1980; Feder, Just, and Zilberman 1985; Lipton and Longhurst 1989). It is argued that the new technology may entail fixed costs in the form of access to information and sources of supply of new inputs and arrangements for finance and marketing, which tend to discourage adoption by small farmers and tenants. However, the review by Knox McCulloch, Meinzen-Dick, and Hazell (1998) of adoption studies suggests that land tenure is not likely to constrain adoption of new crop varieties, because the returns are relatively short term (unlike, for example, planting trees), and the technology itself is not "lumpy," but can be adopted on any size area.

This section reports the findings of the household-level survey on the adoption of improved rice varieties and analyzes what assets influence adoption. The requirement of working capital in cultivating a given amount of land is higher for MVs than for TVs (see below). Farmers who grow MVs need to invest in irrigation equipment, such as tube wells and pumps, or pay water charges to owners of the equipment for the purchase of the services. Unless the government bears the cost of irrigation development, access to capital in the form of accumulated savings or low-cost credit from financial institutions may become an important factor determining the extent of MV adoption. Because small landowners and tenants have little physical capital and limited access to institutional credit, a priori they would adopt modern rice varieties less readily than do large landowner cultivators.

Earlier we reported the findings of the survey regarding the use of irrigation and adoption of MVs by various farm size and land tenure groups (see Table 3.3). Contrary to the a priori hypothesis, the coverage of irrigation was found to be larger in smaller farms during 1987–88, when about 24 percent of the land area was covered by irrigation. The coverage of irrigation has increased substantially since then, reaching about 60 percent of the cultivated land in 2000–01. The inverse relationship between farm size and the coverage of irrigation still persisted. No consistent relationship between the tenure status of the farm and the coverage of irrigation was found. The purely self-farmed and purely tenant-farmed holdings had higher coverage of irrigation than the mixed tenant farms, which tend to be large in size of holding.

How can one explain the above observations? In the early years, irrigation facilities were developed by the government, largely through externally funded projects that benefited cultivators irrespective of farm size. Even with the private ownership of shallow tube wells and power pumps that have expanded greatly since the mid-1980s, the small- and medium-sized farms have a higher probability of having some parcels located within the command area of a tube well compared to the large farmers with greater number of parcels, because of the scattered holdings and the random location of the parcels. The subsistence pressure of producing more rice to meet family needs may also induce small and tenant farmers to take more advantage of the irrigation facilities than larger farmers do.

The coverage of MVs in the villages under study has expanded from 33 percent of rice cropped area in 1987–88 to 70 percent in 2000–01 (see Table 3.3). As reported earlier, the intensity of adoption of MVs is inversely related to the size of farm and is not systematically related to land tenure status, contrary to findings reported in the early green revolution literature. The inverse relationship with farm size was observed in 1987–88 as well as in 2000–01. Similar findings are noted by other microstudies in the Bangladesh context (Hossain 1977a, 1988; Asaduzzaman 1979; Mandal 1980; Hossain et al. 1994).

Factors Affecting Adoption: A Multivariate Analysis

To analyze the relationship between the asset base of the farm households and the intensity of adoption of MVs, a multivariate regression model was estimated with household-level data (Table 3.7). The explanatory variables include both socioeconomic characteristics of the household and the biophysical characteristics of the farm. The dependent variable is measured as the area under modern rice varieties as a percentage of the cultivated area.

The equation was estimated separately for two seasons. For the overlapping *aus* and *boro* seasons (dry season), irrigation is a prerequisite for growing MVs because the rainfall is scanty and the puddling of soil for transplantation of seedlings cannot be done without irrigation. For *aman* rice (wet season), rainfall is plentiful (although farmers face occasional droughts), so MVs can be grown under rain-fed conditions. But physical control is imposed by land elevation, because lowlands remain deeply flooded throughout the monsoon season and are thus unsuitable for growing dwarf MVs, and also by salinity, as saline-resistant MVs have not yet been developed. The model includes land elevation, irrigation coverage, and prevalence of soil salinity (as a village-level dummy variable) in the adoption function to capture the effects of these biophysical and technical factors.

As the observed value of the dependent variable has a limited range, the function was estimated by the TOBIT method using the LIMDEP software (www.limdep.com). Some of the explanatory variables are endogenous (irrigation coverage, tenancy ratio, access to credit, and extension exposure), so the values of the parameters were jointly estimated using the two-stage least squares method.

The estimated parameters of the model are reported in Table 3.7. The relative significance of the variables may be judged by the asymptotic *t*-values of the regression coefficients, because the wider the divergence of the coefficient from the standard error, the larger is the *t*-value. As expected, irrigation is found to be the most significant variable associated with the rate of adoption. For the dry season, the asymptotic *t*-value of the regression coefficient is the highest for

TABLE 3.7 Modern varieties adoption function: Estimates of a tobit model

Factors	Wet season[a] 1987–88	Wet season[a] 2000–01	Dry season[a] 1987	Dry season[a] 2000
Owned land (ha)	1.32 (0.91)	–0.055 (–0.038)	0.29 (0.16)	–3.77* (–3.47)
Tenancy ratio (%)	0.048 (1.69)	0.115* (3.93)	0.065* (2.49)	0.126* (3.83)
Land per worker (ha)	0.259* (4.13)	0.254 (1.56)	0.043 (0.83)	0.515* (3.16)
Nonland fixed assets/cultivated land (US$)	–0.002 (–0.18)	–0.001 (–0.33)	0.010* (3.25)	0.016* (7.58)
Institutional loan/cultivated land (US$)	0.040 (0.33)	–0.035 (–1.39)	0.129 (1.63)	–0.091* (–9.63)
Education of family workers (years of schooling)	0.449 (1.21)	0.031 (1.18)	0.061 (1.78)	0.098* (4.16)
Irrigation coverage (percentage of holding)	0.267* (7.83)	0.277* (9.48)	0.673* (23.49)	0.583* (20.46)
High land (percentage of holding)[b]	0.253* (8.72)	0.290* (10.17)	0.105* (3.72)	–0.237* (7.62)
Low land (percentage of holding)[b]	–0.036 (–0.62)	–0.200* (–4.13)	0.085* (2.19)	–0.070 (–1.39)
Very low land (percentage of holding)[b]	–0.177 (–1.70)	–0.443* (–8.46)	0.178* (4.89)	0.059 (1.12)
Dummy for salinity (villages with saline soil = 1)	5.53 (0.98)	–10.15 (–1.69)	13.4* (3.34)	15.64 (1.91)
Extension contact (dummy; farmer with contact = 1)	10.55* (3.54)	18.43* (7.18)	4.80 (1.86)	6.94* (2.68)
Sigma squared	1082	1427	873	1736
Restricted log-L	–4,061	–5,766	–4,124	–5,788
F-value	7.42	42.2	46.42	50.91
Number of cases	818	1108	808	1,108

SOURCE: IRRI-BIDS sample household survey.

NOTES: Figures within parentheses are asymptotic *t*-values. Highlands are those not flooded during the peak of the monsoon season, lowlands flooded at a depth of 50 cm to 10 cm, and very lowland at a depth of more than 100 cm. The medium highland flooded at a depth of up to 50 cm is used as a control.

* indicates the coefficient is statistically significant at the 5 percent level.

[a] The dependent variable is measured as the area under modern rice variety during the season as a percentage of total cultivated land.

[b] The variables are measured as land under different elevation as a percentage of the total holding.

irrigation compared to other variables included in the model. With wider adoption of MVs in 2000, the effects of the land-elevation variables had become weak and were not significantly associated with the adoption in the dry season except for high land, which had a higher cost of irrigation.

For the wet season, the land elevation is the most important factor affecting adoption. Adoption is less widespread in the low and very low land in the toposequence because the deep flooding of such land is not suitable for growing dwarf MVs. The higher the land elevation the higher is the adoption of MVs in the wet season. The association of adoption with irrigation is relatively weak for the wet season compared to the dry season. If rainfall is favorable irrigation is not required for the adoption of MVs in the wet season. However, access to irrigation may reduce the riskiness of cultivation of MVs. If there is a long dry spell during the monsoon season, supplemental irrigation could help save the crop.

The adoption of MVs does not appear to be significantly related to soil salinity. Indeed, for the dry season, the coefficient is positive and statistically significant in 1987–88. In some costal areas farmers have started sinking tube wells to grow MVs with groundwater that is not saline. The endowment of nonland fixed assets is significantly associated with the adoption for the dry season but not for the wet season. Presumably, it reflects substantial financial working capital needs for the adoption in the dry season, particularly on account of irrigation charges.

The coefficient of the tenancy variable is positive and statistically significant, contrary to the hypothesis that tenants would not have incentives to adopt MVs. In the equation for 2000–01, the coefficient of the landownership is negative, indicating that smaller farmers have a higher rate of adoption. Land per worker has a significant positive effect in the 1987–88 wet and 2000–01 dry seasons. Education of family workers only had a significant positive effect in the 2000–01 dry season.

The farm's contact with extension officials was found to be significantly associated with adoption of MVs. The *t*-values of the regression coefficient indicate that the association is stronger for the wet season than for the dry season, when the access to irrigation subsumes the effect of other factors. The findings also show that the extension contact with farmers has improved in 2000–01 compared to 1987–88. The availability of institutional loans was not a factor influencing adoption of MVs. Indeed, in the 2000–01 dry season the adoption of MVs was lower in households with larger loans from the institutional market.

The association of MV adoption with irrigation coverage at the village level can be seen in Figure 3.3. There are villages at both ends of the adoption scale, and the relationship of adoption with irrigation is very strong. The villages with low levels of adoption are mostly located in the coastal areas or in the depression basins that have a majority of land deeply flooded during the wet season.

FIGURE 3.3 Relationship between modern variety adoption and irrigation coverage, village level, 2000

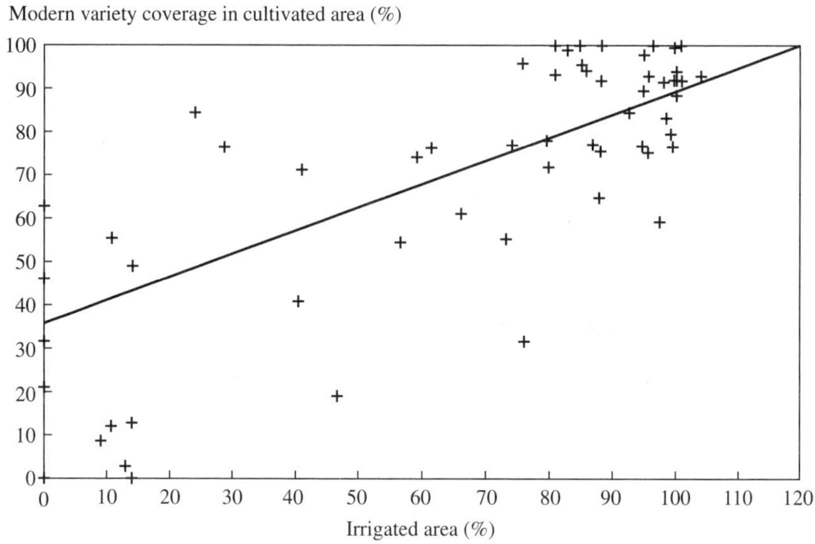

To conclude, the biophysical and technical factors—the availability of irrigation facilities, the elevation of the parcel of land that determines flooding depth, the option to grow nonrice crops, and salinity are more important factors influencing adoption than such socioeconomic factors as farm size, tenancy, endowment of other assets, and access to finance. Subsistence pressure also pushes small farms to adopt the new technology. As noted earlier, MVs have already spread to 70 percent of the land under rice cultivation—almost all the land in the irrigated ecosystem. Farmers now grow traditional varieties only in the flood- and salinity-prone areas for which appropriate MVs have not yet been developed.

Impact of Adoption

This section assesses the impact of the adoption of MVs by estimating (1) the direct effect on farm incomes through changes in the input-output relationships and (2) the indirect benefits accruing to the poor through the operation of different markets, particularly through reduction in the real price of rice.

Effect on Productivity, Unit Cost, and Profitability

As estimated by the survey, the level of input use, yield, and costs and returns for the TVs and MVs are shown in Table 3.8. For ease of comparison, the values are expressed in U.S. dollars at the prevailing exchanges rates of Bangladesh Taka in 1987–88 and 2000–01. The "paid-out cost" includes the costs of seed, fertilizer, manure, irrigation, pesticides, hired labor, and rental of agricultural machinery and draft power. Total cost includes paid-out cost plus the imputed value of family-supplied human and animal labor and the interest charges on working capital. The value of family labor was computed at the wage rate paid to the hired labor. Labor productivity is estimated at the value added (gross value of production minus the cost of material inputs) per day of labor.

The data on input use show that farmers used chemical fertilizers and pesticides in substantially higher amounts in the cultivation of MVs than for TVs. The heavy use of chemicals would have adverse effects on the environment. Pesticides, however, accounted for less than 2 percent of the total cost of production.

The cash cost of MV production per unit of land was almost three times that for TVs in 1987–88, and 120 percent higher in 2000–01. The total cost of production per hectare was about 86 percent higher in MV cultivation in

TABLE 3.8 Costs and returns in the cultivation of traditional and modern rice varieties

	Traditional varieties		Modern varieties		All varieties	
Items	1987–88	2000–01	1987–88	2000–01	1987–88	2000–01
Gross value of production (US$/ha)[a]	325	312	638	625	429	509
Paid-out costs (US$/ha)	106	115	296	251	169	202
Income from rice cultivation (US$/ha)	219	197	342	374	260	307
Total cost (US$/ha)[b]	251	177	467	327	322	272
Operating surplus (US$/ha)	74	135	171	298	107	237
Yield (tons/ha)	1.67	1.98	3.58	4.19	2.30	3.37
Unit cost (US$/ton)	150	89	130	78	140	81
Output price (US$/ton)	174	145	167	141	171	142
Profit (US$/ton)	24	56	37	63	31	61
Labor use (days/ha)	142	110	206	133	163	125
Labor productivity (US$/day)	1.88	2.36	2.11	3.47	2.05	3.05

SOURCE: IRRI-BIDS sample household survey.

NOTES: The paid-out cost includes only out-of-pocket expenses. For 2000–01, the rent paid by tenants is estimated at US$136 per hectare for traditional varieties, US$192 per hectare for modern varieties.

[a] Includes the value of by-products.

[b] Includes imputed costs of family supplied inputs and interest charges on working capital but excludes the land rent.

1987–88 and 84 percent in 2000–01. The numbers clearly show that the MVs are substantially more input-intensive and hence may not favor low-income farmers with limited access to working capital. But the increase in production from the adoption of MVs was much higher than the increase in cost, so the cost per unit of output went down with the adoption of MVs. The unit cost was 13 percent lower in the cultivation of MVs compared to TVs in 1987–88, and 12 percent lower in 2000–01. Thus technological progress contributed to reduction in the unit cost of production, which has helped maintain rice prices at a low level, a major factor behind the improvement in access to food for low-income households (see below).

The increase in the gross value of production from the shift from traditional to MVs was US$313 per hectare in both survey periods. Farmers, however, have reduced the cost by introducing mechanization, which helped reduce use of animal and human labor, and by making more economical use of chemical fertilizers. The labor use in the cultivation of MVs was reduced from 206 days per hectare in 1987–88 to 133 days per hectare in 2000–01, and the use of chemical fertilizers declined from 380 kg/ha (materials) in 1987–88 to 291 kg/ha in 2000–01. Family income from rice farming (gross value of production minus paid out cost) per hectare was lower in 2000–01 than in 1987–88 for TVs, but increased 10 percent for MVs. The net gains from the shift of land from traditional to MVs in fact increased from US$123 per hectare in 1987–88 to US$177 per hectare in 2000–01.

The price of rice declined by US$26 per ton in 2000–01 compared to 1987–88, but the unit cost of production, a measure of total factor productivity, was reduced even more, by US$52. Thus the farmer continued to gain from the adoption of MVs despite the decline in the price of rice. This situation thus benefited both the rice consumer and the producer.

How important are the gains from adoption of MVs in relation to household incomes? The average size of farms in 2000–01 was estimated at 0.67 ha, and the average household income at US$1,151. With two MV rice crops per year, the net gains from adoption for an average household would have been US$237, or 21 percent of total household income.

For an assessment of the effect of MV adoption on household income, we estimated an income determination function with the household-level data. The household income is related to the endowment of different assets—land, worker, physical capital, and education—as well as some location-specific variables, such as access of the village to infrastructure (measured by the availability of electricity in the village). For estimating the effect of MVs, the land under MVs was introduced as an additional explanatory variable. We note in Table 3.7 that the adoption of MVs is strongly influenced by the coverage of irrigation and the elevation of the parcel of land that determined the depth of flooding. These factors would affect household incomes through the adoption of MVs. Because the area under MV is an endogenous variable, the predicted values of the area under MVs were used in the regression model. The area covered by irrigation was used as the instrumental variable for predicting the area under MVs (absolute area, not percentage of area under MVs) for the dry season, and the area with different land elevation was used for predicting the area under MVs for the wet season. A value of zero for this variable was put to households who did not cultivate any land. A village-level dummy variable representing adopter and nonadopter villages was introduced to explore whether, besides the direct benefit from MV adoption, households gain additional income through the expansion of trade and transport activities induced by productivity growth (Mellor 1976; Hazell and Roell 1983; Hazell and Ramasamy 1991). These variables, however, were not found to be statistically significant, and hence were dropped from the model. The parameters of the function were estimated by using the two-stage least squares technique to take into account the possibility that some of the independent variables may be endogenous.

The income function was estimated separately for the landowning (those with more than 0.2 ha) and the functionally landless households (those with up to 0.2 ha). The latter group contained most of the poor households (see Table 3.11). The results of the income function are reported in Table 3.9. The estimated *t*-values of the regression coefficient may be used to delineate the importance of the contribution of the factor to household income, as it shows the distance of the coefficient from the standard error. The results show that the most important factors affecting incomes for the landowning households are the physical capital employed in nonagricultural activities, the number of nonagricultural workers, and the amount of land owned, in that order of significance. For the functionally landless households, the factors were nonagricultural fixed assets, the number of nonagricultural and agricultural workers in the household, and non-land fixed assets employed in agriculture (such as cattle and goats). It is obvious that for the latter group, land was not a significant factor affecting growth in household income. The other variables that were statistically significant for both groups were education of the workers (except for the functionally landless households) and the access to electricity (effect of infrastructure).

The values of the regression coefficient for the MV area at the household level in the equation for the landowning households indicate that incremental income from the adoption of MVs was US$378 per hectare in 1987–88, reduced to US$149 per hectare in 2000–01. For the functionally landless households, some of whom cultivate land rented from others, the estimates are US$336 and US$154 per hectare, respectively, for the two periods. The contribution of MV adoption to household income has declined over time, presumably due to the reduction in rice area and the relative decline in rice prices.

The marginal contribution of land to household income is less when the land is rented than when it is owned, as a substantial portion (from 30 to 50 percent, depending on tenancy contracts) of the gross value of production is paid to the landowner as rent. Still, for landowning households the contribution of

TABLE 3.9 Determinants of household incomes, 1987 and 2000

	Landowning households[a]		Landless households[b]	
Factors	1987–88	2000–01	1987	2000
Dependency ratio (persons per worker)	22.02* (9.78)	−1.61 (−0.86)	19.91* (1.96)	23.7* (2.97)
Land owned (ha)	449* (11.34)	391* (8.24)	173 (0.55)	184 (0.47)
Land rented (ha)	279* (2.88)	298* (3.16)	59 (0.88)	−7 (−0.05)
Predicted area under modern varieties (ha)	378* (4.68)	149 (1.92)	336* (2.54)	154 (1.17)
Agricultural worker (unit)	−40 (−1.05)	62 (1.37)	82* (3.51)	105* (3.82)
Nonagricultural worker (unit)	367* (9.78)	575* (13.43)	144* (7.12)	351* (15.88)
Agricultural fixed assets (US$)	0.250* (2.45)	1.15* (6.05)	0.114 (0.61)	1.262* (6.70)
Nonagricultural fixed assets (US$)	0.512* (13.31)	0.298* (27.96)	0.487* (12.68)	0.73* (18.15)
Education per worker (years of schooling)	19.76* (2.17)	25.82* (2.32)	22.55* (3.72)	2.88 (0.44)
Villages with electricity (dummy)	237* (3.05)	479* (4.67)	63 (1.66)	184* (3.83)
Gender dummy (female = 1)	−74 (−0.60)	−292 (−1.42)	142* (2.42)	−90 (−1.23)
Religion dummy (non-Muslim = 1)	−207 (−1.76)	−56 (−0.37)	−44 (−0.83)	63 (0,92)
Village dummy for flood-prone ecosystem	131 (0.36)	−159 (−1.65)	57 (1.66)	6 (0.15)
Village dummy for saline-prone ecosystem	168* (2.31)	−328* (−2.04)	−53 (−0.83)	−115 (−1.50)
R^2	0.62	0.66	0.37	0.51
F-value	83.1	138	26.01	74.51
Number of cases	653	948	585	948

NOTES: The dependent variable was annual household income measured in U.S. dollars. * indicates the regression coefficient is statistically significant at the 5 percent level.

[a] Includes households that own more than 0.2 ha of land. The income of these households increased from US$1,260 to US$1,745 during 1987–2000.

[b] Includes households that own up to 0.2 ha of land and are defined as "functionally landless" in Bangladesh. The income of these households increased from US$562 to US$724 during 1987–2000.

rented land to household income was US$279 per hectare in 1987–88 and had increased to US$298 per hectare by 2000–01. Thus increased access to land in the tenancy market would increase income. For the functionally landless households (purely tenants), however, the contribution of rented land to household income was insignificant. The effect is presumably captured by higher returns from agricultural workers.

The effect of access to electricity on household income is positive for both periods, but the effect was higher in 2000–01 compared to 1987–88. Higher education seems to have a positive impact on household income through promotion of occupational mobility from agriculture to nonagricultural occupations. Household income was positively associated with the dependency ratio for the land-poor households, indicating that higher subsistence pressure induces poor households to substitute labor for leisure.

Factors Affecting Change in Incomes: Analysis of Panel Data

We also generated panel data on income and its determining factors for households that were common to both surveys, to assess the contribution of different factors to the change in household income over 1987–2000. Among the 1,239 households studied in 1987–88, 217 of them (18 percent) split into 584 households by forming separate households, 148 (12 percent) migrated outside the village, and 874 (70 percent) remained intact. We analyzed the data for the households that remained intact over 1987–2000. We also identified households that were poor in 1987–88 by estimating poverty-line income and counting those with income lower than the poverty line (the head-count index), and analyzed the factors behind the changes in income over 1987–2000 for them. The poverty line was computed as the income needed to have a normative food basket that would give a balanced diet and 2,110 calories per day to an average person as determined by the National Nutrition Council, assuming that at the poverty level, 30 percent of income would be required for meeting nonfood basic needs. The number of poor households was estimated at 59.6 percent in 1987–88 and 41.9 percent for 2000–01. The poverty-gap index was reduced from 23.4 to 16.4 percent and the squared poverty gap from 12.1 to 8.4 percent. Thus our survey data show a substantial reduction in poverty over 1987–2000.

The model used for estimating the factors affecting the change in income over 1987–2000 for the panel households and the estimated values of the parameters can be seen from Table 3.10. Judged by the *t*-values of the regression coefficients, the most important factors affecting the change in income for all households are accumulation of nonagricultural capital, the change in land endowment, mobility from agriculture to nonagricultural occupations, expansion in the area under MVs (technological progress), rural electrification, increase in the level of education of earning members, and increase in the amount of rented land, in decreasing order of importance. For poor households the factors are accumulation of nonagricultural capital, accumulation of land, mobility from agriculture to nonagricultural occupations, rural electrification, access to land in the tenancy market, and accumulation of human capital. It should, however, be mentioned that the effect of education on income would be much higher than indicated by the *t*-value for the regression coefficient of the variable measuring the change in education, because higher education facilitated occupational mobility from lower-productive agricultural to higher-productive nonagricultural jobs. So the effect of education is partly captured by the occupation variable.[5]

The indirect effect of the expansion of MVs for generating employment in rural trade, transport, and processing activities should be captured by village-level dummy variables on technological progress (early adopter and late adopter villages). The value of the regression coefficient for these variables was positive but not statistically significant. Presumably it is difficult to estimate these effects through regression, as the variables representing nonagricultural workers and the nonagricultural capital would also capture the indirect linkage effects on technological progress on the nonfarm economy.

The direct effect of the expansion of MV rice area on household income can be assessed from the estimated coefficient of the change in MV area during 1987–2000 in Table 3.10. The value of the coefficient indicates one hectare of additional area under rice MV would increase average household income for all households by US$339. The mean value of the change in MV area during 1987–2000 was 0.17 ha, which would contribute to an increase in income by US$58 for an average household. The average size for the panel households was 6.12 persons. Therefore the increase in per capita income from the expansion of MVs was US$9.48, only about 7 percent of the poverty level income in 2000–01.

The contribution of MVs is, however, substantial in improving household-level food security for the farm households. Because a majority of rural households are small and marginal farmers (see Table 3.3), a large proportion of farm households has to depend on the market for meeting household food needs. An increase in the productivity of land with the adoption of MVs would reduce farmers' dependency on the market for the provisioning of staple food. We estimate from the 2000–01 survey that only 51 percent of the farm households would meet their rice needs from production on farm, 13 percent were deficit

5. The effect of education is understated by the value of the coefficient, as many household members educated beyond high school migrate out to urban areas in search of better jobs. An analysis of the characteristics of households (for the 1987–88 sample) who migrated out did not find any selectivity of migration with regard to the education of the household head. But it is a general observation in Bangladesh that an individual who migrates to urban areas to receive education at a college level does not return to his/her native village and rejoin the household. Even many household members who graduate from local high schools migrate to urban areas to seek jobs in urban informal sectors.

TABLE 3.10 Factors affecting changes in household incomes, 2000–01 compared to 1987–88

Variables	All households		Households identified as poor in 1987	
	Regression coefficient	*t*-Value	Regression coefficient	*t*-Value
Changes in land owned (ha)	501*	8.75	664*	7.11
Changes in rented-in land (ha)	242*	3.03	282*	3.53
Changes in agricultural worker (unit)	16	0.34	12	0.25
Changes in nonagricultural worker (unit)	319*	6.95	286*	5.62
Changes in agricultural non-land asset (US$)	0.41*	2.94	0.30	1.33
Changes in nonagricultural asset (US$)	0.27*	22.99	0.66*	8.93
Changes in year of schooling for the workers	18.31*	3.27	18.08*	2.57
Changes in modern variety rice area (predicted values)	339*	5.39	145	1.36
Dummy for villages with electricity in 1987	513*	5.24	449*	4.31
Dummy for villages with electricity in 1987–2000	253*	2.49	88	0.84
Dummy for early adopter villages	35	0.32	117	1.01
Dummy for late adopter villages	75	0.76	50	0.48
Constant	–49	–0.58	79	0.86
R^2	0.50	0.41		
F-value	92	40		

SOURCE: IRRI-BIDS sample household survey.

NOTES: The explanatory variables at the household level are measured in changes over the period 1987–2000. * indicates the regression coefficient is statistically significant at the 5 percent level.

for up to six months' rice needs, and another 14 percent would depend on the market for six to nine months' needs. Therefore the adoption of MVs had substantial effect on household rice provisioning from self-production. Farm households who could meet their entire rice needs from self-production were 58 percent for those who adopted MVs in more than half their rice area, compared to 34 percent for households with less than 50 percent of rice area covered by MVs. For households owning up to 0.4 ha of land, the self-sufficiency in rice was 29 percent; 34 percent for those adopting MVs on more than half of the rice area compared to 15 percent for those adopting MVs at less than 50 percent.

Benefits to the Poor: Analysis of Qualitative Data

The wealth-ranking exercise conducted for this study can shed light on the profile of the poor. These numbers are based on the perceptions of the people themselves regarding their status and take into account economic and noneconomic factors and the multidimensional concept of ill-being (Nabi et al. 1999). Table 3.11 shows the association of the incidence of poverty with landholding of the household and the education and occupation of the household head. Twelve percent of rural households considered themselves very poor. Most own less than 0.2 ha of land, and their managers had no formal schooling and provided wage labor to the enterprises of others. Another 31 percent of households considered themselves as moderately poor; 90 percent of them owned land up to 0.4 ha. Households unable to provide three meals a day were reported at 24 percent for all households, mostly concentrated in households owning less than 0.2 ha and having wage labor as the primary occupation. Thus endowment of land and human capital are still predominant correlates of poverty. Since (1) rice production requires land and (2) MVs aim to increase the productivity of land, a pertinent question is how agriculture in general and technological progress in rice in particular can contribute to poverty reduction and improvement of the livelihood of the bottom 50 percent of the rural households that do not own or operate much land.

It is estimated from the 2000–01 survey that an average poor household operated 0.41 ha of land. At prevailing land productivity, this size holding would generate only US$217 per household or US$38 per capita per year. This amount is only 28 percent of the poverty-line income. In that sense, broad-based rural development rather than narrowly focused land-based agricultural development is essential for poverty reduction in Bangladesh.

Poor households may, however, gain indirectly from technological progress, particularly through the operation of different rural markets (Otsuka, Chuma, and Hayami 1992; David and Otsuka 1994; Hossain et al. 2002). The labor market is considered in the literature to be very important in transferring income from landowners to the landless. As landowning households hire labor for conducting farm operations, and MVs require more labor per hectare than TVs, the agricultural labor households could gain from additional employment generated from the adoption of MVs.

But because the proportion of medium and large farmers is very small in Bangladesh (see Table 3.2), the agricultural labor market can generate employment for only a small fraction of the vast number of landless and marginal landowning households. It was noted above that only 22 percent of rural workers had agricultural wage labor as a primary occupation in 1987–88, and the number declined to 12 percent by 2000–01. When the MVs were first introduced, the demand for hired labor increased substantially. But recently, labor use per hectare has declined with the spread of agricultural mechanization in

TABLE 3.11 Association of poverty with asset endowments, perception of respondents, 2000–01

Asset ownership group	Percentage of all households with each asset level[a]	Percentage of households at each asset level who are:			Percentage of households at each asset level who:	
		Very poor	Moderately poor	Not poor	Do not have adequate meals[b]	Have adequate meals
Land ownership						
Households with only homestead	34.5	25.5	52.5	22.0	37.2	62.8
Own ≤0.2 ha	15.8	15.4	44.6	40.0	37.6	62.5
Own 0.21–0.40	15.1	3.5	26.3	70.2	20.4	79.6
Own 0.41–1.00	19.3	0.5	10.2	89.3	8.2	91.9
Own ≥1.01	15.3	0.1	1.0	98.9	1.0	99.0
Schooling of household head						
No formal school	42.9	20.6	39.5	39.9	28.4	71.6
Attended primary school only	27.4	7.4	35.4	57.2	28.2	71.8
Secondary school dropout	15.8	6.0	17.8	76.2	16.7	83.3
Secondary school passed or more	13.9	1.1	7.6	91.3	7.2	92.8
Major source of household income						
Wage labor	22.2	31.3	54.5	14.2	40.0	60.0
Farmer	46.2	4.7	23.6	71.7	18.8	81.2
Service	17.3	11.0	21.2	67.8	17.5	82.5
Business	14.3	6.7	28.5	64.9	21.1	78.9
Total population	100.0	12.0	30.7	57.3	23.6	76.4

[a] Includes household reporting both agriculture and nonagricultural paid labor (for example, transport operation, construction work, agroprocessing) as the major source of income.

[b] Household reporting not having three good meals a day, as a proxy for households with hungry people.

land preparation, irrigation, and postharvest processing.[6] Even full employment in the agricultural labor market cannot provide poverty-escaping income at the prevailing agricultural wage rate of US$1/day. The estimate of the structure of household income from the survey shows that agricultural wage income accounted for 12 percent of the rural household incomes in 1987–88, but declined to 5 percent in 2000–01 (see Table 3.6). So it is the expansion of the tenancy market, rather than the operation of the agricultural labor market, that is more important for poverty-reducing effect of improved technologies.

Marginal landowning households have indeed gained from substantial land transactions that occur through the operation of the tenancy markets. It was noted above that the area under tenancy cultivation increased from 21 percent of the operated land in 1987–88 to 34 percent in 2000–01. The expansion of the tenancy market has occurred mainly for two reasons. First, with rapid rural-to-urban migration taking place, many urban settlers become absentee landowners, having their land cultivated by their relatives still living in rural areas. Second, as higher-productive employment opportunities in the rural nonfarm sector come up with agricultural growth linkage effects (Mellor 1976; Hazell and Roell 1983; Haggblade, Hazell, and Brown 1989) and with developed infrastructure (Ahmed and Hossain 1990), the better educated and capital-rich households find it more economical to rent out land and engage in rural nonfarm occupations. So the rising supply of land in the tenancy market provides greater access to land to the land-poor households. Some households that provided agricultural labor earlier became tenants by renting the land from their former employers. In the focus group discussion, many poor household members mentioned that they learned about MVs when working as paid agricultural workers. Our survey data showed that the increase in the area under tenancy was higher in villages with larger coverage of MVs.

The terms of tenancy have moved in favor of the tenants over time. The sharecropping system, under which the harvest and certain input costs (mostly irrigation charges) are shared between the landowner and the tenants, was the predominant tenancy arrangement in Bangladesh until the 1970s (Hossain 1977b). Since then fixed-rent tenancy both in kind and in cash-rental payments has gained prominence with the spread of cultivation of MVs. The survey data

6. An earlier study (Hossain et al. 2002) based on the same survey data noted that the agricultural labor market has become tight, with many agricultural laborers taking up tenancy cultivation and moving to nonfarm occupations, such as transport operation and road and house construction. As a result the agricultural wage rate has increased substantially, and in response farmers are mechanizing agricultural operations. There has also been a substantial change in the contractual arrangements in the agricultural labor market. The daily wage contract is giving way to piece-rate contracts, particularly in land preparation, transplanting, and harvesting. The wage earning per day of labor was 28 percent higher for piece-rate contracts compared to the daily wage contract in 1987–88, and 32 percent higher in 2000–01. Thus the change in contractual arrangement in the 1990s has benefited the landless households, which are the dominant suppliers of wage labor.

for 2000–01 show that the effective rent paid to the landowner was 31 percent of the gross produce under the fixed-rent tenancy, compared to 50 percent for sharecropping.[7] Thus greater availability of rental land and the increased incidence of fixed-rent tenancy have facilitated redistribution of some benefits of technological progress from land-rich to land-poor households.

Poor households have also gained from expansion of rural nonfarm activities, which can partly be traced to increased land productivity due to technological progress. An impressive development of the rural road network in the 1990s coupled with the increase in the marketed surplus of rice, vegetables, and fruits have created employment opportunities in transport operations and informal small-scale processing and trading activities. The increase in the number of shallow tube wells, pumps, power tillers, rickshaws, and rickshaw vans has created jobs in the operation and maintenance of agricultural machinery and transport equipment. Agricultural growth and the increase in marketed surplus of rice have stimulated jobs in agroprocessing and other business enterprises in rural towns. Many marginal landowning households with some skills for utilizing capital have been able to generate self-employment in livestock and poultry raising, petty trading, and various personal services with the vast increase in microcredit supplied by NGOs.

But the most important way the technological progress in rice cultivation has contributed to reduction in poverty is by keeping rice prices affordable for low-income households. Figure 3.4 shows the trend in the real (deflated by inflation) price of rice in Bangladesh compared to the world market. During 1976–1992 the price in Bangladesh was higher than the price in Thailand, the country with the largest rice exports in the world. Since then the price in Bangladesh has continued to decline sharply, but the price in the world market had an upward trend. The decline in price in recent years has been much more pronounced in Bangladesh than in the world market. Thus although there has been a trend toward liberalization in agricultural trade, the domestic price in Bangladesh is still not aligned perfectly with the world market and is largely influenced by the supply-demand balance within the country. Rice is the dominant staple food in Bangladesh and constitutes a large share of consumer expenditure. The government does not have enough foreign exchange to meet the food gaps through imports. The country also lacks the marketing infrastructure for exporting rice for occasional years when there is some surplus.

7. The sharecropping tenancy is now mostly limited to cultivation of traditional varieties. In earlier times when the MVs were grown under the sharecropping system the landowner used to share the cost of irrigation and chemical fertilizers. These tenancies have now been converted to seasonal fixed-rent tenancies, or to advanced renting of land for a number of years (khai-khalasi). For fixed rent tenancy the tenant pays five maunds (200 kg) of unhusked rice per bigha (1.5 tons/ha) for the boro rice and four maunds (160 kg) per bigha (1.14 tons/ha) for the aman rice crop.

FIGURE 3.4 Trend in real rice prices, Bangladesh and Thailand

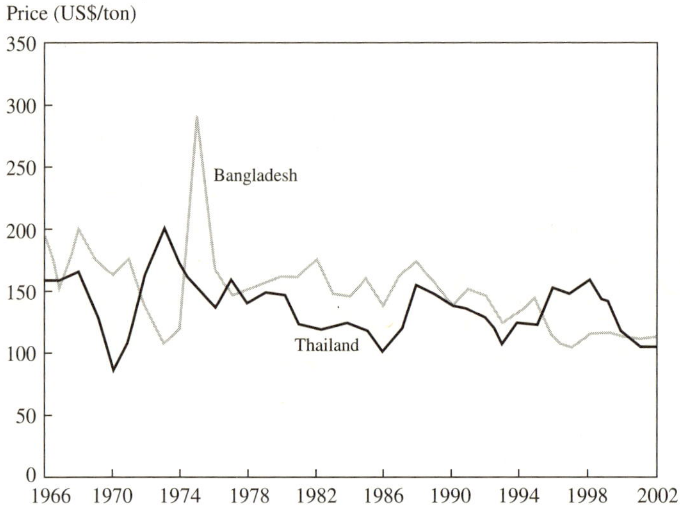

SOURCE: Price data from national statistical documents.
NOTE: Nominal prices deflated by general price index and converted to U.S. dollars using the 2002 exchange rate.

The household income and expenditure survey conducted by the Bangladesh Bureau of Statistics in 2000 found that the bottom 40 percent of households in the per capita income scale spent 68 percent of their income on food —35 percent on rice alone—compared to 44 and 10 percent, respectively, for the top 10 percent in the income scale (BBS 2001). So a reduction in the real price of food relative to other items in the consumption basket benefits poor relatively more than non-poor households. The data from our survey show that the nominal wage rate for agricultural laborers increased from Tk 30 in 1987–88 to Tk 66 per day's labor in 2000–01, while the price of rice increased from Tk 10.91 to Tk 13.07 per kilogram. The rice equivalent wage thus has increased from 2.74 to 5.04 kilograms per day over 1987–2000, a rate of growth of 4.8 percent per year. Thus the relative decline in the price of rice was a major factor behind the increase in food access for the poor.

The indirect benefits from the rapid expansion of area under MVs were also consistently reported by participants in the focus groups (reported in Box 3.3). Higher intensity of crop cultivation, and increased production and marketed surplus of rice were mentioned as important benefits only by the non-poor households. They also mentioned that the increase in rice yield led to reduction

BOX 3.3 Perceptions regarding impact of improved varieties

Non-poor	Moderately poor	Very poor
Rice production increased	Year-round employment	Year-round employment
Cropping intensity increased	Diversified livelihood	Diversified livelihood
Marketed surplus expanded	Rice production increase from tenancy	Wage rate increased
More land available for non-rice crops	Wage rate increased	Affordable rice price increased food entitlement
Capital accumulation for nonfarm activities	Obligation for providing free services to employers reduced	Women's drudgery reduced
Housing conditions improved	Children attending schools	Higher school enrollment of children
More investment in education of children	Housing conditions improved	Higher earnings through migration

in area under rice cultivation, which helped diversification into other crops. The surplus generated by the increase in rice productivity was used for capital accumulation in agriculture, the establishment of nonagricultural business, and investment in children's education, contributing to higher earnings from services and business.

Both the very poor and moderately poor mentioned year-round employment opportunities, diversified livelihood strategies, and increased wage rates as major impacts on livelihoods. The very poor mentioned increased food access from low rice prices and reduced drudgery of women as other important benefits. Increased rice production from tenancy cultivation and reduced obligation to provide services to employers at below-market prices were also mentioned.[8] Both groups mentioned improved housing and increased enrollment of

8. The qualitative data were useful in illustrating these changes. For example, members of the female, very poor group in Patardia said they now benefit from higher wages (which used to depend on the whim of the landowner): "in the past, the landowners sat there comfortably in their shoes, but would not pay us more than Tk 20 a day; now we tell them we won't work for less than Tk 50 and they have no alternative but to agree."

children in schools as important social benefits. At the same time, the spread of the cultivation of MVs has made it possible to free resources, especially land and labor, for other agricultural and nonagricultural uses.

In the focus group discussions, concerns were raised about some negative effects of the expansion of cultivation of MVs. These were reduction in wetlands and common-property resources (such as flood-plain fisheries), reduction in soil fertility, declining stock of cattle due to lack of grazing land, increase in income disparity between the rich and poor, and increased violence. The analysis of the quantitative data from the survey could not adequately capture these dynamic forces.

Impact on Vulnerability

Technological progress in rice cultivation has contributed to farmers' resilience to natural disasters, floods, and droughts. The area under pre-monsoon *aus* crops that are highly susceptible to droughts has been reduced by nearly 2 million ha over the three decades studied: the area has been diverted partly to growing MV *boro* rice and partly to vegetables and fruits. So the loss from the late arrival of the monsoon is now much lower than in the 1970s. Similarly, the area under deepwater broadcast *aman* has declined from 2.2 million to only 0.7 million ha, reducing losses from floods. In the deeply flooded area, farmers now keep the land fallow during the monsoon season and grow high-yielding *boro* rice with irrigation during the dry season. The *boro* area has expanded from 0.5 million to 3.8 million ha, which brings in about 50 percent of the total rice harvest during May–June. Thus losses in *aman* crop from floods or droughts could be recovered within six months, while in the 1970s farmers and consumers had to suffer until the next *aman* harvest in December. The loss of *aman* crop from droughts has also been reduced due to large-scale expansion of shallow tube wells that are used for supplemental irrigation. The seasonal fluctuation in rice prices has been reduced substantially because of two equally important rice harvests during the year. These are some of the reasons why the apprehensions about the severe impact of the disastrous floods in 1988 and 1998 on food insecurity and famine were proved wrong.

Another dimension of vulnerability for the poor is the fragile environment in which low-income households are forced to live. The common-property resources, such as floodplains, are an important source of income for the poor (Knox McCulloch, Meinzen-Dick, and Hazell 1998). There is some evidence from qualitative data that the spread of MVs has contributed to a range of environmental problems, including reduction in fish habitat, contamination of water bodies with pesticides and chemical fertilizers, reduced biodiversity, and declining soil fertility. The loss of previously available wild leafy vegetables was also noted. These developments may impact negatively on the livelihood of the poor in the long run.

Conclusions

Summary of Findings

IRRI has played a major role in developing the rice research capacity in Bangladesh. BRRI has produced large numbers of MVs, two-thirds of them with some IRRI lineage. Farmers have adopted only a few of them, but some remained popular long after their release. MV coverage has now expanded to about 63 percent of the rice area. The technological progress has helped Bangladesh maintain the food-population balance without having to extend rice cultivation to new lands.

The dominance of small farmers and tenants in the Bangladesh agrarian structure did not constrain the adoption of MVs. Indeed MVs are more widely adopted on smaller farms than on larger ones. Technical factors—access to irrigation facilities and the elevation of the land parcel—are the significant determinants of MV adoption. The privatization of minor irrigation equipment (shallow tube wells and power pumps) and reduction in import duties since the late 1980s helped make widespread MV adoption possible in the 1990s, as has the provision of improved infrastructure, such as rural roads and electrification. As a result the general issue of MV adoption is no longer a current one for most farmers, except for the flood- and salinity-prone coastal areas for which appropriate MVs are yet to be developed; in these areas farmers continue to grow TVs.

The quantitative research shows that for the richer 50 percent of households with access to land, there has been direct positive impact from adoption of MVs in the form of increased yields, reduction in unit costs, and increased farm incomes. The productivity increases led to lower output prices, but the unit cost of production was reduced faster, thereby ensuring that MV rice cultivation still remains more profitable than TV cultivation. However, the average farm size has continued to decline due to demographic pressure, and income from rice farming now accounts for only 20 percent of the household income. The direct effect of the MV adoption on overall household income therefore remains small. Nonagricultural income is found to have gained dramatically in importance for rural households. Although not highly profitable, rice contributes to improved food security and provides a springboard for both rich and poor farm households to move into nonfarm income generation and employment. In terms of impact on the poor, MV adoption does not have a substantial direct effect, except for some purely tenant households that were able to gain access to land from the expanding tenancy market. But the focus groups emphasized indirect benefits in the form of (1) stable employment in the expanding nonfarm activities related to growth in rice productivity and marketing of surplus production and agricultural inputs, (2) reduced real price of rice and its seasonal fluctuations having large positive impacts on the food entitlement for the poor, and (3) reduction in the vulnerability from natural disasters. The rice equivalent wage has increased at about 4.8 percent per year from 1987–88 to 2000–01. The poor households mention year-round employment, reduction in women's drudgery, improvement in housing, and increased school attendance of children as major impacts of the expansion of MV cultivation.

The qualitative research highlighted several negative adoption impacts, such as shrinking common-property resources, increased use of pesticides, and declining soil fertility, all of which may increase the long-term vulnerability of the poor. It also throws light on the processes involved in technology dissemination that, after initial release and demonstration on a small scale by BRRI and extension agents, has taken place primarily through informal farmer-to-farmer learning. The focus group discussions revealed the low levels of confidence in the public sector agencies and highlighted the highly variable performance of NGOs engaged in providing credit.

Reflections on Methodology

The linking of the quantitative and the qualitative methodologies proved instructive in broadening the ability of the project researchers to examine the relationship between technology adoption and poverty, particularly in picking up dynamic processes. Whereas the quantitative survey data speak to changes in household structure, landholding, and change in occupation and income, the qualitative data provide insights on the nonincome dimensions of poverty and social and institutional processes that impact on poverty and vulnerability, with particular implications for the poor: the prioritization of assets; the importance of health, trust, and social networks; and the complexity of gender issues.

What was less satisfactory was the approach taken in the study to "bolt-on" the qualitative component to ongoing quantitative research. Although this procedure was necessary to conserve resources, and the longitudinal quantitative data with a large sample covering wider geographical area certainly added to the quality of the research, limitations became evident because the research was not designed to integrate both approaches. With both qualitative and quantitative research it can be difficult to separate the impact of one component of change (such as technological progress) from the overall development interventions on the changes in livelihood systems. Designing studies so that the quantitative and qualitative data can be collected together, and the members of the project team leading each component have time to work together in the field and during analysis, would help reduce these problems. New capacities will be needed among researchers of all persuasions to ensure that the synthesis of a large volume of diverse forms of data (such as statistics, perceptions, and observations) can take place in a transparent way that builds meaning and avoids bias brought about by researcher loyalty to one research methodology or another.

Implications for Future Agricultural Research

The research confirms the relevance of this particular CGIAR technology to poverty reduction, but it raises a set of issues and questions about the future direction of agriculture-related research. For mainstream technical research, the findings could point to the need for rice varieties that require less water to reduce pressure on groundwater, particularly given the current crisis of arsenic contamination in many areas of Bangladesh. There may also be a need to understand the complexity of vulnerability to develop specific technological solutions to suit less favorable or more unpredictable conditions, such as in the flood-prone areas and saline-affected coastal areas. New varieties of rice may be needed in the medium term that are relevant to risk-prone lands, suit a diversified crop portfolio, and are amenable to sustainable crop management techniques.

The study also shows the need for agricultural researchers to recognize important changes in the economic landscape of rural Bangladesh. It has long been known that more than half the rural population of Bangladesh is functionally landless and is therefore dependent on a combination of various forms of agricultural tenancy, laboring, and nonagricultural livelihoods. But the growing importance of nonagricultural income among the better-off households now means that very few people are full-time farmers who rely on agriculture as the main source of income.

There are now limits to the indirect benefits available to the poor from this technology in the form of trickle-down effects of higher employment and lower prices. Research and development need to take into account the livelihood strategies of the poorest more directly—by connecting rice research with research on other agricultural enterprises, such as vegetable cultivation, fisheries, and poultry raising, which provide more opportunities for the poor (for example, research on developing quickly maturing varieties for accommodating nonrice crops in rice-based systems, and nonchemical pest control in rice for expansion of fisheries in the flood-plains).

Acknowledgments

The authors acknowledge financial and institutional support for the study from the Rockefeller Foundation, DFID UK, the Bangladesh Institute of Development Studies, the International Rice Research Institute, and the International Food Policy Research Institute. They are grateful to numerous rural households in Bangladesh for their cooperation in responding to the enumerators and sharing personal information with them during several visits for data collection. They also acknowledge support received from Uttam Kumar Deb, Noel Magor, B. A. A. Mustafi, Hossain Zillur Rahman, Mahfuzur Rahman, and Binayak Sen, the staff members of Socioconsult Ltd., in developing the research protocol, implementing the surveys, and analyzing the data. The final version of the chapter benefited from comments from Ruth Meinzen-Dick, Robert Chambers, and four anonymous referees. The authors are, however, solely responsible for any remaining errors.

References

Abdullah, A. A., and Q. Shahabuddin. 1997. Critical issues in agriculture: Policy response and unfinished agenda. In *The Bangladesh economy in transition,* M. G. Quibria, ed. Delhi: Oxford University Press.

Ahmed, R., and M. Hossain. 1990. *Development impact of rural infrastructure in Bangladesh.* IFPRI Research Report 83. Washington, D.C.: International Food Policy Research Institute.

Alauddin, M., and C. Tisdell. 1986. Market analysis, technical change and income distribution in semi-subsistence agriculture: The case of Bangladesh. *Agricultural Economics* 1 (1): 1–18.

Asaduzzaman, M. 1979. Adoption of HYV rice in Bangladesh. *Bangladesh Development Studies* 7 (3): 23–49.

BBS (Bangladesh Bureau of Statistics). 1999. *Census of agriculture—1996.* National series, vol. 1. Dhaka: Bangladesh Bureau of Statistics, Ministry of Planning.

———. 2001. *Report of the Household Income and Expenditure Survey 2000.* Dhaka: Bangladesh Bureau of Statistics, Ministry of Planning.

———. 2002. *The yearbook of statistics 2002.* Dhaka: Bangladesh Bureau of Statistics, Ministry of Planning.

Bebbington, A. J., and T. Perreault. 1999. Social capital development and access to resources in highland Ecuador. *Economic Geography* 75(4): 395–418.

Bhaduri, A. 1973. A study of agricultural backwardness under semi-feudalism. *Economic Journal* 83 (329): 120–137.

Boyce, J. K. 1987. *Agrarian impasse in Bengal: Institutional constraints to technological change?* Oxford: Oxford University Press.

Chuta, E., and C. Liedholm. 1979. *Rural nonfarm employment: A review of the state of art.* Rural Development Paper 4. East Lansing: Department of Agriculture, Michigan State University.

David, C. C., and K. Otsuka. 1994. *Modern rice technology and income distribution in Asia.* Boulder, Colo., and London: Lynne Rienner.

Evans, P. 1996. Government action, social capital and development: Reviewing the evidence on synergy. *World Development* 24 (6): 1119–1132.

Evenson, R. E., and D. Gollin. 1997. Genetic resources, international organizations, and improvement in rice varieties. *Economic Development and Cultural Change* 45 (3): 471–500.

Feder, G., R. F. Just, and D. Zilberman 1985. Adoption of agricultural innovations in developing countries: A survey. *Economic Development and Cultural Change* 32 (2): 225–298.

Geertz, C. 1963. *Agricultural involution.* Berkeley: University of California Press.

Griffin, K. 1974. *The political economy of agrarian change: An essay on the green revolution.* Cambridge, Mass.: Harvard University Press.

Haggblade, S., P. Hazell, and J. Brown. 1989. Farm-nonfarm linkages in rural Sub-Saharan Africa. *World Development* 17 (8): 1173–1201.

Hazell, P. B., and C. Ramasamy. 1991. *Green revolution reconsidered: The impact of high-yielding rice varieties in South India.* Baltimore: Johns Hopkins University Press.

Hazell, P. B., and A. Roell. 1983. *Rural growth linkage: Household expenditure patterns in Malaysia and Nigeria.* IFPRI Research Report 41. Washington, D.C.: International Food Policy Research Institute.

Hossain, M. 1977a. Farm size, tenancy and land productivity: An analysis of farm level data in Bangladesh agriculture. *Bangladesh Development Studies* 5 (3): 285–348.

———. 1977b. Nature of tenancy markets in Bangladesh agriculture. *Journal of Social Studies* 3 (1): 1–24.

———. 1988. *Nature and impact of green revolution in Bangladesh.* IFPRI and BIDS Research Report 67. Washington, D.C., and Dhaka: International Food Policy Research Institute and Bangladesh Institute of Development Studies.

———. 1996. Agricultural policies in Bangladesh: Evolution and impact on crop production. In *State, market and development: Essays in honor of Rehman Sobhan,* A. A. Abdullah and A. R. Khan, eds. Dhaka: University Press.

Hossain, M., and B. Sen. 1992. Rural poverty in Bangladesh: Trends and determinants. *Asian Development Review* 10 (1): 1–34.

Hossain, M., B. Sen, and H. Z. Rahman. 2000. Growth and distribution of rural income in Bangladesh: Analysis based on panel survey data. *Economic and Political Weekly* 35 (52–53): 4630–4637.

Hossain, M., M. A. Quasem, M. A. Jabbar, and M. M. Akash. 1994. Production environments, modern variety adoption and income distribution in Bangladesh. In *Modern rice technology and income distribution in Asia,* C. C. David and K. Otsuka, eds. Boulder, Colo.: Lynne Rienner.

Hossain, M., A. Janaiah, M. Husain, and F. Naher. 2001. The rice seed delivery system in Bangladesh: Institutional and policy issues. *Bangladesh Development Studies* 17 (4): 1–40.

Hossain, M., M. L. Bose, A. Chowdhury, and R. Meinzen-Dick. 2002. Changes in agrarian relations and livelihoods in rural Bangladesh: Insights from repeat village studies. In *Agrarian studies: Essays on agrarian relations in less developed countries,* V. K. Ramachandran and M. Swaminathan, eds. New Delhi, India: Tulika Books.

Hossain, M., D. Gollin, V. Cabanilla, E. Cabrera, N. Johnson, G. S. Khush, and G. McLaren. 2003. International research and genetic improvement in rice: Evidence from Asia and Latin America. In *Crop variety improvement and its effect on productivity: Impact of international agricultural research,* R. E. Evenson and D. Gollin, eds. Wallingford and Oxford: CAB International.

Jahangir, B. K. 1979. *Differentiation, polarization and confrontation in rural Bangladesh.* Dhaka: Center for Social Studies.

Jansen, E. G. 1987. *Rural Bangladesh: Competition for scarce resources.* Dhaka: Dhaka University Press.

Januzzi, F. T., and J. T. Peach. 1980. *Agrarian structure in Bangladesh: An impediment to development.* Boulder, Colo.: Westview Press.

Khan, A. R. 1990. Poverty in Bangladesh: A consequence of and a constraint on growth. *Bangladesh Development Studies* 18 (3): 19–34.

Khan, A. R., and B. Sen. 2001. Inequality and its sources in Bangladesh, 1991/92 to 1995/96: An analysis based on household expenditure surveys. *Bangladesh Development Studies* 27 (1): 1–49.

Knox McCulloch, A., R. Meinzen-Dick, and P. Hazell. 1998. *Property rights, collective action, and technologies for natural resource management: A conceptual framework.* CAPRi Working Paper 1. CGIAR System-Wide Program on Property Rights and Collective Action. Washington, D.C.: International Food Policy Research Institute.

Lipton, M., and R. Longhurst. 1989. *New seeds and poor people.* London: Unwin Hyman.

Magor, N. 1996. Empowering marginal farm families in Bangladesh. Ph.D. dissertation. Adelaide, Adelaide University, South Australia.

Mandal, M. A. S. 1980. Farm size, tenancy and productivity in an area of Bangladesh. *Bangladesh Journal of Agricultural Economics* 3 (2): 21–42.

———. 1989. Market for and returns from groundwater irrigation in Bangladesh. In *Bangladesh agriculture sector review,* Vol. III: *Land, water and irrigation.* Dhaka: United Nations Development Program.

Mellor, J. W. 1976. *The new economics of growth: A strategy for India and the developing world.* Ithaca, N.Y.: Cornell University Press.

Muqtada, M. 1986. Poverty and inequality: Trends and causes. In *Bangladesh: Selected issues in employment and development,* R. Islam, ed. New Delhi: International Labor Organization.

Nabi, R., D. Datta, S. Chakrabarty, M. Begum, and N. J. Chaudhury. 1999. Consultation with the poor: Participatory poverty assessment in Bangladesh. Dhaka: NGO Working Group of the World Bank.

Osmani, S. R. 1990. Notes on some recent estimates of poverty in Bangladesh. *Bangladesh Development Studies* 18 (3): 75–87.

Osmani, S. R., and M. A. Quasem. 1990. *Pricing and subsidy policy for Bangladesh agriculture.* BIDS Research Monograph 11. Dhaka: Bangladesh Institute of Development Studies.

Otsuka, K., H. Chuma, and Y. Hayami. 1992. Land and labor contracts in agrarian economics: Theories and facts. *Journal of Economic Literature* 39 (4): 1965–2018.

Pearse, A. 1980. *Seeds of plenty, seeds of want: Social and economic implications of the green revolution.* Oxford: Clarendon.

Rahman, A., and T. Haque. 1988. *Poverty and inequality in Bangladesh in the 1980s.* Research Report 91. Dhaka: Bangladesh Institute of Development Studies.

Rahman, H. Z., and M. Hossain. 1995. *Rethinking rural poverty: Bangladesh as a case study.* New Delhi: Sage.

Ranis, G., and F. Stewart. 1993. Rural nonagricultural activities in development. Theory and application. *Journal of Development Economics* 40: 75–101.

Ravallion, M., and B. Sen. 1996. When method matters: Monitoring poverty in Bangladesh. *Economic Development and Cultural Change* 44 (4): 761–792.

Reardon, T., J. Berdegué, and G. Escobar. 2001. Rural nonfarm employment and incomes in Latin America: Overview and policy implications. *World Development* 29 (3): 395–409.

Rosegrant, M. W., and P. Hazell. 2000. *Transforming the rural Asian economy: The unfinished revolution.* Manila: Asian Development Bank.

Shand, R. 1986. *Off-farm employment in the development of rural Asia.* Canberra: Australian National University.

Sidhu, S. S., and C. A. Baanante. 1984. *Agricultural production, fertilizer use and equity considerations: Results and analysis of farm survey data.* Muscle Shoals, Ala.: International Fertilizer Development Corporation.

Van Schendel, W. 1981. *Peasant mobility: The odds of peasant life in Bangladesh.* Assen, The Netherlands: Van Gorcum.

Woolcock, M. 1998. Social capital and economic development: Toward a theoretical synthesis and policy framework. *Theory and Society* 27 (2): 151–208.

4 Assessing the Impact of Vegetable and Fishpond Technologies on Poverty in Rural Bangladesh

KELLY HALLMAN, DAVID LEWIS,
AND SURAIYA BEGUM

Poverty reduction is the central policy challenge for Bangladesh, one of the poorest countries in the world. According to the World Bank (1999), around 36 percent of the population was very poor and 53 percent poor in 1995–96: the period just before the survey data for this study were collected.

There is a strong gender dimension to poverty in Bangladesh (Kabeer 1994). The distribution of consumption within households favors men. Of 43 global studies reviewed by Haddad et al. (1996), pro-male bias in nutrient allocations appears to be most prevalent in the South Asian region;[1] boys in this region are also more favored in the distribution of nonfood health inputs, such as health care (Haddad et al. 1996; see also Mitra et al. 1994, 1997; Filmer, King, and Pritichett 1998). Furthermore, this is the only region of the world where girls have higher child mortality rates than boys. Food for Education and other relatively recent incentive programs for female education have, however, raised girls' school enrollment and educational attainment.

Rural poverty is still a pervasive problem in Bangladesh. Recent reductions in poverty in the 1990s were larger in urban than in rural areas (World Bank 1999). Ninety-three percent of very poor households and 89 percent of poor households are in rural areas. Rural poor employed in the nonfarm sector tend to be better off than those whose primary employment is in the farm sector. This situation implies that the promotion of off-farm income sources—such as fisheries, livestock, and forestry—constitute a potentially attractive policy option for addressing rural poverty (World Bank 1999).

This case study seeks to integrate economic and social analyses to assess the impact of new vegetable and fish technologies on poverty and vulnerability

1. One careful study asserts that men both receive more nutrients than women and expend more energy; that is, they are nutritionally taxed more than women (Pitt, Rosenzweig, and Hassan 1990). This explanation still does not account, however, for the pro-male bias found in intrahousehold distributions of nutrients for children. Walker and Ryan (1990) show that in Indian villages the daily wage premium that working men with higher body mass indices receive does not nearly cover the costs of extra food consumption required to sustain their additional energy requirements for work.

in Bangladesh. The study is distinctive in that it draws on data that include both traditional economic measures (for example, household income, profits from farm production, nutrition outcomes, food expenditures) and those that are more social in nature (for example, social connectedness, empowerment, institutional structures). Drawing on both types of information provides a more integrated and holistic view of rural livelihoods.

The research combines survey data collected in 1996–97 with focus group discussions and semistructured interviews conducted in 2001. Elements of the sustainable livelihoods (SL) framework were used to help frame specific research questions, devise a qualitative data-collection strategy that would generate new insights into the existing data, and orient the collection of supplementary data on a range of new issues, such as technology dissemination pathways and the wider social context.

This study follows on earlier research undertaken by the International Food Policy Research Institute (IFPRI) and its partners in Bangladesh. Data were collected in 1996–97 to examine the effects of the adoption of new vegetable varieties and polyculture fishpond management technologies on household resource allocation, incomes, and nutrition. Data were also collected for individuals within households, therefore allowing analysis of gender-related issues.

Households were surveyed in three sites where nongovernmental organizations (NGOs) were active in disseminating technologies developed by international agricultural research institutions. These sites were (1) Saturia *thana,* Manikganj district (referred to below as Saturia); (2) Gaffargaon *thana,* Mymensingh district, and Pakundia and Kishoreganj Sadar *thanas,* Kishoreganj district (referred to below collectively as Mymensingh); and (3) Jessore Sadar *thana,* Jessore district (referred to below as Jessore). The technologies and the modes by which they were disseminated differ by site, as indicated in Table 4.1.

Vegetable Intervention

In Saturia, credit and training in small-scale vegetable technology is provided to women who grow vegetables on small plots on or near the household compound. The vegetable varieties were initially developed at the Asian Vegetable Research and Development Center (AVRDC), based in Taiwan, and then adapted to Bangladesh conditions at the Bangladesh Agricultural Research Institute (BARI). The vegetables were introduced through the small NGO Gono Kallayan Trust (GKT). GKT has been operating in Saturia since 1987. In March 1994 GKT added vegetable production using AVRDC/BARI seeds to their portfolio of income-generating programs. Selected GKT extension agents received training at AVRDC sites outside Bangladesh. The improved vegetables introduced include tomato, okra, Indian spinach (*pui shak*), red amaranth (*lal shak*), radish, eggplant, amaranth (*data*), kangkong (*kalmi shak*), mung bean, and sweet gourd.

TABLE 4.1 Study sites, technologies, and approaches

Site	Saturia (five cases, five comparison villages)	Mymensingh (fourteen cases, seven comparison villages)	Jessore (eight cases, eight comparison villages)
Community characteristics	Less than two hours northwest of Dhaka; some access to Dhaka markets; high levels of NGO activity; low-lying flood-prone area	Four to five hours north of Dhaka; remote and socially conservative; little NGO activity; not flood-prone; some water shortages in dry season	Close to western border with India; less socially conservative but politically volatile
Agricultural technology	Privately grown vegetables	Privately operated polyculture fishponds	Group-operated polyculture fishponds
Institution originating technology	AVRDC	ICLARM	ICLARM
Dissemination approach	Training and credit to all adopters	Training to all adopters; credit to poor adopters	Training to some members of each adopter group; credit to all group members
Type of disseminating institution	Small local NGO	Government ministry extension program	Medium-sized local NGO
Target group	Women NGO members in households with marginal landholdings	Individual or joint pond owners	Poor women, NGO members, predominately landless

NOTES: AVRDC, Asian Vegetable Research and Development Center; ICLARM, International Center for Living Aquatic Resource Management; NGO, nongovernmental organization.

Fish Interventions

The International Center for Living Aquatic Resource Management (ICLARM—now known as WorldFish Center), with headquarters in the Philippines, began providing technical advice to the Fisheries Research Institute (FRI) in Mymensingh in 1988 in regard to polyculture fish production and other fish culture technologies. Seven fish species were promoted: silver fish, carp (*katla*), rohu (*rui*), mrigel, mirror carp, *sharputi,* grass carp; black fish (*kalibouch*), shrimp, and tilapia are also cultivated.

In Mymensingh, polyculture fish production was undertaken in privately held, single-owner fishponds. The Mymensingh Aquaculture Extension Project (MAEP) began operating in July 1990 and was jointly implemented through MAEP extension agents and Ministry of Fisheries extension agents. They provided training to better-off households and training with credit to poorer households. The intervention was directed at both men and women, but in practice to men more often than women.

In Jessore, the NGO Banchte Shekha arranged long-term leases of ponds, which were managed by groups of women who received credit and training in polyculture fish production methods. Banchte Shekha extension agents received training from both ICLARM and FRI personnel in pond management for polyculture fish production starting in 1993.

Table 4.2 shows results of a census of households in each site on the extent of adoption of the target technology just before the household surveys began. Although the percentages are not negligible, the time experienced using the technologies was short when the survey began in both the vegetable and the group fishpond sites (Saturia and Jessore, respectively). They had only been available to the disseminating institutions for two to three years in these sites; thus experience for any particular adopting household would have been for an even shorter period.

The household survey collected data in three sites at four different months and covered one complete agricultural cycle in 1996–97. The study design included two village types in each site: (1) case villages (those where the improved agricultural technology had already been introduced by the disseminating institution), and (2) comparison villages (those where the technology had not yet been introduced but where the disseminating institution planned to eventually introduce it).

In both types of villages, the disseminating institution delivered all the same supporting services (mainly microfinance). It is important to note that in each site, case and comparison households in both village types were affiliated with the same NGO and undertook the same agricultural activities, but those in comparison villages did not have access to the improved technologies. Although the interventions were not randomized to villages, a comparison of

TABLE 4.2 Study sites and extent of adoption

Site	Saturia	Mymensingh	Jessore
Technology	Vegetables	Private ponds	Group ponds
Adopters as percent of households in case villages	40	50	16
Year technology introduced	1994	1990	1993
Survey inception year	1996	1996	1996
Elapsed time between introduction of technology and beginning of household survey (years)	2	6	3

village characteristics indicated few significant differences between case and comparison villages in infrastructure and access to services (see IFPRI-BIDS-INFS 1998).

The survey sampling methodology was designed so that households with and without access to the technologies, but otherwise similar in nature, could be compared. Access to the technologies was determined at the village level, so that all NGO members in case villages, whereas none of the NGO members in comparison villages, had access to the technologies. Differences in incomes and various other livelihood outcomes between these two groups—"with access and adopting" and "without access but wishing to adopt"—indicate the impact of access and adoption (IFPRI-BIDS-INFS 1998). This sampling design involved careful selection of comparison households in comparison villages: each expressed a demand for the technology and had similar physical capital (land, buildings, and livestock) and human capital (education and experience) as case households in case villages. The selection process involved a census of households in both village types. In case villages with access to the technology, adopting and nonadopting households were interviewed. In comparison villages, households were divided into two groups based on answers recorded in the censuses: those likely to adopt (all NGO members likely to adopt) and those not likely to adopt (non-NGO members plus NGO members not likely to adopt). Likely-adopter households were randomly selected from the first group. Site-specific conditions required unique sample selection methodologies in each case (see IFPRI-BIDS-INFS 1998 for more details). Sampling weights were calculated and used in the analysis to account for each household's probability of being selected for the survey.[2]

2. Site-specific details of the sampling methodology are available upon request from www.ifpri.org.

For each type of household, the study collected detailed information on production and other income-generating activities by individual household members; expenditure on food, health, and other items; food and nutrient intakes for each household member; and time-allocation patterns and health and nutritional status of individual household members (Table 4.3). Survey data were supplemented with qualitative research undertaken between survey rounds 3 and 4 on factors affecting intrahousehold bargaining and outcomes. Insights from the

TABLE 4.3 Topics covered by survey questionnaires

Topic	Explanation
General household information	Demographics, education, migration
Parcels of land	Ownership, tenure relations
Agricultural production	Steps in production, record input use, output, post-harvest processing, disposition of output including revenues from sales, loans, past production history
Agricultural wage labor by household member	By crop, by task
Other sources of income by household member	Nonagricultural employment and transfers
Backyard livestock and vegetable production	Livestock, fruits, vegetables
Asset ownership by family member, dowry, inheritance	History of assets at marriage, current assets
Women's autonomy, mobility, decisionmaking	
Credit use	
Food expenditures	One-month recall
Nonfood expenditures	Four-month recall
Source of water; food preparation; preschool feeding practices	
Reproductive history	
Health services; nutritional knowledge	
Time allocation of head man and woman and children under 10 years of age	Twenty-four-hour recall
Anthropometrics and recent morbidity	Two-week recall
Individual food intake	Twenty-four-hour recall
Blood analysis; clinical signs of micronutrient malnutrition	
Chronic illness history; use of health infrastructure	

qualitative analysis fed into formulation of questions in the fourth survey round on dowry, assets brought to marriage, and bargaining power.

The richness of the existing data and the variety of agricultural technologies and dissemination strategies provide an excellent base for further study—supplemented by additional qualitative data collection—of the linkages among agricultural production, livelihood strategies, nutritional status, vulnerability to shocks, and empowerment of women.

Previous analysis of these data (IFPRI-BIDS-INFS 1998) revealed that although vegetables and fishpond production both gave higher rates of return than rice production, the production of high-yielding varieties of rice, rearing of livestock, and off-farm activities were larger sources of income than were vegetable or fishpond production. Vegetable production (both AVRDC and others) in adopter and likely-adopter households in Saturia contributed only 2.5 percent and 2.1 percent, respectively, of total household income. The marginal effect of adoption of AVRDC-improved seeds compared with other improved and local seeds would seem therefore to be less than 1 percent of total household income. Several possible explanations are put forth in IFPRI-BIDS-INFS (1998): (1) rice can be grown virtually all year (subject to availability of water), whereas vegetables do not grow well during periods of heavy rain and high temperature; (2) the risk of growing vegetables was high in the project area, as it is highly subject to flooding; (3) rice can be stored whereas vegetables must be marketed immediately, and because vegetable prices are more variable, their production is riskier. The report also found that producers of AVRDC vegetables do not consume larger amounts of these and other vegetables in total compared with likely-adopter households. Determining the constraints to expanded vegetable production in this demonstration project was identified as a question meriting further inquiry.

In Mymensingh, fishpond production accounted for 9.9 percent and 5.4 percent of total household income in adopter and likely-adopter households, respectively. If the case-comparison sampling scheme is valid, the difference between the two figures, 4.5 percent of income, represents a rough estimate of the marginal effect of applying the polyculture management technology to existing fishponds. The adoption of the polyculture technology was associated with greater consumption of large fish, but not of fish consumption in total—there was apparently a one-for-one substitution of large fish for small ones. Although the magnitude of this substitution is small (4–5 calories per day per capita increase in large fish consumption and a corresponding decrease in small fish consumption), small fish are more nutritious gram-for-gram than large fish, so that the impact on dietary quality was negative.

In Jessore, only five of nine group fishponds surveyed were operated as intended under the program. In two of the four cases of nonoperation, excavation of ponds was not undertaken or excavation was inadequate. Two groups leased out their ponds as a consequence of intragroup disagreements on how to operate the pond and share the output. Cash profit per acre over the 16 months of the surveys for the five group-operated ponds was about Tk 17,500 (US$168 per hectare; marginally better than fishpond profits of likely-adopter households in the private fishpond site of Mymensingh).[3] However, average cash profit per month per group member for these five ponds was estimated to be only Tk 16 (US$0.38), a modest sum due to the large size of the groups. This was, however, income earned directly by the women themselves, and presumably more under their own control. Mean own-consumption of fish among adopting households in this site did not differ statistically from that of likely-adopter households. These findings raise questions of why there was not more apparent impact or greater uptake of the technologies. It is possible that certain non-income gains (such as increases in social capital or empowerment) were realized but not identified in the original evaluation studies. Another possibility is that at the time surveys were fielded, the adopting households may not have had a sufficiently long enough period to comfortably incorporate the new technologies into their livelihood portfolios. Yet another scenario is that the technologies were not well suited to the target households' vulnerability or political/institutional contexts. To investigate these issues in more depth, the current study seeks to examine how the technologies were introduced and how they fit into overall livelihood portfolios, especially of poor households. In particular: Were the new technologies perceived as being riskier? Did they increase the vulnerability of households? Did the additional labor the technologies required conflict with the pattern of other livelihood activities?

Research Questions

The SL framework was used in the study to help organize the main research questions. It was applied as a means of widening the understanding of poverty and drawing together the various perspectives of social and economic analyses to undertake a broader poverty impact assessment. Given the overall concern with understanding the effects of technology on poverty, and the factors structuring these effects, the research focused on the following questions:

- How do we understand the overall vulnerability context, and what is the relationship between adoption of the new technologies and household vulnerability?
- What are the relationships between household assets, technology adoption, and livelihood strategies? What are the constraints on adoption?

3. Cash profit per acre was calculated as revenue minus the costs of hired labor and purchased inputs; imputed value of family labor and own inputs were not included in this definition (IFPRI 1998).

- What are the transforming effects of intervening organizations and institutions? How do the dissemination approaches of the NGOs and public sector agencies involved affect livelihood strategies?
- How are decisions taken within households concerning choices of livelihood strategies, and how do the agricultural technologies fit with these strategies?
- What outcomes can be used to assess the direct and indirect effects of technology adoption on adopting and nonadopting households?

Vulnerability Context

In rural Bangladesh, poverty is pervasive and associated with high vulnerability. A number of factors influence the vulnerability of households in our study sites. These include features of the natural environment (lowland flooding versus seasonal water shortages), lack of access to existing natural resources because of poverty or social isolation, lack of availability of agricultural technologies and the inputs to effectively use them, food shortages during lean months coupled with already low nutritional status, lack of access to insurance mechanisms due to weak social networks or lack of physical assets to use as collateral, and lack of decisionmaking power among women.

These factors, along with the method and quality by which the technologies are introduced and supported, influence whether and which agricultural technologies are used. In particular, new technologies may be unproven in the local context and thus perceived to be so risky as to increase the vulnerability of already at-risk households, or the risks may co-vary with other existing sources of risk. Adoption may also influence vulnerability by changing assets controlled by households or by certain types of individuals within them. Successful adoption may, however, increase physical and financial assets, so that food and livelihood security are enhanced. Other types of assets could also be affected. For example, increased human and social capital may result from the technical training and group involvement that delivery of interventions often entails. Less successful adoption could result in loss of physical or financial assets and even negative social capital if conflicts arise in the delivery or application of the technology. The introduction and use of technologies may affect vulnerability by changing the transforming structures and processes that influence access to various assets and livelihood strategies.

Assets and Technology Choice

The asset situation of households influences adoption and choice of technologies by changing access to resources. If the poor lack the ability to obtain the inputs required to use a technology, it is unlikely they will adopt it. Direct ownership of such assets as land and agricultural equipment is a key component; however, other types of assets that one does not necessarily own may bolster access to needed inputs. Membership of poor women in an NGO that arranges leases of fishponds and organization of women into groups to manage these fishponds are examples of social assets influencing the adoption and successful utilization of the technology. The research investigates how various assets influence adoption of the different technologies.

Transforming Structures

Research and extension systems that have inadequate information flows, adverse (for example, top down, nonparticipatory) incentive structures, and overly complex organizational structures can thwart the effective design and implementation of even technically sound interventions (Lewis 1991). In this study we investigate the effectiveness of alternative pathways of dissemination (government and NGOs) in reaching the poorest households. In particular, are NGOs more effective than government in service delivery?

We also investigate how well programs are targeted. For example, did the strategy of organizing fishpond groups reach the poor more effectively than targeting households with sole ownership of fishponds? Particular attention is given to the role of gender relations in the adoption of technology, and how the effects of the program vary by gender.

Class and caste relations, market organization, and governance are also relevant, but they are not analyzed in detail here. Our focus is rather on those structures more immediately subject to the influence of external agents.

How Technologies Fit into Livelihood Strategies

That off-farm activities and production of high-yielding varieties of rice and livestock were more important sources of income (in that both had higher per unit profitability) than vegetable or fishpond production raises questions about how the different agricultural technologies fit into the overall livelihood portfolio, especially of poor households. In particular, do patterns of time allocation shift and conflict with other activities? What are the gender implications of such shifts? What do households give up or gain by adopting the technologies?

Outcomes

We investigate a series of outcomes that adoption of the technology is expected to influence. Household survey data from 1996–97 are used to assess the effects of technology adoption on income, consumption, nutritional status, and empowerment of women. We also use qualitative data collected in 2001 within the context of focus groups. Separate focus groups were held for men and women from different wealth levels. Information on household background characteristics from the 1996–97 survey was used to recruit particular types of individuals to participate in these focus groups. Using this method for focus group recruitment enabled us to attribute particular qualitative information to individuals from particular types of households.

Although adoption of technologies can have important indirect influences on outcomes for both adopting and nonadopting households, we do not expect to find indirect effects in the survey data because (1) the prevalence of adoption and length of experience with applying the technologies in the case villages was still rather short at the time of the household surveys; and (2) even for adopting households, the contribution of the technologies to their overall income portfolio was quite small. The focus groups, however, provide more opportunity to probe for indirect effects of the technology, as they were conducted several years after the household surveys.

Methods

This research combines an existing quantitative study with the collection of new qualitative data in the three study sites. We found that the existing household survey data could only go so far on certain issues, creating a need to follow up with more qualitative data collection.

Methods for Further Analysis of the Existing Household Survey Data

Two primary issues were addressed using the existing survey data: (1) the effects of assets on technology adoption; and (2) the relationships among technology adoption and expenditure, income, empowerment of women, and nutritional status.

To explore associations between technology adoption and livelihood outcomes, we first examine whether villages that received the technology differ from the comparison villages. We then investigate possible differences in long-term livelihood assets of adopter and likely-adopter households. If there are few observed differences found at either the village or household level, then comparing mean livelihood outcomes of adopter (case) and likely-adopter (comparison) households may give a reasonable indication of the impacts of adoption of the technology. If, however, case and comparison villages differ or if adopter and likely-adopter households differ, then multivariate methods to control for having access to the technology at the village level, and using the technology at the household level, are necessary. These issues are explored below.

Methods for Supplementing Existing Quantitative Data with New Qualitative Data Collection

Because the original study did not focus on the broader concepts of poverty or livelihood strategies, the existing data are supplemented using data collection that is both qualitative and participatory, making use of selected participatory rural appraisal (PRA) techniques. For example, at the time the survey was done, women adopting the new vegetable technologies in Saturia reported that they were working the same number of hours as before they adopted the technology

(IFPRI-BIDS-INFS 1998). Did this mean that the new vegetable types required little additional labor input, or did it mean that some other kind of activity was being displaced? There were also issues that were clearly important but could not be addressed with the quantitative data in hand, such as the need to record the views of local people on local services. A particular institutional process on which we needed more information, given the context of the study, was the issue of dissemination pathways, such as the question of the effectiveness of NGOs and government agencies as technology providers.

The qualitative data collection employed focus groups as its main approach, which combined discussion around a common set of questions with the selective use of certain group-based PRA techniques—chiefly those of seasonality mapping and the ranking of priorities. In each location, the focus group data were followed up where necessary with semistructured interviews with key informants, such as NGO staff. Some researchers view qualitative and PRA as the same; however, for us the PRA techniques yielded both quantitative and qualitative data, as in the case of the ranking exercises.

A pretest for the participatory data collection was organized in Saturia in January 2001, in which one focus group discussion was held. This event provided an opportunity to refine the questions further and to train the fieldwork team.

Working with the SL Framework

The main strength of the SL framework is that it allows systematic analysis of the range of social and economic forces affecting how members pursue livelihood improvements. However, there are also certain limitations to the SL framework. First, its conceptual inclusiveness and complexity can make it difficult to operationalize, particularly at the level of practice. Second, understanding power relations remains difficult within this framework (and in many others). Third, linking the global and the local in understanding how wider polices and economic forces—such as export policies—can affect household-level strategies remains a challenge (Kanji and Barrientos 2002).

The open-ended nature of the SL framework meant that clear lines had to be drawn around the types of data we would collect and the level at which we would collect them. As described below, we assessed which study questions and elements of the SL framework we could address using the information already available in the survey. The types of information missing were mainly on the vulnerability environment and process and institutional factors. The data collected in the focus groups addressed how the agricultural technologies affected vulnerability, fit into livelihood strategies, and affected selected livelihood outcomes. We did not use supplementary participatory analysis to examine how the technologies affected livelihood assets or transforming structures and processes, because the time frame for the participatory data collection was not long enough to investigate these issues.

Drawing on the World Bank NGO Working Group Participatory Poverty Assessments (PPA) (Nabi et al. 1999) and other related studies, such as the Poverty Alleviation through Rice Research Assistance study, we developed three categories of households for comparative qualitative data collection. Household characteristics reported in the PPA helped us to roughly stratify households in the 1996–97 survey data. We defined poor (that is, the category termed "social poor" in the PPA) and very poor (those people termed "helpless poor" and "bottom poor" in the PPA). We have also included a single "non-poor" category (those termed as "rich" and "middle" in the PPA) in the study, so that we can examine the position and perspectives of better-off households for a comparison with those that are poor. The rough breakdowns are given in Appendix Table 4A.1. Because the household survey was choice-based and designed to oversample technology adopters and likely-adopters, the profiles of sample households in each site reflect site-specific targeting priorities and differences in livelihood assets needed to adopt the different technologies offered. Hence households in the group fishpond site of Jessore are more likely to be poor, whereas those in the private fishpond site of Mymensingh are less poor.

What New Data Were Needed?

As well as the problem of the SL framework's open-ended nature, there was the additional challenge that the framework needed to be grafted onto an existing study. We took the view that this challenge represented an opportunity rather than a constraint. The combination of an existing quantitative study and the conceptual insights generated by the SL framework provided the means to generate a set of new research questions (to both supplement and complement the quantitative data) that could be addressed through further qualitative research. Table 4.4 illustrates the ways in which different types of data and data collection methods used here addressed the research questions. This approach provided a framework in which the integration of new qualitative data and existing and new quantitative data could take place.

In each site, three sample villages were drawn for further study. One was chosen randomly from among remote villages as far from the main road system as possible. A second was chosen randomly from among accessible villages, close to the road and with good communications. The third one was randomly selected from the "middle ground." This sample allowed us to compare villages with different levels of infrastructure, information, and market access. The focus groups were held in the case villages (where adoption of technology will be relatively advanced) but not in the comparison villages.[4]

4. Unfortunately, we could not explore when and to what extent the technologies had in fact been disseminated in comparison villages after the survey. Without a better handle on those factors, it seemed difficult to be able to study comparison villages; hence focus groups were done only in case villages.

TABLE 4.4 Matching data sources to research questions

Vulnerability	Assets	Strategies	Dissemination pathways	Outcomes
Qualitative* (focus groups)	Qualitative	Qualitative* (focus groups)	Qualitative* (mainly semistructured interviewing)	Qualitative* (in terms of people's perceptions drawing on PRA)
Quantitative	Quantitative* (strong survey data)	Quantitative	—	Quantitative* (strong income and nutrition data)

NOTES: * indicates strong data, in terms of the relative strengths of quantitative and qualitative data methods in relation to livelihoods. —, Not available; PRA, Participatory Rural Appraisal.

Each focus group consisted of 6–10 people, and each contained a uniform group structured by well-being category and gender. Households that had participated in the 1996–97 survey were classified into the well-being categories described above using the survey data: non-poor, poor, and very poor. These classifications served as the basis for contacting households to participate in focus group discussions. Their members and members from similar types of households in terms of well-being status were invited to participate in a gender-specific focus group discussion. There were therefore six types of focus groups in each village studied, for a total of 54 focus groups (3 sites × 3 types of case villages per site × 6 focus groups per village).

The approach was generally successful, although fieldwork took longer than expected due to the problem of *hartal* (a general political strike) stoppages. There was also variation in accessibility and openness to research across the three districts, with Mymensingh being the most socially conservative.

There were several important methodological and practical lessons to be learned from the research experience. The first was that we underestimated the logistical complexity of convening focus groups of this kind, where different categories of busy people in frequently remote village locations had to be convened. Keeping a focus group discussion within the broad range of issues we had targeted was challenging for organizers, especially when such discussions sometimes attracted interest from other villagers and passersby. It was also complicated by the reliance on local consultants—who were better acquainted with administering quantitative data collection than with the requirements of this kind of qualitative research—to identify and convene the groups. A tendency among some of the collaborators to simply equate qualitative research with PRA was also a complication in the qualitative research process. We were interested in participatory approaches to all kinds of data collection in the study, but we also wanted to combine as innovatively as possible certain PRA skills with more formal research methods, such as the use of semistructured interviewing guides.

This study in part built on ongoing research work, and the addition of new agendas and activities inevitably generated problems that might have been avoided if the study had been conceived as a new stand-alone activity. For example, there was a time lag between the original quantitative data collection work and the design and implementation of focus groups and semistructured interview work. In the case of Mymensingh and Jessore, the lag was four years. This delay meant that we were unable to examine in detail the changes that may have been implemented by ICLARM in other areas based on lessons learned from some of our cases. The time lag between the quantitative and qualitative studies also reduced some of the potential synergies from integrating the methods.

Findings on the Vulnerability Context

Qualitative Findings

The qualitative data helped reveal both material and nonmaterial aspects of vulnerability. The focus groups highlighted the importance of a range of broad aspects of vulnerability in addition to the obvious lack of material assets, such as land or cash, or vulnerability to fluctuating markets. These include:

- Female subordination or dependence on male household members (for example, for the sale of products they have produced or for permission to participate in fish production training);
- Lack of technical knowledge about vegetable or fish cultivation, creating perceptions of high risk or disappointing yields;
- Law-and-order problems (threats of violence to minority households at times of social tension, which can lead to forced sale of land; theft of fish from ponds; or malpractice by officials, staff, or group leaders);
- Low levels of trust in a government or NGO service (sometimes after evidence of malpractice) or in fellow members (as in the case of some of the fisheries groups); and
- Lack of access to justice (the non-poor may forcibly prevent poor from taking part in certain activities, or they may take over profitable activities).

There was wide variation in the general vulnerability context among the three study sites. Saturia is the poorest overall of the three areas, despite being closest to Dhaka. Mymensingh is relatively well off, with agriculture supplemented by business and services. However, there is a severe water shortage during the dry season and a relatively high degree of social conflict over such issues as land and marriage. Compared with Saturia, village women in Mymensingh are less mobile and *purdah* is observed more strictly. The research team therefore found it difficult to get permission from husbands and religious leaders for non-poor women to participate in the focus groups. Jessore is the least conservative area of our study and is reasonably prosperous. Despite this advantage, there is a high level of social and class tension, which produces a high level of fear and insecurity among the poor.

Although these general village-level variations in vulnerability context were significant, the variations in vulnerability among different social categories within each site were greater. Therefore we break vulnerability down into two aspects, as elaborated in the following sections.

Social and Political Dimensions of Vulnerability

The non-poor households in Saturia reported less vulnerability because they have access to cash and extensive kin support networks to assist with cultivation. It was also reported that some poor and very poor adopters of vegetables distribute produce to family and neighbors as a way of building and maintaining social solidarity. One very poor woman in Saturia remarked, "we distributed vegetables among our family and other relatives, and we also gave them to those among our neighbors who have not grown vegetables." This observation was a key insight from the qualitative research that was not apparent from the quantitative data. It was interesting to note gendered differences within focus group discussions on this issue, where men and women can be seen to place different values on goods and transactions. A male poor group member did not see value in distributing vegetables to relatives: "How can we give things of low status as gifts? Vegetables should not be given to the father-in-law's house."

Vulnerability in this area is also a function of membership in the wider community. In this area of Saturia, there is a substantial Hindu minority, some of whom reported discrimination. Many Hindus are found among the poor and very poor categories.

There is an important gender dimension to vulnerability. For women in all sites, movement between private and public space is problematic. Poor women make an important distinction between outside work (*bairer kaj*) as paid, and inside work (*ghorer kaj*) as household work, which is unpaid. The women combine a range of activities, such as paddy husking, producing and small-scale trading of *mourri,* and sewing the traditional *katha* (a Jessore local specialty)—all hard work for small returns.

There is a perception among the poor that their plight is ignored by those who are better off. One of the male poor group members said that many people have a good economic situation but that few of the rich ever help the poor. "In this area the overall situation is not so bad. But he who has has, he who has not has not. Because of self-interest, the rich do not bother to uplift the poor."

The poor are disproportionately affected by law-and-order problems. There is the perception of an increasing crime problem. We were told that although people may know who the criminals are, there is a culture of fear, and it is dangerous to try to do anything about law-and-order problems. The poor women explained that nobody speaks out unless they want more trouble.

Adoption and Vulnerability

Adoption of agricultural technologies can reduce vulnerability through increased income, strengthening of social relationships, and strengthening of self-confidence and problem-solving capabilities at the individual level. The non-material side to vulnerability is also useful in highlighting the ways in which very poor and non-poor can successfully use the technologies to build social relationships (for example, distributing vegetables to friends, neighbors, and patrons) to build both horizontal and vertical ties that can reduce their vulnerability.

The study found that economic and social empowerment generally follows adoption of these technologies, but that vulnerability can be reduced or increased independently of increases in income. For some people who do adopt, new forms of vulnerability can arise related to the technology. This is particularly true in relation to fish culture. In Mymensingh, for example, it was reported that fish polyculture carries some distinct vulnerability problems of its own. Fish can be stolen, poisoned, or suddenly stricken by disease. They are highly perishable and need to be sold quickly if they are grown in seasonal ponds. In Jessore, group-operated fish production was found to be subject to the same problems, with added social dimensions of mistrust within groups, and "principle and agent"-type incentive problems between poor groups and non-poor pond owner-leasers. This problem of postadoption vulnerability can disadvantage women adopters, who may find themselves working harder to produce vegetables or fish but have no direct access to the market or control of the cash profits.

In the case of vegetables, the study revealed that the technology was relatively easy to adopt and unlikely to increase vulnerability, because these were cultivated on homestead land where security was easy to ensure and access did not bring a time cost. Nor was there likely to be a displacement of other crops, as homestead land tends to be unutilized for cultivation. Failure of vegetables does not, therefore, imply the loss of other income-earning opportunities. Moreover, the ability to produce vegetables within the homestead was deemed attractive to women and their families because this activity brought less vulnerability to harassment and loss of reputation than working outside.

Adoption can therefore both increase and reduce vulnerability—but the general picture is that vegetables were relatively easy to adopt (compared to fish) and were unlikely to increase vulnerability. In general, we found that it was a difficult task to collect a wide range of qualitative data on the vulnerability context. It was easier for people to talk about the impact of the 1998 floods (in the sense of vulnerability to natural hazards) but more difficult to discuss social vulnerability in the focus groups. Focus groups may not be the most effective means for the collection of this kind of information, due to its sensitivity.[5] Also, this difficulty perhaps reflected a tendency for people to recall only dramatic episodes and events of vulnerability rather than systemic or the everyday experience of it.

Assets, Adoption, and Organization

Quantitative Findings Using the Survey Data

Purposive placement of interventions is a concern when assessing the impact of programs such as these. If technologies are disseminated to areas that are either more prepared to benefit from their availability or are in greater need of them, comparing outcomes for areas with and without the technology may result in misleading conclusions about program impact. There are two predominant approaches to dealing with this potential problem: (1) a fixed-effects (difference-in-difference) estimator or (2) an instrumental variables approach. Both depend on data availability.

The first approach (for example, Pitt, Rosenzweig, and Gibbons 1995) tests whether changes in outcomes are greater in areas where there are greater changes in program coverage net of changes in individual-, household-, and community-level factors. This approach could not be used because we do not have information on differential program exposure at two points in time. Comparison villages were without the technology for the entire survey period, and changes in program exposure in case villages over the survey period were not measured (because the surveys covered only a single agricultural year). The second approach uses instrumental variables methods where particular variables are hypothesized to influence the outcome of interest only indirectly via their effects on program exposure. We experiment with this method, using distance from the office of the technology-disseminating institution to each village as the instrumental variable in the first-stage technology recipient probit equation presented in Appendix Table 4A.3, the hypothesis being that villages closer to the office may have been chosen for earlier introduction since they were easier to access. The predicted value of village-level technology recipient status from this probit equation will be used in the livelihood outcome equations pre-

5. Although we were not able to test this hypothesis conclusively, evidence from other recent research (for example, Kaplowitz and Hoehm 2001) suggests that individual interviews and focus groups tend to produce different, often complementary, types of information.

sented in Appendix Tables 4A.7 and 4A.9 (with standard errors corrected using the bootstrap method).

Appendix Table 4A.2 presents a comparison of long-term characteristics of case (technology-recipient) and comparison (technology-pending) villages at the time of the survey. Although the number of villages is small for this type of comparison, we find no statistically significant differences by village type in distance from the technology-disseminating institution. Among other longer-run village characteristics—unlikely to be due to the effects of the technologies—there are a few significant differences, and most of these appear to be driven by differences in Mymensingh, the private-owner fishpond site.

The determinants of whether a village was a case (technology-recipient) versus a comparison (technology-pending) village were further investigated with a probit regression, where village characteristics described above were used as explanatory variables. Given the small number of villages per site, site-specific village regressions could not be undertaken. As shown in Appendix Table 4A.3, neither the distance from the village to the technology-disseminating institution office nor any of the other village characteristics were found to be significant at the 5 percent level or better. With few systematic differences between the villages that were and were not receiving the technologies at the time of the surveys, concerns that nonrandom village-level placement of the technology could bias our comparisons of otherwise similar households are allayed. Even with these results, however, we still present means comparisons of households with and without access to the technology, as well as multivariate results where predicted village status is included as a regressor.

To more fully ensure that households with and without technology access (adopters and likely adopters) are in fact similar, we compared their asset positions at the first survey round. As presented in Appendix Table 4A.4, adopter and likely-adopter households have similar asset holdings. Where differences exist the patterns of advantage are not unidirectional. Using these assets as multivariate determinants of being an adopter versus a likely-adopter household, as presented in Appendix Table 4A.5, reveals that over all sites, households with higher valued houses and better educated adult males are more likely to have access to the technology. Similar to findings at the village level, these effects are driven by the Mymensingh site. In Saturia, there are no statistically significant differences, whereas in Jessore the only significant differences between actual and likely adopters are that the former group is more likely to have adult males with some university education.

Qualitative Findings Based on the Focus Groups

The focus groups suggest that membership in NGOs and other organizations is weighted toward the poor, but that asset ownership/power also allows some non-poor households (but not the wealthiest) to become NGO members. At the same time, there are some very poor households who find themselves excluded from NGO membership because they are asset-poor (for example, some report that they may be unable to keep up with loan repayments or do not have necessary collateral assets or documentation).

It was also reported that lack of social connections contributes to isolation for the very poor, which makes it difficult to become part of an organization. Lack of education can also make poor people unconfident about joining an organization. In the case of government extension, status issues make it harder for very poor and poor, and the women in particular from those groups, to gain access to public services. NGOs in general are better at overcoming these barriers.

A certain level of material and nonmaterial assets is a precondition for adoption. It was striking that the poor tended to have the widest range of livelihood strategies, whereas the very poor and the non-poor had fewer.[6] Lack of access to financial resources is, as might be expected, a key element of vulnerability. The male very poor in Jessore said that they could not easily reduce their vulnerability without access to cash. Although money cannot solve all problems, it can solve many of them, they said. Credit is therefore very useful. They said that if they cannot maintain their basic household expenditure, how can they be expected to expand into fish production? First, money is needed, then advice and information.

For many of the poor, financial vulnerability makes it unlikely that they will be able to adopt new technology. This barrier was apparent in focus group discussions regarding microcredit services from NGOs. For very poor people, the pressure of taking a loan that has to be strictly repaid in weekly installments and that demand regular group meetings can act as a disincentive to adopt technology.

Other recent studies (for example, Hulme and Mosley 1996) have pointed out that these technologies—and NGOs/credit services in general—cater most effectively to the poor rather than to the very poor. This finding is supported by the focus group data.

However, lack of adoption is also attributed to other factors, including lack of access to an NGO *samity* or group (due either to lack of availability or a reluctance to join); for women a reluctance to go outside the household; and lack of access to land or a pond. The problem of a lack of control over irrigation water was cited by members of a male poor group in Saturia because the government Power Development Board that controlled the local tube well cut the water supply after the rice-growing season ended. This policy made it difficult to obtain water for vegetable growing.

A lack of both material and nonmaterial assets was shown to be significant and interrelated in constraining household choices. For example, one very poor

6. This pattern is consistent with that noted by Reardon, Berdegué, and Escobar (2001) for Latin America.

group member in Saturia said, "we have no land so we can't do anything. If we had some land, then we would cultivate vegetables." Another female very poor group member from Jessore said, "since we were very poor at the time it started, we could not get involved with the *samity.*"

Disseminating Institutions and Targeting

The main finding is that in all three communities the poor generally held a more positive view of nongovernmental actors than of governmental ones; the latter were seen as remote and sympathetic only to the interests of the rich.[7] However, people saw a marked difference among various NGOs and observed that NGOs vary considerably in competence, integrity, and operating style. NGOs disseminating technology for adoption by individual households met with more success than those promoting group-based or collective adoption. In relation to targeting, it was found that NGOs do reach the poor relatively effectively. However, many of the very poor tend to be excluded due to lack of resources, and there are many cases of non-poor members participating in NGO groups.

Saturia: Vegetable Production and Sales

In Saturia, recent infrastructure improvements have made vegetable sales more profitable, with new roads reported by several informants. However, there was almost no contact reported in the groups between villagers and any government offices or programs in support of agricultural development, only with NGOs. The AVRDC seeds were originally disseminated by GKT, but this NGO is now seen primarily as a source of credit and only secondarily as a source of vegetable technology, which is also available more widely. Many villagers are now producing and storing their own seeds instead of buying them from GKT, although there are reports that seed quality varies. The consensus seemed to be that although GKT had initially done a very good job of promoting the technology in the early 1990s, it is now less effective. Many people have withdrawn from the GKT program. Some people reported being coerced into taking seeds when they only wanted credit. Others complained they had been forced to contribute to pension savings schemes. There were many complaints about the lack of timeliness in the delivery of seeds and credit. Some informants complained of "rough treatment" or lack of attention from NGO staff. In later interviews with NGO staff, we were told that many of these problems were localized and had been addressed through staff changes. According to some informants, other NGOs, such as the Bangladesh Rural Advancement Committee (BRAC), are now providing better credit and seed services in the area. These observations illustrate the dynamic

7. The exception here was the government family-planning workers, who were regarded with more respect than most other government workers.

quality of NGO service provision over time and the range of perspectives on the effectiveness and responsiveness to local needs of such provision.

Mymensingh: Fish Cultivation by Individuals

In Mymensingh, adopters do not refer to the government MAEP project at all, but instead perceive Danida (the Danish aid agency supporting the program) as the organization that is introducing the technology. Most people do not have much respect for the government's extension services. Even non-poor men say that the *thana* fisheries officer does not provide any services or visit the village:

> There is no government hatchery. There is a government fisheries organization in the district, but it is not active.
>
> The government people are there but they just exist on the record, not for us.
>
> The government officers are just there for their own interests. They sit in their offices but they don't come to us.

Information about fish culture is also gained informally from people involved in the fish business, such as hatchery owners, fishermen, and fish traders outside the Danida project. In this way, some fisheries technology information is being extended informally through private-sector sources.

Jessore: Group-Based Fish Cultivation

In Jessore, the services received from the NGO were adequate at first, but problems had arisen among the group members. The poor women's focus groups reported that these organizational problems made the technology unsustainable, not the technology itself. That the NGO Banchte Shekha leased the pond and then provided training and advice was seen as a good strategy. However, the problem reported by the focus groups was that the staff of Banchte Shekha did not supervise the groups after the initial training and the groups tended to fall apart. As a result the group leaders were able to misappropriate the group funds and exploit members—they were not held accountable to the NGO. The group members then stopped participating. Non-poor men reported that credit, training, and irrigation facilities were all necessary services for modern fisheries. The non-poor women's explanation for group failure was that members did not take the group seriously, only 5 of 21 members were given training by Banchte Shekha, and the group was too big and could not be easily united or cooperative. The different explanations are illustrative of different social categories leading to very different perspectives on technological change; these perceptions influence adoption behavior.

Perceptions of Service Delivery and Targeting by Dissemination Agencies

In general, people are more positive about the role of NGOs than that of government services. In the case of government extension, status issues make it harder for very poor and poor, and females in particular from those groups, to gain access to public services. NGOs in general are better at overcoming these barriers and reaching the poor, but many non-poor households also become members. Furthermore, many very poor households are excluded because of social exclusion, lack of confidence to participate in groups by those with low education, or lack of assets, which makes it difficult to keep up with loan repayments.

Can Agencies Empower the Poor?

Adoption of the technology, where successful, brings empowerment for women in the sense that earning money can increase their decisionmaking power within the household, and—in some localities—create opportunities to move into public space, such as the market, to sell produce. The gains in confidence reported by women NGO group members arises from the solidarity of the group and the added status of being part of an outside organization. There is also a strong demand from the community for more training and other services from NGOs. However, some women report that joining an NGO may have political/social/factional implications and that the NGOs (like the government) are not neutral.[8] Members of very poor female groups reported that "they [NGOs] don't treat us all equally," and "they only give seeds and loans to people with whom they have a good relationship." This unequal treatment may be disempowering. In the group-operated fishponds, lack of adequate NGO supervision is given as a reason for failure and this failure contributed to disempowerment.

Findings on How Technologies Fit into Livelihood Strategies

The main finding is that because the poor are engaged in multiple income-earning strategies, technology adoption needs to take account a range of activities within an overall livelihoods portfolio. Questions of technology adoption therefore need to be understood in relation to their overall fit within these multiple strategies—especially for the very poor, who tend to have the most diversified livelihoods. A second key finding is that adoption is time consuming, but adopters perceive that the return from adoption outweighs the burden of the extra work.

Saturia

In Saturia, where vegetables have been introduced, the fit with women's livelihood strategies is generally a good one among all wealth categories. For small-scale homestead vegetable cultivation, the vegetable technology requires very little land, no real need to operate beyond the homestead, low levels of cash in-

---

8. Given the well-documented broader political context of NGOs, the government, and civil society in Bangladesh, this nonneutrality of many NGOs is probably not surprising, but it is rarely acknowledged by project planners, agricultural research institutions (such as AVRDC or ICLARM), or donors (Lewis 2004).

vestment, and flexibly timed labor inputs and offers a high level of nutritional benefits. Vegetable cultivation can be coordinated with all the many other household tasks relatively easily. But adopters who wish to undertake the cultivation of vegetables on farmland beyond the homestead, and the sale of vegetables by women in the market, are definitely constrained by the public/private space dichotomy. However, there are cases where this dichotomy is being challenged (see below).

Seasonal commitments vary widely among the groups. Poorer women have no savings and therefore need to work steadily to secure income throughout the year. They tend to be less busy in July–August when there is less work available (the rainy season). Compared to poor women, non-poor women have a shorter busy time (October–March) when they are concerned with pre- and post-harvest rice work. The non-poor tend to cultivate a smaller range of vegetables than do the poor, because they do not bother with vegetables that they can easily buy from the market (such as chilies). Instead, they give more importance to their wider household-related work during this period, such as paddy husking, seed preservation, and *kata* sewing for winter.

Mymensingh

In Mymensingh, agriculture used to be the main occupation in the village, but now it has been joined by the new fish polyculture technology as the second most important source of income. Fish cultivation has become a business, providing a source of cash when needed; therefore, it is a source of security for some households. Now that fish production has become a commercial business—even among the relatively few poor men who have adopted it—it is no longer just for consumption.

However, those who cultivate shared ponds (as opposed to ponds owned by one individual) have less access to fish for consumption on a regular basis. Apportionment has to be negotiated with other members of the group, who may decide that in different months only specific individuals can consume fish. This extra layer of negotiation is an important difference between sole- and shared-access cultivation identified by very poor women.

There are strong status reasons why husbands do not want their wives involved in these aspects of fish production. Women would be willing to get more involved if there were not such social pressures, which makes them vulnerable. One poor woman said, "fish cultivation is related to the market, so this is dominated by men, and women cannot talk with the men." Another said if she did not have a husband, she would go to the market, but other villagers would criticize her.

Jessore

In Jessore, the collective fish technology has been less successful, mainly due to the failure of organizational arrangements and lack of trust—at times justified—in relations between NGO staff and beneficiaries (see the section on disseminating institutions and targeting above). The public/private space dichotomy is another constraint on women's room to maneuver, and hence distance to ponds was an important constraint on adoption. One reason for group failure was because there were always group members who were unable to go to the pond. Younger women were compelled to send older household members—such as the mother—to feed the fish and visit the pond.

Although they are aware of the technology, there are still non-poor households that continue with traditional "extensive" fish cultivation for consumption. There is a belief that if modern varieties and techniques are used, the fish do not taste as good. Because they have other sources of income, some non-poor males say they do not therefore have adequate incentives to move into commercial fish production.

Findings on Outcomes

Empowerment

QUANTITATIVE FINDINGS. In the fourth round of the household survey, one year after households had first been interviewed and familiarity had been established between respondents and the survey teams, a module on intra-household decisionmaking was included. Questions were formulated based on qualitative work done between survey rounds 3 and 4 by Naved (2000). Multiple dimensions of male-female bargaining power and interactions were measured: family background, assets brought to marriage, current asset ownership, individual contributions to household income, household expenditure patterns, mobility, and decisionmaking. Hallman (2000), Quisumbing and de la Brière (2000), and Quisumbing and Maluccio (2000) analyze a number of these outcomes. Here we focus on physical mobility, control over resources, experience of domestic violence, and political knowledge and activity.

As shown in Appendix Table 4A.6, there are a number of significant differences between women NGO members who have access to and adopt the technology (adopters) and those who have similar livelihood assets but no access to the technology (likely adopters). In every site except Mymensingh (the private-owner fishpond site where the technology reached men in practice) women in adopting households have similar or more favorable outcomes than women in likely-adopter households. For all sites pooled, women from adopting households were more likely to have visited friends or relatives outside the village and attended NGO training or programs in the year before the survey; they were more able to correctly name political leaders and less likely to report having been beaten by their husbands in the past year. By site, women in adopting households in Saturia, the vegetable technology site, reported having more mobility and were more politically aware than women in likely-adopter households. Saturia was the only site where the technology was both targeted at and successfully delivered to women. In Mymensingh, even though the technology was officially targeted at women, it was often men who operated the ponds in practice. The ponds were located largely outside the household compound, making it difficult for women here to physically have any involvement in their operation. Mymensingh is the most culturally conservative of the three sites, and mobility of women is very limited. Women in adopting households here report a greater ability to save for their own expenses, but it also appears that such savings may increase their vulnerability to some degree. Women in adopter households report having their money and assets taken against their will more often than women in likely-to-adopt households.

In Jessore, the group fishpond site, there are also statistically significant differences between women in adopter and likely-adopter households. The former were more likely to have attended NGO training or programs, less likely to have been beaten by a husband or family member, and more likely to have chosen whom to vote for the last time they did vote.

Next we explore multivariate regressions using predicted case (technology-recipient) village versus comparison (technology-pending) village status from the first-stage probit regression (Appendix Table 4A.3). These findings should be considered in conjunction with the means comparisons and interpreted with a degree of caution, given that (1) site-specific technology-recipient regressions were not possible because of the small number of villages in each study site, and (2) the one instrumental variable in the technology placement probit village-level regression (distance from village to office of disseminating organization) is weak and insignificant.

For all sites pooled there are no statistically significant differences in female empowerment outcomes by village technology access status, as shown in Appendix Table 4A.7. Site-specific results indicate a few significant effects of having access to and adopting the technology after other factors are controlled for. In Saturia, technology adoption was associated with women visiting friends and going to the market more, but unlike the means comparison results, adoption is associated here with attending NGO training sessions less often. In Mymensingh, although women in adopting households could more easily name the prime minister of their country, they were more influenced by others in their voting decisions. They are also found to be less likely than women in likely-adopter households to report having assets forcibly taken by husband or relatives in the year before the 1997 fourth round of surveys—contrary to the means comparison results. In Jessore, adoption of the technology was related to women attending NGO training sessions and programs more often—reflecting the group-based nature of the technology delivery there and the location of the ponds away from the household compound. Controlling for other factors, women in adopting households in Jessore had a higher likelihood of working for pay than did women in likely-adopter households.

FOCUS GROUP FINDINGS. In general, people are more positive about the role of NGOs than they are about government services, and women NGO group members report gains in confidence from NGO membership. There is a strong demand for more training and other services from NGOs. Some women report that joining an NGO may have political, social, or factional implications and that NGOs are not neutral in their treatment of all groups. In the group-operated fishponds, lack of adequate NGO supervision is given as a reason for failure that contributed to disempowerment.

For women who have gained direct access to cash income (in general through vegetable production rather than fish culture), some women from the poor groups reported empowerment through an improved understanding of personal finance and enhanced status: "if you have money, then you have status."

Higher female status is given as an outcome of adoption by women's groups. "Now women give money to their husbands from their own earnings. Once husbands would have been angry about this, but they don't say anything now." Several of the groups reported changing norms subsequent to adoption—for example, if women go outside the home in pairs or groups "no one complains nowadays."

Education level is also improving after adoption: "If I didn't grow fish I could not educate my children" (from a member of the female very poor group). Although the additional income may be negligible in monetary terms, it is likely that this woman is reporting the empowerment effects of managing the new income, which is contributing to stronger intrahousehold bargaining power.

Overall Impacts of Technologies on Well-Being

QUANTITATIVE FINDINGS. In this section household-level expenditure and income and individual nutritional outcomes are described for adopter households with technology access and likely-adopter households without such access. As with the empowerment outcomes, we present pooled and site-specific bivariate evidence, followed by multivariate regression results, where predicted technology-recipient village status (from the probit regression in Appendix Table 4A.3) is the regressor of interest.

Appendix Table 4A.8 indicates that average monthly per capita expenditure levels and the percentage of the household budget spent on food over the study year did not differ significantly between adopter and likely-adopter households. For the sites pooled, the only significant difference in incomes is profit from ponds, which is driven by households in the Mymensingh site. In Jessore total household income and household off-farm income are much greater for adopter households. Rental income from leased-out land differed statistically by adopter status within each site, but no pattern is apparent.

Means for individual-level nutritional outcomes show variation by technology-access status. Aggregated across the sites, child weight-for-height *z*-scores do not differ between adopter and likely-adopter households. By site, children in likely adopter households in Jessore have better scores than do those in adopter households.

The multivariate results, presented in Appendix Table 4A.9, should be interpreted with similar caution as for those presented above for women's empowerment. Access to the technologies does not significantly affect household expenditure, nor did it influence the incomes of households in Saturia; the latter is probably due to the small scale and short duration of the vegetable technology dissemination at the time of the survey. In Mymensingh having access to the improved fishpond technology was associated with higher farm incomes, in the form of greater crop and pond profits. In Jessore the technology was associated with higher total and off-farm income, which may be due in part to the greater likelihood that adopting women worked for pay, as was found in the empowerment outcome regressions.

Examining the multivariate results for individual nutritional status reveals no significant effects of access to the technologies in the pooled sample. Adoption of the vegetable technology appears, however, to be negatively associated with preschooler weight-for-height *z*-scores in Saturia. The evidence that the negative association in Saturia is due to the technology is not convincing, as the technologies were relatively new and appeared to have few other effects. Among adults, there are no statistically significant effects of access to the technologies, and hence these are not reported.

FOCUS GROUP FINDINGS. The most positive stories are from the vegetable-growing site of Saturia, and from the individual-pond site of Mymensingh (for example, "before we could only eat fish—now we can sell it as well and solve some of our problems"). The group pond work in Jessore seems to be the least successful—many people here are left embittered with the failure of the collective action and are suspicious of the NGO concerned.

These findings have implications for future scaling up of adoption projects. With regard to the vegetable case study, there is clearly scope for this technology to have wider impact in terms of poverty reduction. Saturia is of course known to be one of the centers of vegetable production in view of its high land, rich soil, and proximity to Dhaka markets. However, the nonlumpy character of this technology and the potential nutrition, gender empowerment, and social network benefits to poorer groups from even very small-scale adoption is apparent from the study. But the dedication and commitment by GKT has clearly played a key role, and care would have to be taken in the selection and training of other NGOs that might undertake this type of work.[9] One of the benefits of this technology is that it can remain small-scale and household based, although it could also lose its gender benefits in contexts where growers can connect with markets and export potential.

Conclusions

Poverty, Impact, and Vulnerability

The study found varying effects of the agricultural technologies. Each site was unique in terms of the nature of the group being targeted, the technology de-

9. It was also clear from the focus group discussions that GKT itself has been through a difficult period, during which relations had broken down in some communities due to inappropriate behavior by some field staff. Some staff members were subsequently dismissed.

livered, the length of exposure of program participants to the technology, and the mode and effectiveness of delivery by the disseminating institutions. Even though half the ponds in the female group fishpond site were not excavated and the groups reported experiencing collective action problems, women in adopting households in this site seemed to fare best relative to their likely adopter peers with regard to having greater mobility, a higher likelihood of working off the farm, and higher levels of household income. In the private-owner fishpond site, where the technology reached better-off households and was in practice targeted at men, pond and crop profits and hence farm incomes saw large increases; changes in female empowerment associated with the technology were mixed in this site, however. In the case of the vegetable technology, which is targeted at women in households with relatively small amounts of land and is essentially a nonlumpy technology that requires a very low level of investment, analysis of the survey data revealed no effects on household income, but apparent negative impacts on preschooler nutritional status and on women's attendance of NGO programs in the year before the survey. The focus group results from this site, however, indicate benefits of this technology in the form of network building and reciprocity among women.

It was also found that the technologies had the capacity to increase vulnerability in a number of ways, such as through the theft of fish or through intrahousehold inequalities that lead to coercion. Women who begin to gain income may be compelled to pass on resources to their husbands or in-laws, particularly in areas where restrictions of female freedom, mobility, and decisionmaking are strong. Institutional factors may also contribute to increased vulnerability, as in the case of the collective action problems that contributed to group fishpond failures. The qualitative element of the research showed a higher level of trust for NGO as opposed to government services, but it also highlighted the variable performance of NGOs. Political dimensions to NGO activity were also shown to be important and are perceived by some sections of the community to affect the dissemination of technologies and extension support services for the technologies.[10]

Research Methods

Quantitative and qualitative data were found to complement each other well in the research across a range of issues. For example, the survey addressed female empowerment of adopters in terms of measuring the frequency of women's visits outside the home, attendance at meetings, knowledge of local politics, while

---

10. These issues of politics and power may constrain the room for maneuver of the poor, as shown by the well-known 1983 study of the rural power structure in Bangladesh by BRAC, in which the idea of "the net" was developed. However, recent research (for example, Lewis and Hossain 2005) suggests that the rigidity of the local power structure may be loosening in small ways that can present opportunities for NGOs, local elites, and the poor to benefit from win-win negotiated outcomes. Such opportunities include support by local leaders and NGOs for groups of local poor to secure improved access to land rights in return for political support.

the focus groups revealed interesting material on the nonmonetary exchange of vegetables among households to build and maintain social networks in the attempt to reduce vulnerability. However, the time lag between the quantitative and the qualitative data collection was a weakness of the study, because sometimes earlier findings were out of date by the time of the focus group meetings. Nevertheless, the approach was found to be useful, and overall the SL framework helped sharpen understanding of the different entry points through which technology can affect household well-being and vulnerability.

A particular strength of combining the social and economic approaches here is that questions that cannot easily be answered by a quantitative survey (even such a thorough one as the one used) were informed by a series of qualitative studies with households, groups, and institutions in the survey areas. These included such issues as the perceptions of poverty, livelihoods strategies, the institutional setting, and technology dissemination pathways.

Another potential weakness of the research approach is that the period of time between introduction of the technologies and the household surveys in two of the three study sites was rather short. This brevity may not have allowed adopters sufficient time to comfortably incorporate the new technologies into their livelihood portfolios. In particular, only two years had elapsed between technology introduction and the survey in the vegetable study site. The technology with the largest productivity effects—the private fishpond experiment—was not only the one targeted most heavily to the non-poor but had also been in operation for the longest period at the time of the household survey—six years. Thus the timing of project evaluations should be carefully considered.

Wider Implications

What lessons might be drawn from this research in relation to wider questions of the relevance of technological research to poverty reduction issues? First, understanding the reality of poor people helps agricultural research to reach and benefit this most important of clienteles. Technologies that build on the assets of the poor (for example, underused homestead land to grow vegetables) are more likely to be adopted by and benefit poor households and the individuals within them; these benefits may come in the form of higher profit or other livelihood outcomes, such as reduced social vulnerability. Conversely, technologies that require high threshold levels of certain assets, such as land or financial capital, are likely to exclude the poor unless programs find other arrangements to work around the assets they lack (for example, group-leased fishponds for those without ponds of their own). An important lesson from this study is that program approaches intended to overcome the low asset stocks of the poor—and therefore allow them to adopt technologies—may be extremely difficult to design and tailor in practice. In this instance collective action difficulties were too great an obstacle to overcome in the group fishpond experiment. More careful attention should be paid to the design of such potential solutions in future research. It is possible that more in-depth participatory research before the introduction of the group-based scheme may have helped program designers foresee and avoid the problems that arose.

We also found that even when technologies are delivered in a technically efficient manner, they may have the potential to increase vulnerability, especially among the poor. Research into how the operational aspects of agricultural technologies impact upon different types of households and individuals within them should be emphasized. Some technologies may be inherently riskier for the poor or for women.

It is not only the technology that matters, but also how it is disseminated. Special efforts to reach poor households, and especially the women within those households, were key to achieving poverty reduction. Untargeted dissemination is more likely to benefit men and better-off households. Reaching women with the technologies provided empowerment effects that led to welfare increases greater than the income effects alone might indicate. The disseminating institutions—whether government, NGO, or social networks—also play a pivotal role in building trust, both with the technology and within the community. Hence the technical competence and the general approach of the disseminators are both important. In the case of the fish polyculture technology, many of the problems raised in the focus groups had more to do with the failure of broader institutional arrangements than with the specific technology itself. Decontextualizing technologies from their institutional and political settings should be avoided: research must focus in a more integrated way on holistic approaches based on sound contextual information.[11]

More generally, the SL framework has been sometimes criticized for paying insufficient attention to issues of power and structure in relation to transforming structures and institutions. The research findings presented here have implications beyond the question of transforming structures simply as dissemination pathways and offers insights into the wider conditions of structural change. The adoption of new technologies is of course influenced in part by prevailing structures of rural gender relations, but it also influences them in the form of changed gender outcomes in which poor women can build greater room for maneuver. The relationship between technology and structure is therefore a dynamic one, and the importance of gender relations as a transforming structure needs to be recognized both within the SL framework and in the design of pro-poor, technology-based interventions in rural Bangladesh.

11. For example, Lewis (1998) argues from data collected in the early and mid-1990s that the constraints on the poor using fish technology in Bangladesh have tended to be presented in terms of a technical problem instead of more accurately as institutional and political.

Appendix

APPENDIX TABLE 4A.1 Rural well-being categories

World Bank PPA category	Characteristics reported in World Bank PPA	Study well-being category and percentage of households by category in 1996–97 survey by site	WB PPA category and percentage of study households by category in 1996–97 survey by site
Rich	Large landowners (approximately five or more acres); own cattle and draft power and agricultural equipment; able to hire laborers; generate surplus income for savings; no food deficits; good quality house structure; have tube well and latrine; can afford to send children to school and use health care; women seldom work outside home; dominate local community power structures	Non-poor: Vegetable: 55 Private pond: 64 Group pond: 32	Rich: Vegetable: 4 Private pond: 11 Group pond: 2
Middle	Medium landowners (approximately 1.0–4.5 acres) with some draft power and agricultural equipment; may take on sharecroppers; may have non-agricultural income sources; expenditures equal income; no food deficits; no housing problem; can afford to send children to school and use health care; take investment but not consumption credit; women do not generally work outside home; have two sets of clothes per year; some take credit from NGOs		Middle: Vegetable: 51 Private pond: 53 Group pond: 30
Social poor	Food deficits experienced but ability to somehow manage two regular meals per day during the slack season; small landholdings (0.3–0.6 acres), profits from which can meet one to two months' needs; adopt multiple livelihood strategies; occasionally work as wage laborers or in factories;	Poor: Vegetable: 27 Private pond: 20 Group pond: 35	Social poor: Vegetable: 27 Private pond: 20 Group pond: 35

	women may work outside the home; own some homestead land but not high quality house; have no or poor water and sanitation facilities; very little to spend on clothing; trusted in community due to interaction as laborers with middle and rich; can borrow for consumption and repay; many are NGO members; express opinions in community but do not take leadership positions		
Helpless poor	Landless; often live on others' land in dilapidated structures; wage labor in combination with sharecropping and fishing; accept low wages during lean periods; suffer from food deficits, especially children; women work as wage laborers; illness of a household member, particularly a wage earner, can have devastating effects; do not have any assets for fallback on during crises; often are NGO members; have very poor clothing; cannot afford dowry for their daughters; cannot afford to entertain guests	Very poor: Vegetable: 19 Private pond: 16 Group pond: 32	Helpless poor: Vegetable: 16 Private pond: 13 Group pond: 22
Bottom poor	Landless; households often headed by women or aged men and do not have able-bodied income earner; going hungry is constant and not a seasonal occurrence; always working to eat; often forage for food and collect fuel to save on expenditures; begging is a source of livelihood; receive clothes donated at Eid festivals; high prevalence of illness; cannot afford health care; cannot take consumption loans because of inability to pay; rarely able to join NGOs; have low social interaction with other groups; attend feasts uninvited		Bottom poor: Vegetable: 3 Private pond: 3 Group pond: 10

SOURCE: Nabi et al. (1999).

NOTES: NGO, nongovernmental organization; PPA, Participatory Poverty Assessments.

APPENDIX TABLE 4A.2 Characteristics of sample villages

	All sites		
Characteristic	Case	Comparison	Means test (*p*-value)
Number of villages	**27**	**20**	
Instrumental variable			
Distance to office of disseminating organization (miles)	4.45	5.05	0.44
Physical			
Distance to nearest paved road (kilometers)	1.44	1.26	0.67
Any household in village has electricity	0.63	0.85	0.10
Village has a market (1 = yes, 0 = no)	0.44	0.20	0.08
Minutes to nearest phone (= 0 if in village)	34.63	35.90	0.87
Minutes to nearest post office (= 0 if in village)	12.33	20.45	**0.05**
Minutes to nearest bus stop (= 0 if in village)	19.89	21.95	0.74
Political			
Village has a Union Parishad representative (current or in past five years; 1 = yes, 0 = no)	0.74	0.90	0.18
Social			
Number of mosques	3.04	1.72	**0.03**
Village has a youth organization (1 = yes, 0 = no)	0.81	0.70	0.37
Number of NGOs with members in this village	3.48	2.90	0.21
Human			
Minutes to *thana* health center (wet season; = 0 if in village)	49.63	67.75	**0.03**
ORS available in village (1 = yes, 0 = no)	0.93	0.75	0.10
Has a BRAC school (1 = yes, 0 = no)	0.74	0.55	0.18
Has a primary school (1 = yes, 0 = no)	0.74	0.50	0.09
Has a secondary school (1 = yes, 0 = no)	0.33	0.05	**0.02**
Has a madrasa school (1 = yes, 0 = no)	0.33	0.20	0.32
Has adult education classes (1 = yes, 0 = no)	0.33	0.40	0.65
Natural			
Logarithm of value per decimal irrigated upland (1996 taka)	7.53	7.71	0.45
Logarithm of value per decimal irrigated lowland (1996 taka)	7.60	7.59	0.96
Tube well is primary source for drinking water (1 = yes, 0 = no)	0.96	1.00	0.40
Number of tube wells in the village	45.19	48.85	0.68
Other			
Village perceived to be richer than neighboring villages (1 = yes, 0 = no)	0.41	0.30	0.46

NOTES: BRAC, Bangladesh Rural Advancement Committee; NGO, nongovernmental organization; ORS, oral rehydration solution. na, not applicable. Boldface indicates significance at the 5 percent probability level.

Saturia: Vegetables			Mymensingh: Individual fishponds			Jessore: Group fishponds		
Case	Comparison	Means test (*p*-value)	Case	Comparison	Means test (*p*-value)	Case	Comparison	Means test (*p*-value)
5	**5**		**14**	**7**		**8**	**8**	
3.10	4.30	0.43	3.61	3.00	0.37	6.75	7.31	0.67
1.65	0.60	0.24	1.53	2.11	0.45	1.15	0.94	0.73
0.40	1.00	**0.04**	0.64	0.71	0.74	0.75	0.88	0.52
0.40	0.20	0.49	0.57	0.00	**0.01**	0.25	0.38	0.59
23.00	28.00	0.62	28.21	26.86	0.82	53.13	48.75	0.81
7.80	14.00	0.11	12.79	15.57	0.70	14.38	28.75	**0.05**
13.40	17.00	0.76	13.57	18.86	0.40	35.00	27.75	0.60
0.60	0.80	0.49	0.93	0.86	0.60	0.50	1.00	**0.02**
0.80	1.90	**0.04**	4.57	1.43	**0.00**	1.75	1.88	0.79
1.00	0.80	0.29	0.93	0.57	**0.05**	0.50	0.75	0.30
5.00	4.80	0.74	3.21	1.57	**0.00**	3.00	2.88	0.88
44.00	50.00	0.71	36.43	66.43	0.07	61.25	80.00	0.26
1.00	1.00	na	0.93	0.71	0.19	0.88	0.63	0.25
0.80	1.00	0.29	0.71	0.29	0.06	0.75	0.50	0.30
0.40	0.60	0.53	0.86	0.43	**0.04**	0.75	0.50	0.30
0.20	0.00	0.29	0.50	0.14	0.11	0.13	0.00	0.30
0.20	0.00	0.29	0.50	0.00	**0.02**	0.13	0.50	0.11
0.80	0.60	0.49	0.07	0.00	0.47	0.50	0.63	0.61
7.78	7.94	0.43	7.20	7.36	0.30	7.95	7.89	0.92
7.79	7.98	0.37	7.39	7.30	0.68	7.84	7.60	0.72
1.00	1.00	na	1.00	1.00	na	0.88	1.00	0.30
54.40	59.80	0.81	42.71	22.86	0.11	43.75	64.75	0.17
0.40	0.60	0.53	0.43	0.14	0.19	0.38	0.25	0.59

APPENDIX TABLE 4A.3 Determinants of case versus comparison village status, probit marginal effects

Variable	*dF/dx*	*z*
Instrumental variable		
Distance to office of disseminating organization (miles)	0.05	0.65
Physical		
Distance to nearest paved road (km)	0.12	1.13
Any household in village has electricity	–0.35	–1.73
Village has a market	0.24	1.27
Minutes to nearest phone[a]		
Minutes to nearest post office	–0.01	–1.16
Minutes to nearest bus stop	0.00	–0.46
Political		
Village has a Union Parishad representative (current or in past five years)	–0.30	–1.64
Social		
Number of mosques	0.15	1.21
Village has a youth organization	–0.08	–0.30
Number of 12 local NGOs with members in this village	0.00	0.06
Human		
Minutes to *thana* health center (wet season)	–0.01	–1.89
ORS available in village[b]		
Has a BRAC school	0.04	0.13
Has a primary school	0.10	0.42

Has a secondary school	0.14	0.42
Has a madrasa school	0.23	1.07
Has adult education classes	0.21	1.01
Natural		
Value per decimal irrigated upland (1996 Tk)	–0.05	–0.24
Value per decimal irrigated lowland (1996 Tk)[c]		
Tube well as primary source for drinking water[d]		
Number of tube wells in the village	0.00	–0.05
Other		
Village perceived to be richer than neighboring villages	0.13	0.47
Number of observations	47	
Likelihood ratio test $\chi^2(19)$	31.38	
Probability χ^2	0.0367	
Pseudo R^2	0.4894	
Test all parameters equal zero		
$\chi^2(19)$	12.32	
Probability χ^2	0.8716	

NOTES: For a village status of case, technology recipient = 1; for comparison villages, technology pending = 0. BRAC, Bangladesh Rural Advancement Committee; NGO, nongovernmental organization; ORS, oral rehydration solution.

[a] Variable was dropped, as it is highly correlated with minutes to bus stop.

[b] Variable was dropped, as it is highly correlated with minutes to *thana* health center.

[c] Variable was dropped, as it is highly correlated with value of irrigated upland.

[d] Variable was dropped, as there was no variation with respect to it.

APPENDIX TABLE 4A.4 Household livelihood assets

Asset	All sites		
	Case	Comparison	Means test (p-value)
Number of households	**321**	**318**	
Physical capital			
Value of wife's assets at marriage (1996 Tk)	2,433	3,280	0.36
Value of husband's assets at marriage (1996 Tk)	86,668	80,288	0.67
Wife's share of current household assets	0.06	0.06	0.97
Total value household assets (1996 Tk)	203,794	191,370	0.51
Value of durables (1996 Tk)	14,591	13,675	0.66
Value of house (1996 Tk)	2,899	1,655	**0.02**
Value of land (1996 Tk)	180,212	169,766	0.56
Value of livestock (1996 Tk)	6,422	6,274	0.79
Land owned (acres)	1.55	1.44	0.48
Homestead area (acres)	0.16	0.15	0.16
Cultivable crop area (acres)	1.26	1.20	0.68
Cultivable pond area (acres)	0.17	0.21	0.33
Land area leased in (acres)	0.12	0.19	0.11
Sanitation score index (0–10)[a]	5.49	5.17	**0.02**
Human capital			
Adult male height (centimeters; age, 19–45 years)	162.7	162.6	0.83
Adult female height (centimeters; age, 19–45 years; nonpregnant, nonlactating)	150.9	150.3	0.18
Household size	5.80	5.60	0.45
Number of prime-aged male earners	1.19	1.19	0.98
Household female headed	0.04	0.04	0.82
Highest level of adult male education in household (years)	5.51	4.79	0.06
Highest level of adult female education in household (years)	3.39	2.91	0.10
Dependency ratio	0.49	0.49	0.94
Financial capital			
Number of loans taken between survey rounds (zero to three)[a]	2.10	2.00	0.26
Loan amount, round 1 (1996 Tk)[a]	7,689	7,293	0.65
Social capital			
No food gifts given during year[a]	0.50	0.48	0.50
No food or income gifts received during year[a]	0.35	0.38	0.36

NOTE: Boldface indicates significance at the 5 percent probability level.

[a] Variable may more appropriately be considered a livelihood outcome instead of a livelihood asset.

Saturia			Mymensingh			Jessore		
Case	Comparison	Means test (*p*-value)	Case	Comparison	Means test (*p*-value)	Case	Comparison	Means test (*p*-value)
106	**103**		**106**	**106**		**109**	**109**	
3,421	5,788	0.35	2,320	2,275	0.96	1,497	1,824	0.65
12,437	137,311	0.50	89,937	56,835	0.10	55,339	47,837	0.62
0.06	0.04	0.22	0.02	0.02	0.18	0.09	0.12	0.42
69,818	172,799	0.91	321,239	311,377	0.80	122,623	92,215	0.19
11,474	10,629	0.67	22,860	21,476	0.76	9,533	8,967	0.87
2,469	1,747	0.38	4,175	1,155	**0.00**	2,077	2,054	0.98
50,141	153,173	0.91	285,913	282,424	0.93	105,982	75,888	0.18
5,733	7,250	0.09	8,290	6,321	0.05	5,275	5,306	0.97
1.10	1.10	0.98	2.62	2.49	0.69	0.95	0.75	0.28
0.14	0.15	0.89	0.23	0.19	0.17	0.12	0.10	0.35
0.90	0.93	0.86	2.12	2.05	0.79	0.77	0.64	0.46
0.01	0.02	0.66	0.37	0.46	0.36	0.12	0.13	0.77
0.00	0.04	0.02	0.18	0.31	0.10	0.18	0.21	0.68
5.06	4.61	0.06	5.54	5.08	0.08	5.85	5.79	0.75
162.3	162.3	0.91	162.7	162.6	0.84	163.1	162.9	0.80
151.3	151.7	0.63	149.4	148.5	0.32	151.8	150.5	0.11
5.58	5.64	0.85	6.81	6.25	0.12	5.00	5.04	0.90
1.20	1.28	0.45	1.33	1.25	0.51	1.04	1.05	0.91
0.04	0.03	0.73	0.00	0.00	0.32	0.07	0.08	0.80
3.97	3.76	0.73	7.79	6.93	0.18	4.71	3.58	0.06
2.33	1.90	0.34	4.86	4.20	0.22	2.92	2.61	0.50
0.50	0.51	0.46	0.53	0.48	**0.05**	0.44	0.47	0.13
2.77	2.71	0.38	1.38	1.41	0.83	2.10	1.89	0.10
12,262	14,234	0.22	6,496	4,405	0.20	4,389	3,543	0.31
0.72	0.61	0.11	0.25	0.20	0.33	0.54	0.63	0.22
0.20	0.30	0.09	0.66	0.70	0.56	0.19	0.16	0.48

APPENDIX TABLE 4A.5 Determinants of adopter versus likely-adopter household status,

Variable	All sites	
	dF/dx	*z*
Household size	0.00	0.45
Value durables/1000 (1996 taka)	0.00	0.40
Value house/1000 (1996 taka)	0.01	**2.81**
Value livestock/1000 (1996 taka)	0.00	0.05
Value land/1000 (1996 taka)	0.00	0.00
Male has some primary education (versus none)	0.17	**2.80**
Male has some secondary education (versus none)	0.03	0.49
Male has some university education (versus none)	0.28	**3.22**
Female has some primary education (versus none)	0.05	0.90
Female has some secondary education (versus none)	0.01	0.22
Mymensingh dummy (versus Saturia)	–0.23	**–4.25**
Jessore dummy (versus Saturia)	–0.06	–1.27
Constant		
Number of observations	638	
Wald χ^2	39.96	
Probability χ^2	0.000	
Pseudo R^2	0.0574	

NOTES: na, not applicable. Boldface indicates significance at the 5 percent probability level.

APPENDIX TABLE 4A.6 Female empowerment outcomes (percentages)

Outcome	All sites		
	Case	Comparison	Means test (*p*-value)
Visited friends/relatives outside of village in past year	95	90	**0.05**
Gone to *haat*/bazaar in past year	19	19	0.84
Attended NGO training or programs in past year	31	17	**0.00**
Husband/family member verbally abused you in past year	66	71	0.21
Husband/family member beat you in past year	23	33	**0.01**
Woman knows name of UP chairman	82	74	**0.02**
Woman knows name of her representative in parliament	47	35	**0.00**
Woman knows name of prime minister	88	81	**0.02**
Woman has ever voted	89	87	0.45
For last vote, woman chose who she voted for	32	26	0.15
Worked for pay in past year	70	67	0.38
Ever decides alone about family expenditures	45	50	0.20
Keeps money on own for expenses or security	73	69	0.34
Husband or family member took money from woman against her will in past year	22	18	0.35
Husband or family member took asset from woman against her will in past year	11	7	0.12

NOTES: NGO, nongovernmental organization; UP, Union Parishad. Boldface indicates significance at the 5 percent probability level.

probit marginal effects

Saturia		Mymensingh		Jessore	
dF/dx	*z*	*dF/dx*	*z*	*dF/dx*	*z*
0.00	0.00	0.01	0.74	–0.01	–0.31
0.00	0.19	0.00	0.49	0.00	–0.15
0.01	1.22	0.01	**2.35**	0.00	0.00
–0.01	–1.92	0.01	1.50	0.00	–0.19
0.00	0.63	0.00	–0.83	0.00	1.43
0.17	1.78	0.31	**2.52**	0.03	0.31
0.03	0.31	0.09	0.73	0.01	0.09
–0.01	–0.06	0.33	**2.36**	0.42	**2.72**
0.06	0.55	0.07	0.75	0.00	0.02
0.07	0.59	0.04	0.39	–0.08	–0.78
na		na		na	
na		na		na	
209		212		217	
9.15		29.92		10.03	
0.5178		0.0009		0.4376	
0.0326		0.1175		0.0412	

Saturia: Vegetables			Mymensingh: Private fishponds			Jessore: Group fishponds		
Case	Comparison	Means test (*p*-value)	Case	Comparison	Means test (*p*-value)	Case	Comparison	Means test (*p*-value)
93	84	**0.05**	92	91	0.88	99	96	0.17
20	22	0.80	11	10	0.77	25	26	0.83
40	19	**0.00**	18	12	0.23	33	18	**0.02**
64	72	0.23	65	67	0.85	68	73	0.45
22	29	0.27	26	31	0.40	22	38	**0.02**
86	72	**0.02**	73	70	0.63	86	79	0.23
51	29	**0.00**	61	53	0.24	29	24	0.43
81	62	**0.00**	94	95	0.85	89	86	0.48
94	91	0.49	81	73	0.25	94	97	0.31
44	41	0.74	19	21	0.77	30	15	**0.01**
60	60	0.98	63	53	0.14	89	88	0.80
22	29	0.26	18	23	0.42	98	99	0.57
59	60	0.86	70	55	**0.03**	90	92	0.62
13	15	0.62	37	17	**0.00**	15	22	0.24
11	8	0.47	9	1	**0.03**	12	11	0.80

APPENDIX TABLE 4A.7 Marginal effects of predicted case village (technology-recipient)

Outcome	All sites
Visited friends or relatives outside of village in past year	0.00
Gone to *haat*/bazaar in past year	0.01
Attended NGO training or programs in past year	0.03
Husband or family member verbally abused woman in past year	–0.12
Husband or family member beat woman in past year	–0.05
Woman knows name of UP chairman	–0.09
Woman knows name of her representative in Parliament	–0.06
Woman knows name of prime minister	0.05
Woman has ever voted	0.03
For last vote, woman chose who she voted for	–0.07
Worked for pay in past year	0.12
Ever decides alone about family expenditures	–0.02
Keeps money on own for expenses or security	0.12
Husband or family member took money from woman against her will in past year	–0.06
Husband or family member took asset from woman against her will in past year	0.02

NOTES: NGO, nongovernmental organization; UP, Union Parishad. *, **, and *** indicate statistical significance at the 5 percent, 1 percent, and .1 percent levels, respectively. A negative coefficient implies that residing in a case village and adopting the technology (versus residing in a comparison village and wishing to adopt the technology) has a negative effect on the outcome of interest. In all regressions sampling weights are applied, standard errors are corrected for intracluster correlation, and survey round dummies are included. All site-pooled regressions include site dummies. Regressions include adopter and likely-adopter households only.

APPENDIX TABLE 4A.8 Livelihood outcomes: Expenditure, income, and nutritional status

	All sites		
Outcome	Case	Comparison	Means test (p-value)
Average monthly per capita household expenditure	712	697	0.62
Total annual household income (all sources)	38,559	35,655	0.32
Total annual household off-farm income	18,144	16,103	0.19
Total annual household farm income	16,380	14,529	0.20
Total annual crop profit	11,309	10,382	0.43
Total annual pond profit	2,527	1,365	**0.00**
Total annual livestock profit	1,796	2,069	0.36
Weight-for-height z-score (children aged five years and younger)	–1.14	–1.10	0.49

NOTE: Boldface indicates significance at the 5 percent probability level.

status versus comparison village (technology-pending) status on female empowerment outcomes

Saturia	Mymensingh	Jessore
0.09	0.03	0.00
0.14	0.02	–0.01
–0.25*	0.04	0.02***
–0.16	–0.05	–0.29
0.14	–0.13	–0.16
0.14	–0.02	–0.18
–0.22	–0.05	–0.06
0.06	0.02*	–0.02
–0.03	0.09	0.07
–0.10	–0.15**	–0.02
–0.06	0.12	0.11**
0.14	0.04	—[a]
0.20	0.12	0.10
–0.08	–0.09	0.01
0.01	–0.06*	0.05

Each marginal-effect cell in the table is from a separate regression. Complete regression results available upon request from the authors. Each regression also includes wife's share of assets brought to the marriage, her age and age squared, her education, husband's education, total value of household assets, and household size and demographic composition.

[a] In Jessore all women reported "yes."

Saturia			Mymensingh			Jessore		
Case	Comparison	Means test (*p*-value)	Case	Comparison	Means test (*p*-value)	Case	Comparison	Means test (*p*-value)
722	695	0.62	668	707	0.32	743	690	0.33
43,806	43,944	0.98	41,344	42,475	0.85	30,748	21,189	**0.00**
19,677	19,882	0.92	15,007	15,743	0.83	19,701	12,884	**0.00**
18,036	18,641	0.84	22,192	19,091	0.20	9,117	6,206	0.06
13,888	14,031	0.95	13,645	13,463	0.92	6,531	3,939	0.06
178	292	0.66	6,268	3,097	**0.00**	1,173	695	0.38
2,611	3,055	0.53	1,895	2,145	0.63	908	1,064	0.45
–1.33	–1.32	0.96	–0.92	–0.98	0.46	–1.22	–0.99	**0.01**

APPENDIX TABLE 4A.9 Effects of predicted case village (technology-recipient) versus comparison village status (technology-pending) on expenditure, income, and nutritional status

Outcome	Coefficient			
	All sites	Saturia	Mymensingh	Jessore
Average monthly per capita household expenditure	–8.71	9.55	–27.58	11.34
Total annual household income (all sources)	3,928.99	–236.84	4,895.34	8,240.22*
Total annual household off-farm income	3,526.34	2468.58	2,432.26	5,985.30*
Total annual household farm income	4,701.87***	3561.09	7,254.59***	2,973.97
Total annual crop profit	3,467.90**	3,266.42	3,886.13*	3,720.40
Total annual pond profit	1,277.88**	–256.92	3,163.33***	–266.69
Total annual livestock profit	17.62*	176.29	218.39	–359.67
Weight-for-height *z*-score (children aged five years and younger)	0.10	–0.64*	0.12	0.10

NOTES:*, **, and *** indicate statistical significance at the 5-percent, 1-percent, and .1-percent levels, respectively. Each row represents four separate regressions. In each regression sampling weights are applied, standard errors are corrected for intracluster correlation, and survey round dummies are included. All site-pooled regressions include site dummies. Regressions include adopter and likely-adopter households only. Each marginal effect cell in the table is from a separate regression. Complete regression results available upon request from the authors. Each household-level regression includes wife's share of assets brought to marriage; wife's and husband's age, age squared, and education; total value of household assets; and household size and demographic composition. Each child-level regression includes child's age, age squared, sex, and sex-specific birth order; mother's share of assets brought to marriage; whether child's mother and father are head female and male in the household; mother's and father's ages, age squared, and education; mother's and father's heights; and household assets, size, and demographic composition.

Acknowledgments

The economic data used in this chapter were originally collected under a grant funded by the United States Agency for International Development, Office of Women in Development: FAO-0100-G-00-5020-00, on "Strengthening Development Policy through Gender Analysis: An Integrated Multicountry Research Program." Thanks go to Howarth Bouis, Agnes Quisumbing, and Benedicte de la Briere for designing the survey, and to Data Analysis and Technical Assistance (DATA) of Dhaka, Bangladesh, for collecting the survey data. In Bangladesh, we thank Sonia Sultana Ashrafee for her excellent contribution to qualitative data collection and analysis. Lawrence Haddad, Jere Berhman, Robert Chambers, Tony Bebbington, and four anonymous reviewers are thanked for their helpful comments. Ruth Meinzen-Dick is gratefully acknowledged for her continuous support and encouragement. An immeasurable debt is owed to the local survey respondents and focus group participants for their willingness to give their time and share details of their lives.

References

Filmer, D., E. M. King, and L. Pritchett. 1998. *Gender disparity in South Asia: Comparisons between and within countries.* Policy Research Working Paper 1867. Washington, D.C.: Development Research Group, Poverty and Human Resources, World Bank.

Haddad, L., C. Peña, C. Nishida, A. Quisumbing, and A. Slack. 1996. *Food security and nutrition implications of intrahousehold bias: A review of the literature.* Food Consumption and Nutrition Division Discussion Paper 19. Washington, D.C.: International Food Policy Research Institute.

Hallman, K. 2000. *Mother-father resource control, marriage payments, and girl-boy health in rural Bangladesh.* Food Consumption and Nutrition Division Discussion Paper 93. Washington, D.C.: International Food Policy Research Institute.

Hulme, D., and P. Mosley. 1996. *Finance against poverty,* Vol. 1. London: Routledge.

IFPRI (International Food Policy Research Institute). 1998. *Commercial vegetable and polyculture fish production in Bangladesh: Their impacts on income, household resource allocation, and nutrition.* Food Consumption and Nutrition Division. Washington, D.C.: International Food Policy Research Institute.

IFPRI-BIDS-INFS (International Food Policy Research Institute–Bangladesh Institute of Development Studies–Institute of Nutrition and Food Science). 1998. *Commercial vegetable and polyculture fish production in Bangladesh: Their impacts on income, household resource allocation, and nutrition.* Donor report. Washington, D.C.: International Food Policy Research Institute.

Kabeer, N. 1994. *Reversed realities.* London: Verso.

Kanji, N., and S. Barrientos. 2002. *Trade liberalisation, poverty and livelihoods: Understanding the linkages.* IDS Working Paper 159. Brighton, U.K.: Institute of Development Studies.

Kaplowitz, M. D., and J. P. Hoehm. 2001. Do focus groups and individual interviews reveal the same information for natural resource valuation? *Ecological Economics* 36: 237–247.

Lewis, D. 1991. *Technologies and transactions: A study of the interaction between new technology and agrarian structure in Bangladesh.* Dhaka: Center for Social Studies.

———. 1998. Partnership as process: Building an institutional ethnography of an inter-agency aquaculture project in Bangladesh. In *Development as process: Concepts and methods for working with complexity,* D. Mosse, J. Farrington, and A. Rew, eds. London: Routledge.

———. 2004. On the difficulty of studying "civil society": Reflections on NGOs, state and democracy in Bangladesh. *Contributions to Indian Sociology* 38 (3): 299–322.

Lewis, D., and A. Hossain. 2005. An analysis of the local power structure in Bangladesh with an emphasis on Faridpur and Rajbari districts. Report to SIDA. Dhaka.

Mitra, S. N., M. Nawab Ali, S. Islam, A. R. Cross, and T. Saha. 1994. *Bangladesh demographic and health survey, 1993–1994.* Calverton, Md.: National Institute of Population Research and Training, Mitra and Associates, and Macro International.

Mitra, S. N., A. Al-Sabir, A. R. Cross, and K. Jamil. 1997. *Bangladesh demographic and health survey, 1996–1997.* Calverton, Md.: National Institute of Population Research and Training, Mitra and Associates, and Macro International.

Nabi, R., D. Datta, S. Chakrabarty, M. Begum, and N. J. Chaudhury. 1999. *Consultation with the poor: Participatory poverty assessment in Bangladesh.* Dhaka: NGO Working Group of the World Bank.

Naved, R. T. 2000. *Intrahousehold impact of the transfer of modern agricultural technology: A gender perspective.* Food Consumption and Nutrition Division Discussion Paper 85. Washington, D.C.: International Food Policy Research Institute.

Pitt, M., M. Rosenzweig, and D. Gibbons. 1995. The determinants and consequences of the placement of government programs in Indonesia. In *Public spending and the poor: Theory and evidence,* D. van de Walle and K. Nead, eds. Washington, D.C., and Baltimore: World Bank and Johns Hopkins University Press.

Pitt, M., M. R. Rosenzweig, and M. N. Hassan. 1990. Productivity, health and inequality in the intrahousehold distribution of food in low-income countries. *American Economic Review* 80 (5): 1139–1156.

Quisumbing, A. R., and B. de la Brière. 2000. *Women's assets and intrahousehold allocation in rural Bangladesh: Testing measures of bargaining power.* Food Consumption and Nutrition Division Discussion Paper 86. Washington, D.C.: International Food Policy Research Institute.

Quisumbing, A. R., and J. A. Maluccio. 2000. *Intrahousehold allocation and gender relations: New empirical evidence from four developing countries.* Food Consumption and Nutrition Division Discussion Paper 84. Washington, D.C.: International Food Policy Research Institute.

Reardon, T., J. Berdegué, and G. Escobar. 2001. Rural nonfarm employment and incomes in Latin America: Overview and policy implications. *World Development* 29 (3): 395–410.

Walker, T., and J. Ryan. 1990. *Village and household economies in India's semi-arid tropics.* Baltimore: Johns Hopkins University Press.

World Bank. 1999. Bangladesh: From counting the poor to making the poor count. World Bank Country Study. Washington, D.C.: World Bank.

5 Impacts of Agroforestry-Based Soil Fertility Replenishment Practices on the Poor in Western Kenya

FRANK PLACE, MICHELLE ADATO, PAUL HEBINCK, AND MARY OMOSA

This study examines the impact of tree-based improved fallow and biomass transfer systems on the rural poor in western Kenya. Western Kenya is one of the most densely populated areas in Africa—densities of more than 1,000 people per square kilometer are not uncommon. Much of western Kenya is considered to have good potential for agriculture, with medium elevation (1,100–1,600 meters), deep, well-drained soils, and relatively high rainfall (1,200–1,800 millimeters per year) that permits two growing seasons. The history of farming in the area, however, is characterized by low input–low output farming. Recent studies have found that crop productivity is very low (less than 1 ton of maize per hectare per year) and that nutrient balances are seriously in deficit. As a result, along with a swelling population and decreasing farm sizes (now around 1 hectare per household), poverty is rampant in the region. Fifty percent or more of the population is estimated to be in poverty in several of the districts, which is significantly higher than in other good agricultural potential regions of Kenya.

The World Agroforestry Centre (ICRAF), the Kenya Agricultural Research Institute (KARI), and the Kenya Forestry Research Institute (KEFRI) developed an agroforestry research program that had as one of its pillar systems the improvement of welfare through soil fertility replenishment (SFR). Initially, several systems were tested, including alley farming, but the two that appeared most promising were improved fallows and biomass transfer. "Improved fallow" refers to the intentional planting of a fallow species. Improved fallows are more efficient than natural fallows and can normally achieve the same effect on crop productivity in a much shorter time. In western Kenya, two main fallow species are being used by farmers. *Crotalaria* spp. and *Tephrosia* spp. are shrub species that develop a good canopy and leaf biomass in a short time, and both fix nitrogen from the atmosphere. They are left for fallow for one season, normally the short rainy season (October–December), after having been planted toward the end (April–May) of the long rainy season. Farmers then plant their crop (normally maize) the following long rainy season and may continue to cultivate the crop for more seasons, using the residual fertility effect from the fallows.

Biomass transfer systems are those in which organic nutrient sources are grown in one place and then transferred to crops in another. In western Kenya, the most popular shrub species used is *Tithonia diversifolia.* This species was selected among many locally found species because it is a prolific grower (found throughout the region), is easy to establish and work with, and its leaves contain high concentrations of nutrients, especially nitrogen. Farmers gather the leaves off-farm or plant the shrub on boundaries or contours on their own farms. They then incorporate the leaves into the soil at planting and sometimes use new leaf growth as a mulch later in the season. This system allows farmers to grow crops continuously, which is an advantage over the improved fallow system, but the available space for producing organic nutrient sources on-farm is limited. As a result, farmers are using biomass transfer systems significantly and increasingly on high-value crops, such as kales and tomatoes, rather than on maize.

This study of SFR technology is one of a set of Consultative Group on International Agricultural Research (CGIAR) case studies examining the impact of agricultural research on poverty. This particular technology was selected for the study because it is an example of natural resource management research as opposed to the more common crop variety research. Because the agroforestry technologies offered an affordable option for soil fertility improvement, it was expected that rates of use and adoption would be relatively high among the poor. Further, there was some question whether the non-poor would perceive any benefits of agroforestry compared to fertilizer. Even if there were such a perception, it is still an empirical issue as to whether the poor actually can substantially benefit from the use of agroforestry technologies. This benefit would seem to depend on their understanding of how to effectively manage the systems as well as their capacity and willingness to increase their land and labor investment in these systems.

The study was also unique in its focus on comparing approaches to technology dissemination. Exploring dissemination processes speaks to debates around social capital, empowerment, and participatory development. Technology is mediated by social processes and the social relationships into which it is introduced. In addition to examining how these processes unfold, the study explores several hypotheses related to the use of local organizations and other forms of participation for dissemination: that social capital will be enhanced, that social divisions will emerge, that farmers will be newly empowered in certain dimensions, and that existing power relationships will be reinforced. Although these consequences may appear contradictory, it is hypothesized that they will occur simultaneously, with a range of effects on different groups. Also, in disaggregating the dissemination study focus groups into women and men, and poor and less poor farmers, a hypothesis suggesting socially differentiated effects is implicit. The findings have implications for policy and program choices related to forms of farmer participation in technology development and dissemination, suggesting the importance of understanding social dynamics when designing program interventions. This study demonstrated the feasibility and value of using social science methods (ethnographic or rapid, depending on the time and budget available) to achieve at least a basic understanding of social dynamics in different geographical areas among different social, cultural, and economic groups before undertaking major new dissemination initiatives. Although not every aspect of a microregion will be understood, sufficient information can be gathered to provide insight into which methods are more likely to be helpful to different social groups. Partnerships with local nongovernmental organizations (NGOs) and research institutions can be developed for this work, given their local knowledge and likely interest.

The range of issues covered in the study required the use of a mix of research methods and interdisciplinary perspectives. Issues pertaining to local perceptions of poverty; the mediation of technology by social processes; and the role of gender, power, and other social constructs in understanding adoption and impact could only be explored using qualitative research methods and sociological perspectives. These were combined with quantitative measures of adoption and impact and economic analysis. The quantitative analyses proved valuable in identifying the prevalence of patterns of adoption and impact relationships among the general population and the poor. We also drew on long-term knowledge of the region, based on work by members of our study team and others. The study was designed using the sustainable livelihoods (SL) conceptual framework as a starting point, drawing on concepts of vulnerability, access to and limitations on combinations of assets (for example, natural, human, and social capital), and the importance of institutions and processes. However, other constructs from economics and sociology were introduced as required. This expansion was done not so much to promote a particular alternative paradigm as to handle important research questions raised by the entire team, composed of individuals from diverse backgrounds and experiences.

The next section presents the methods used in this study, and explains the sampling procedures and outcomes. We then provide a contextual background for the study areas, including an overview of people's livelihood strategies; the reasons behind these strategies; and concepts of poverty from official, researcher, and local perspectives. Next we focus on adoption of soil fertility practices, describing the process in the pilot and nonpilot villages. This section also explores in depth patterns of adoption across different types of households, including poor and non-poor. The section on livelihoods impact examines the extent to which various productivity and welfare impacts occurred as a result of adoption of soil fertility replenishment practices. We then explore approaches to dissemination taken by different institutions in the study areas and their effects, covering methods, knowledge acquisition, sustainability, and the implications of and for social capital and social relationships. Finally, we summarize the methodological and empirical findings and suggest considerations for future poverty alleviation programs in western Kenya.

Research Methods

This chapter synthesizes results from the application of a range of analytical methods in several sites. Sites are selected from Luo and Luhya ethnic group communities. An important distinction is made between those communities within an initial pilot project area (covering parts of Vihiga and Siaya districts) and those from outside. Communities within the pilot project area received high levels of technical support from project staff from 1997 through early 2000. In addition, all researcher-managed trials were conducted with farmers in the pilot villages, and all tree seed purchased from farmers was again from these villages. Qualitative analyses covered four villages in the pilot project area and four villages outside the pilot project area, both Luo and Luhya areas. The quantitative analysis covered 17 villages within the pilot project area and eight villages outside the area.

Qualitative and quantitative methods have been used in conjunction throughout the study to address issues of impact. Qualitative wealth-ranking exercises were undertaken several years ago to help understand the types of indicators local people use to assess relative poverty or wealth; these indicators are important to measure the effect of any research program. Quantitative surveys were then implemented to capture such wealth indicators over wide areas. These surveys formed the basis for stratification and sampling of households in both the qualitative and quantitative impact assessment studies within the pilot project area. Qualitative analyses used mainly techniques of focus group and case studies. Focus group discussions were used to discuss the concepts of poverty and livelihoods and served as the main research tool for assessing the effectiveness of alternative dissemination methods in reaching the poor. Case study methods were used for 40 individuals and their households and/or families. Field assistants interacted with the individuals over a six-month period. Data were analyzed with the assistance of qualitative data analysis software. Dissemination approaches used by different organizations in western Kenya were later studied through 24 focus groups across six villages, disaggregated by gender and wealth status. In addition to discussions, these studies involved some selected visual (participatory rural appraisal [PRA]–type) exercises. These exercises were designed to yield information directly relevant to the research questions and generated both qualitative and quantitative data. Table 5.1 shows the overlap between the quantitative and qualitative methods across villages.

Quantitative analyses relied on data collected from surveys. Within the pilot project area, adoption was analyzed from annual monitoring data on the use of agroforestry from more than 1,600 households (1997–2001). Impact was analyzed from a cohort of 120 households that were part of a baseline survey on assets, expenditure, and food consumption done in 1999–2000 and revisited in 2002. In the nonpilot project sites, no baseline was collected, so all the data come from a 2002 survey that included 360 households, stratified on the basis of use of agroforestry (thus use rates are not representative of the communities

TABLE 5.1 Villages studied and research methods[a]

Site	Type of dissemination approach	Disseminating organizations[a]	Survey	Focus group discussions	Case studies
Luo villages					
Sarika	Village approach	**ICRAF,** KEFRI	X		X
Muhanda Arude	TRACE approach	**CARE-Kenya,** ICRAF, MoARD	X	X	X
Sauri	Village approach	**ICRAF,** KEFRI, KARI, MoARD	X	X	X
Gongo	Catchment area approach	**MoARD,** ICRAF		X	
West Kanyaluo	Sub-chief visited Maseno	Local leaders	X		
Ugunja	Local CBO through ICRAF	ICRAF	X		
Luhya villages					
Eshikhuyu	Village approach	**ICRAF,** KEFRI	X		X
Mwitubi	Catchment area approach	**MoARD,** ICRAF	X	X	X
Mutsulio	PLAR	**KARI,** MoARD, ICRAF, KIT	X	X	
Bukhalalire	Umbrella group approach	**KWAP,** MoARD, ICRAF	X	X	

NOTES: Survey analysis was undertaken in about 14 other pilot villages not listed here. Other organizations were also active with projects that may have included soil fertility in some of these villages, but were not the main soil fertility replenishment interventions of interest to this study. CBO, community-based organization; ICRAF, World Agroforestry Centre; KARI, Kenya Agricultural Research Institute; KEFRI, Kenya Forestry Research Institute; KIT, Royal Dutch Institute for Tropical Agriculture; KWAP, Kenya Woodfuel Agroforestry Programme; MoARD, Ministry of Agriculture and Rural Development; PLAR, Participatory Learning and Action Research; TRACE, Training of Resource Persons in Agriculture for Community Extension.

[a] The main disseminating organization is in bold. The rest joined in after the approach was in place and used it to reach farmers.

sampled). From this survey, issues of dissemination, adoption, and impact were analyzed. In all cases, analyses employed descriptive and econometric analytical methods.

The empirical evidence of both the qualitative and quantitative parts of the study suggests that lessons can be context-specific, differing across communities and households. The inherent heterogeneity of activities and strategies is part of a livelihood perspective that aims to capture processes of change and that perceives development as intrinsically nonlinear. Nevertheless, some findings were sufficiently widespread to enable the distillation of certain patterns of outcomes related to combinations or limitations on assets, institutional environments, vulnerability contexts, and other factors. These patterns emerge at different stages in this chapter and in our conclusions.

Context of the Research

Livelihood Context

Generally, rural households pursue and combine several livelihood strategies, both on- and off-farm. In Siaya and Vihiga districts of western Kenya, most households interviewed pursued at least one of the following sources of livelihood: rain-fed farming, livestock rearing, business, employment, and remittances from family members. In the pilot areas, most family members were children who were mainly occupied as students and depended on adults for their subsistence. As the case studies show, they also devoted time to assisting their parents or as caretakers in some kind of productive activity, such as weeding, planting, and herding, as well as activities related to housekeeping. The adults of the immediate household were all engaged in productive work on and off the farm. In terms of allocation of time, farming as farmers or as farmworkers is an important activity, but difficult to quantify. Among the adults, women are slightly more likely than men to be farmers. Men, however, are more likely than women to have nonagricultural casual jobs. Among other categories, there are hardly any distinctive differences according to gender. Full-time work off-farm was an important livelihood, mentioned by nearly one-third of surveyed households. Casual labor, although common and, in terms of monetary income, very important, was not often cited as a major livelihood source. Remittances and pensions were mentioned as important in only a couple of the case studies.

For those who rely on agriculture as a source of livelihood, maize and bean production dominates throughout, though some of the nonpilot villages are in drier zones where sorghum or millet—and, increasingly, sweet potatoes—are more common. Among the higher-value crops, vegetables are also important sources of livelihoods, but there are hardly any "industrial crops," such as tea, coffee, or sugarcane, grown in these villages. It must be emphasized that there are different types of farmers and farming systems and these are not static, but have changed over time.

One of the main features of the livelihood strategies pursued in rural Kenya is that several strategies are applied sometimes in combination, whereas others are applied in succession, with the possibility of making reversals. The question, therefore, is how these strategies are applied and under what circumstances. Many of the case studies show that households attempt to engage in several livelihood strategies. In some cases, the livelihood strategies are contradictory and therefore interfere with the success of individual strategies. For example, those efforts that combine off-farm with farm-related livelihood activities compete for the limited amount of labor available. The labor required, for instance, for agricultural production is often lacking—particularly for those households whose adult labor force is partly in town, and where the ability to access outside labor is problematic due to complex kinship relations. In other situations, however, these livelihood activities complement one another to the extent that many of them cannot be pursued in isolation. For example, not all off-farm activities compete for resources; they may also render resources that are used to strengthen the use of agricultural resources (financial resources generated from urban employment have benefited agricultural investment for some households).

The set of livelihood strategies pursued and the importance of any particular one may also change over time, for a number of reasons. In spite of investments already made in terms of farming knowledge and skills, some farmers easily shift labor from their own farm to take up casual employment. It is also apparent that the types and combinations of livelihood strategies that households are able to manage are often dependent on the availability of labor. Labor can be hired from outside the household, but poor households generally lack financial resources with which to employ such help.

Among the findings that therefore emerged as central in this impact assessment is the need to understand the driving forces behind the choices people make and why they sometimes persist with strategies that seem unprofitable. The various case study accounts suggest that, generally, choices depend on the resources at hand, perceptions of incentives (rewards and costs), the desire to belong (and the fear of isolation), and how events unfold both for the individual and his or her networks. Some of the issues that help to explain choices among rural households include people's notion of good farming and how this influences the type of strategies they pursue to earn a livelihood, and the nature of investments that they put in place, including the SFR technologies adopted. But intentions are easily thwarted by lack of resources or competition for them. Whereas the rural poor may be in a position to appreciate the dynamic changes around them, they are often unable to take full advantage of opportunities they perceive as potentially beneficial. Research and development organizations must thus be aware of the costs and trade-offs associated with different strategies, including the adoption of a new technology. In trying to make an impact on different types of households, it is important for researchers to understand the issues that drive decisions, and to make strategic choices in technology design that target households and individuals facing different conditions. Although not every circumstance and event can be accounted for, of course, this study demonstrates how quantitative and qualitative research can be used to understand livelihood strategies, resource constraints, social and economic trends, and beliefs and cultural issues. Such knowledge can be used to determine those technologies and dissemination methods most responsive to local conditions.

Local notions about good farming are based on people's aspirations, and these largely hinge on output and recognition from neighbors and friends. Hence people's livelihoods need to be interpreted as culturally embedded. Other evidence, however, suggests that reluctance to use mineral fertilizer goes beyond finances to include how people perceive the dangers associated with these technologies. For example, we noted the view that "fertilizers are spoiling the soil." In some cases, fertilizers were observed to have a negative effect on yields when rains were below normal. Other farmers claimed that once they used some fertilizer, they had to continue to do so just to maintain a reasonable yield. This view probably also has social and institutional dimensions: fertilizer is usually not affordable and often not available at the appropriate times. The choices that people make regarding farming activities—and which end up characterizing their farming styles—depend on how they perceive the practices that they engage in, irrespective of expert opinion.

We also explored the question of different types of farmers, and found that, generally, the livelihood strategies that households pursue vary with the gender of household members. Often, women diversify much more than men do, which means that their time available to invest in new methods or technologies for any one enterprise is limited.

Nevertheless, all these strategies and livelihood approaches are linked because they constitute identity and therefore provide a sense of belonging. Almost all the people tend to want to do that which is commonly undertaken in their home area. The view that a "good farmer is one who feeds his family first" is widely shared and may explain why a seemingly quite unprofitable, subsistence farming has persisted; sometimes just because people wonder about what the neighbors will say. Hence, the struggle to belong and the continued search for identity forces some people to undertake operations that they would otherwise gladly put aside. These shared notions are not static, but they change (though slowly) in the region. The relatively progressive farmers may still produce hybrid maize, often combined with local maize varieties to prevent an accusation of diverging from local cultural repertoires. For other farmers, custom is a reason to mainly plant local varieties (Mango and Hebinck 2004). Local maize is often combined with natural fallowing as was customary, even though seasonal fallows are not effective in raising yields. Again, these issues will affect uptake of new technologies, and it helps to understand them in assessing the potential and design of new interventions and dissemination strategies.

Concept and Measurement of Poverty

As expected by the small farm sizes and low agricultural production, most of the study sites include many poor households. Indeed, national statistics show that more than 50 percent of the population in many of the districts in western Kenya live below the poverty line of 1,240 Kenya shillings per adult per month (about US$16). In our study sites, qualitative and quantitative methods were used to better understand the local concept of poverty and to estimate the extent of poverty among study households.

Poverty or being poor appears to be a very slippery concept. Most commonly heard was the perception that "nobody is poor." The notion of "poor" or "poverty" is often not accepted and rather represents a rejected kind of personality, a deformed person. This opinion comes through in such statements as "poor are people who are handicapped. The poor stay and beg in towns, as they do not have land and shelter. At least I have a shelter."

Generally, most people are reluctant to classify themselves or others as poor. Instead, these people described their condition as a result of lacking various things and, according to most of the people interviewed, this condition is different from that of being poor. Poverty is perceived as a state of being unable to engage in productive activities, and this, it was felt, can only be the case for those people that are physically disabled. Poverty is also associated with lacking income both from employment and business. Poor people have small pieces of land, grass-thatched houses, and large families with children walking in tattered clothes and who have fallen out of school.

Although poverty is real and evident, there is an attempt by all people, the poor included, to run away from both the state of and the reference to being poor. But the main descriptors of poverty in the words of respondents are:

- Having insufficient land;
- Having no daughter or son on the farm;
- Being unable to feed one's family;
- Being unable to pay for education, health care, or the like;
- Wearing tattered clothes;
- Having unemployed children;
- Being physically disabled; and
- Having a house with a leaky roof.

The perception is that poor people engage in casual jobs and must always buy food but never have enough. Diseases like mental disorders are very common among the poor because they are unable to manage the condition. Physical disability is seen as a cause of poverty, and many people in this village who are physically challenged engage in begging at the market. Sometimes laziness and drunkenness are cited as causes of poverty within the community.

It is equally challenging to uncover many rich people. The "rich" are those that have something extra. Yet nobody admits to having anything extra. The notion of being rich is not favored or used at all in everyday discussions. Notwithstanding the murkily defined notions of "rich" and "poor," it is widely acknowledged that the rich have more land (to leave fallow and thus profit more from improved fallow technologies), and they have more capital with which to hire labor.

One may question whether rural poverty is best reduced through improvements in agriculture. In densely populated areas with small average farm sizes, there are obviously limitations as to how much income and production can be generated by a household. However, nonfarm job growth has been slow and itself is driven, especially in rural areas, by the success of the agricultural economy. In the short run, there appears to be no alternative for poverty reduction than through intervening in agriculture. A model for this approach exists within Kenya itself. The Central Kenyan highlands, with similar small farm sizes of 1–2 hectares, boasts extremely high productivity and income, and the lowest poverty rates in all of Kenya. This has come about despite lack of good roads, but with some strategic public investments in high-value enterprise sectors and dynamic private investment by farmers.

In terms of measuring the degree and extent of poverty, we used different methods, ranging from quantification of assets, expenditures, and consumption to using farmers' perceptions of their own poverty to enumerator assessments. Though the different measurements were not always strongly associated, each resulted in the estimation of a large number of poor households. In the pilot villages 50–66 percent of households are classified as very poor, depending on the measure used. Similarly, in the nonpilot villages 41–50 percent of households were classified as very poor. Table 5.2 shows the results of a comparison among three different poverty assessments in the nonpilot villages: one based on value of assets owned, a second on farmers own relative ranking, and a third on enumerator evaluation of the asset and welfare conditions of the household. Although the percentages of poorer, middle, or wealthier households are similar across poverty measure, the different methods of poverty measurement classified different households into each of these groups. For example, only 28 percent of households were classified into the same wealth group by all three measures—that is, 72 percent of households could be classified into more than one wealth group depending on the method of poverty measurement. Moreover, 13 percent of households were simultaneously classified as wealthier and poorer, depending on the method used. Poverty is thus not an easy condition to identify—and it will be differently measured depending on how and by whom it is defined.

Adoption of Improved Fallows and Biomass Transfer

There are distinctive patterns of adoption inside and outside the pilot area. Inside the pilot villages, there was a rapid surge of users between 1997 and 1999. The use rates reached about one-quarter of households for each technology (Figure 5.1). A significant decline in use occurred in 2000 followed by a recovery in 2001. In 2001, 16.7 percent of households were using improved fallows, and 15.2 percent were using biomass transfer. A likely interpretation is

TABLE 5.2 Distribution of poverty in nonpilot villages of western Kenya using alternative classifications

	Months sustained by assets	Farmer relative ranking	Enumerator evaluation
Wealthier group	15.0	10.0	9.4
Middle group	43.8	40.1	48.8
Poorer group	41.3	49.9	41.8

NOTE: Entries are percentages of 360 households.

that considerable technical support—along with the bandwagon effect—may have led to early high rates of testing. Subsequently those who did not receive sufficient benefits (including those who were mainly interested in selling seed) or were unable to manage after ICRAF and partners reduced backstopping efforts (see below) abandoned the new technology. Finally, in 2001, when the villages adjusted to being weaned from ICRAF support, some early testers retried the systems and new testers surfaced.

Outside the pilot villages, the dynamics were much different, with steady increases found over time for both technologies (and other SFR technologies as well).[1] Starting with just about 5 percent of households using agroforestry in 1997, by 2001, 12.4 percent were using improved fallows and 21.6 percent were using biomass transfer (but not necessarily adopting). There appears to be a leveling off of interest in improved fallows, whereas the trend for biomass transfer continues upward. In both pilot and nonpilot areas, there are a number of new testers of the technologies, so, for example, many of those contributing to the 2001 rebound in the pilot villages are new testers.

Further differences are noted in the use of SFR within and outside the pilot villages. Within the pilot villages, 54 percent of those who use agroforestry use both the improved fallows and biomass transfer. Outside the pilot area, only 38 percent of users are using both systems. Thus, when households have less interaction with researchers, they more often than not prefer only one system. Where ICRAF technicians were present, perhaps some households used both systems to please the scientists.[2] In the pilot villages, 88 percent of new testers are trying just one of the systems, which further supports this hypothesis.

1. Although use of agroforestry was monitored annually for all households within the pilot villages, the data for the nonpilot villages were based on recall from 2001. Moreover, the nonpilot village sample was stratified on the basis of use of agroforestry, so actual percentages of users from the sampled farmers are not necessarily representative. A separate census was done that indicates the overall use rates (see below).

2. Thus it might be better to dissociate ICRAF technicians from the dissemination of technologies and the collection of data. Among social scientists this effect is known as the "courtesy bias."

FIGURE 5.1 Adoption patterns of improved fallows and biomass transfer in the pilot villages, 1997–2001

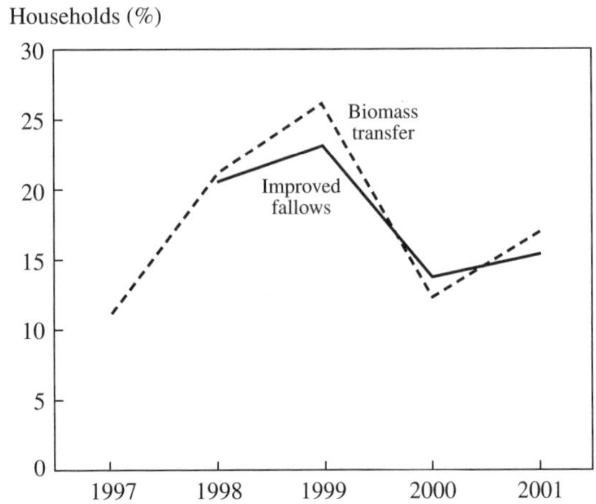

NOTE: Percentage of 1,630 households.

Because of reasonably lengthy exposure to agroforestry within pilot villages, it is possible to classify households into different categories of adoption. The adoption dynamics for each technology were summarized in four mutually exclusive outcomes:

1. Households that never used the technology (nonusers);
2. Households that used the technology early on but never again (disadopters);
3. Households that did not use the technology early on but used it recently (recent testers); and
4. Households that used the technology throughout the period (adopters).

As indicated in Figure 5.2, the highest proportion of pilot village households had not tried either technology as of 2001, about 60 percent in each case. A greater percentage of households have adopted improved fallows (22.0 percent) than biomass transfer (15.0 percent). However, about twice as many households have recently tested biomass transfer than have tried improved fallows (14.6 versus 7.6 percent). For each technology, about 10 percent of households tried and then dropped the practice. This rejection may be due to disappointment with the performance of the technology, the realization that the investment required is too much to bear, or the farmer was initially using the technology for other purposes, such as to sell tree seed to ICRAF or develop closer ties to external organizations.

FIGURE 5.2 Adoption patterns of improved fallows and biomass transfer in the pilot villages by 2001

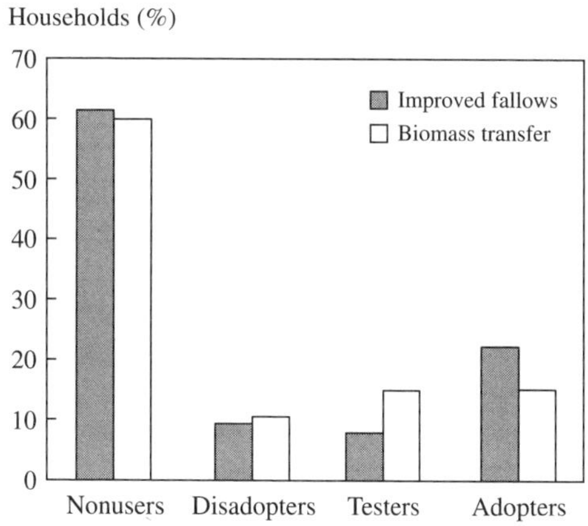

NOTE: Percentage of 1,630 households.

Outside the pilot villages, censuses were done for six different sites (about 1,000 households). Because the rates of use are expected to be relatively high compared to other nonpilot villages (indeed, it is one of the variables used to select these villages), these should not be taken to be representative of dissemination success. Rates of use are very high in five of the six sites (about 24 to 59 percent), which is encouraging, given that direct technical support from the project in these sites has been relatively low. However, these villages have received significant support from other intermediaries. ICRAF's long-term plans in the pilot villages were not clear to the farmers, and many expected continued long-term support from an international organization. Thus, when ICRAF reduced its local support in favor of a broader outreach strategy, the number of users declined. The limitations of involvement of the intermediating organizations in the nonpilot villages should have been clarified for the farmers.

Average fallow area in the pilot area was highest in 1998, dropping to a low in 1999, and recovered somewhat in 2000 and 2001. Fallow size was reduced in 1999, partly due to lower rainfall and seed supply constraints and partly because ICRAF began diverting some attention from the pilot areas to scale-out information to other places. In 2001, the mean fallow size (among practitioners) was 0.04 hectare. Although seemingly not much, the average farm size in the pilot villages is 0.73 hectare, of which perhaps 0.3–0.4 hectare is under maize. Further, the fallow system calls for a rotation of a fallow followed by three seasons of maize. If this pattern is followed, one would expect only one-fourth of the maize area to be under fallow at one time—or 0.075–0.1 hectare. Viewed this way, adoption intensity among those using fallows would be quite high.

Planting *Tithonia* to provide the organic matter for biomass transfer systems is perceived as increased investment in the system. It also reduces the subsequent labor required for collection of the material off-farm. Considerable planting occurred in 1998, 1999, and 2001, when more than 11 percent of households planted in each year. Curiously, the percentage of households planting in 2000 was much lower (4.2 percent). Whether the decline is a sign of saturation or an anomaly is unknown.

Who Adopts Agroforestry Systems: Qualitative Findings Related to Decisionmaking and Social Context

The issue of who decides to adopt SFR technologies is complicated. Among the Luo and Luhya, the husband customarily makes such decisions. In our study sites, women have to ask their husbands' permission to attend seminars and meetings called by ICRAF and other agencies to disseminate SFR. This observation does not imply directly that women do not participate or that they have no say in such matters. In fact, in some of the households, women farmers took the lead in acquiring information about and testing agroforestry on their farms. Decisions appear to be made at the nuclear household levels in many cases. For instance, one co-wife may adopt and another may not. One son in a Luo compound may adopt and the others may not. Nonetheless, women hardly mention that they decide what to do and prefer to give their husbands the credit. It is also evident from some of the case studies that the decision to adopt or not to adopt has brought about disagreements, some of them at the level of the family unit.

An interesting difference occurs at the level of pilot versus nonpilot villages. Women are active adopters of the new SFR technologies in the pilot villages, but not in nonpilot villages, where men were more often mentioned as the main adopters. One explanation is that in nonpilot villages people would have to search for new knowledge from a distance, whereas in the pilot villages, the information is locally available and women are able to attend the dissemination activities more easily. In fact, there were significant efforts to provide information to all social groups and individuals within the pilot villages.

The men who had adopted in the nonpilot villages were those who were known to have connections with people from outside the village, and they have been exposed to other development work through exchange visits by different organizations within and outside the village. This difference is due to two factors. First, looking at their schedules of activities and chores around the household, men have a considerable amount of time that they can be away from home to acquire new knowledge. Second, due to their larger "social space," they can easily interact with other people and attend exchange visits and other meetings that are not yet well known to many people in the village.

Education was not found to play a major role in farmers' decision to adopt. Women, who had less education, excelled in the uptake of the new technologies as long as the explanations were given in the simplest terms possible. The qualitative data on knowledge before, during, and after the project suggest that women understood SFR technologies better than did the men.

Generally, most participants appreciate the work of the ICRAF-KEFRI-KARI staff based at Maseno, but mixed feelings also exist. They "love the Maseno people" for their inputs, but do not like the "agents." The term "agents," or more specifically "ICRAF agents," refers to certain villagers who were the focus of attention from ICRAF-KEFRI-KARI as these organizations introduced and disseminated agroforestry technologies. These agents were not selected by the villagers but were selected by ICRAF staff. Because of the perception of bias, some farmers stopped attending meetings and workshops organized by ICRAF. The farmers blame the ICRAF staff for heavily relying on the agents to choose people to attend seminars and workshops. Questions were raised about how the agents maneuvered themselves into such strategic positions. In both the Luo and Luhya villages clan and political party affiliations appeared to play an important role.

In the pilot villages the different modes of interaction with villagers (in some cases, researchers rented land to conduct experiments) created confusion and led to perceived social differentiation among villagers. A related issue is the ownership of trees: often trees are perceived by the farmers as "CARE trees" or "ICRAF trees" rather than their own trees.

We need to interpret such images and views as well as favoritism as part of the social relationships that emerge over time between such institutions as ICRAF and individual farmers and communities. Rightly or wrongly, such perceptions of favoritism have shaped some negative views of agroforestry-based technologies.

The SFR-project generated money, particularly in the beginning when seeds from improved fallow trees could be sold at high prices. Quite a few people took advantage of this situation and made money from the early seed market. Some farmers decided to adopt after ICRAF promised to purchase the seeds; others adopted with the hope that the many visitors who used to come to the demonstration site could give them money and farm implements. When the market for seed stagnated, many farmers abandoned the technology, as their expectations were not met.

At this point we can elaborate in more detail the typology of agroforestry users. *Seed adopters* are those who saw the opportunities that the seed market presented at the start of the project. Fetching the relatively high prices of seeds stimulated these adopters to grow the seeds that were collected by ICRAF. Most seed adopters dropped out of seed provision as the prices for seeds went down and the seeds were no longer collected by formal organizations. *NGO networkers* are individuals who, through their early involvement with agroforestry and ICRAF, maneuvered themselves into strategic positions to gain access to resources distributed by NGOs and other projects or programs. Their involvement with agroforestry in the capacity of village elder or secretary of a community committee made them known to other agencies. *Keeners* are those who perceive agroforestry as a good addition to the many ways to replenish soil fertility. They are keen on agroforestry, as it increases yields and reduces monetary costs of maintaining soil fertility. *Nonusers* and *disadopters* are, respectively, those who never tried agroforestry and those who stopped using agroforestry because of various reasons including labor or land shortages. The major factors behind nonuse or disadoption are explored in more depth in the quantitative analysis.

Quantitative Findings from Pilot Villages

In both the pilot and nonpilot study areas, regressions were run to examine the effect of several explanatory variables on the different classifications of use or nonuse of improved fallows or biomass transfer (examined in separate models). The explanatory variables pertain mainly to household-level factors, such as household structure and resource levels. All the explanatory variables included in the model are predetermined in relation to the adoption variable, as they were collected at the beginning of the study period. In the pilot villages, multinomial logit regressions tried to identify factors that distinguished nonusers, disadopters, recent testers, and adopters from one another. In the nonpilot villages, a similar multinomial logit analysis was used to identify factors that distinguished nonusers, occasional users, and frequent users from one another. The regression results can be found in Appendix Tables 5A.1–5A.4.

IMPROVED FALLOWS. In the pilot villages wealth was not related to use of the improved fallows, suggesting that the different use patterns are neutral with respect to wealth—the poor are as likely to adopt as the wealthy (Appendix Table 5A.1). Household type was also not related to adopting improved fallows—the technology is being adopted by female-headed and other nontraditional household structures as frequently as by the more common male-headed monogamous household. A final variable linked to poverty shows a different pattern. Nonusers of fallows have smaller farm sizes than do disadopters and adopters (although it should be noted that 97.6 percent of households in the pilot area have farms of 2 hectares or less).[3] Somewhat encouraging is that households who are newly trying improved fallows tend to have farm sizes indistinguishable in size from nonusers. Using the land/adult labor ratio in an alternative regression, it is found that greater ratios are positively related to the

3. Note that farm size is not always identified by rural households as a key criterion for wealth differentiation among households. For example, in the pilot villages, the correlation between farm size per capita and number of cattle is –.03.

adoption of fallows (though not significant for disadoption or recent testing). Thus, for adoption of improved fallows, land is a more important household constraint than labor.

Among other variables, being in one of the focal pilot villages (10 of 17 villages in the pilot area that received more technical assistance) was instrumental in testing fallows at an early date, regardless of whether the practice was continued. However, location is not important for recent testers—suggesting that recent testing is less related to technical backstopping, other external motivations, and to the sheer number of existing users. Early use was similarly higher among Luos compared to Luhyas. However, as was the case with the pilot location variable, new testers are equally likely to be Luhyas or Luos.

Education levels and age of the household head were not related to adoption of improved fallows (or to early testers). But older household heads and those with a secondary education were less likely to have disadopted fallows rather than having never used one. In other words, comparing disadopters to nonusers, the former tend to be younger than the latter.

BIOMASS TRANSFER. For biomass transfer in the pilot villages (Appendix Table 5A.2), the wealth index variable was again not related to adoption of biomass transfer compared to nonusers (thus wealth is not linked to the adoption of either agroforestry practice).[4] However, wealthier households are more likely to have disadopted or recently tried biomass transfer. Thus if the recent testers become adopters, those who adopt will be described as more wealthy. The structure of household is not related to the pattern of use of biomass transfer, so that the technology is completely neutral with respect to household decisionmaking structures. The size of farm is positively related to the adoption of biomass transfer, though not to decisions to disadopt or test in recent times. However, the supply of labor is also very important in the use of biomass transfer (for all three outcomes involving use). When the land-labor ratio is used as a regressor (rather than the two variables independently), it is not significantly related to any of the outcomes, implying that neither land nor labor dominates as a constraint.

Luo ethnicity and being in a focal pilot village are positively related to adoption. The lowest adoption rates are among Luhya households in nonfocal villages. New testers are likely to follow these patterns. Because external assistance has largely been withdrawn from these sites, the emergence of new testers may indicate that there has been significant farmer-to-farmer learning in which large concentrations of early users leads to large concentrations of new testers. The reason for higher use of biomass transfer among the Luo is not clear.

---

4. A wealth index was created by using a principal components analysis of about 10 wealth indicators identified by farmers in the pilot villages. Among them are number of cattle, hiring in or out of labor, purchase of fertilizers, nonfarm jobs, and quality of dwelling.

One hypothesis is that their strong subclan affiliation may lead to increased use among clusters of households. But we find only partial support for this hypothesis, with very high or low rates of adoption in about half the Luo villages, but moderate levels in the other half.

Education and age play a stronger role in use of biomass transfer than they do for improved fallows. Better educated household heads are more likely to have adopted biomass transfer than are uneducated heads. Similarly, there is some support for the hypothesis that more education leads to less disadoption than nonuse. Age of household head is not statistically related to adoption, but younger heads are more likely to be recent testers as well as to be disadopters, compared to those who had never tried biomass transfer. So younger household heads seem to show great interest in biomass transfer but have not always had sustained interest or the ability to maintain the practice.

Quantitative Findings from Nonpilot Villages

In the nonpilot villages, there are hardly any statistically significant results among the household variables, in contrast to the results from the pilot villages (see Appendix Tables 5A.3 and 5A.4 for the nonpilot villages).[5] One possible reason is that the number of observations is about 20 percent of those in the pilot villages, and standard errors of estimates will be higher, all else being equal.

The only household variable that was linked to the frequent use of improved fallows was one of the wealth variables (farmer perception of relative wealth), in which case the wealthier households were more likely to be frequent users as opposed to nonusing households. The same variable was positively related to infrequent use, and the enumerator evaluation of household wealth was also positively related to infrequent use. So, although not all the wealth variables are producing similar results, there are indications that wealth is important in the use of improved fallows. The only other significant result in the fallow regression was that Luhya households were much more likely to be infrequent users than were the Luo. Unlike the pilot villages, farm size is not significantly related to use of improved fallows, which is surprising, given that both the mean and variance of farm size is greater outside than within the pilot area. A similar pattern emerges for biomass transfer. Only the wealth variables are related to the use of biomass transfer. In particular, the asset and farmer measures are positively related to frequent use of biomass transfer. The farmer measure is also related to infrequent use, and the enumerator evaluation of wealth is weakly positively related to infrequent use. No other household variables were statistically significant in the regressions. When the wealth variables are omitted altogether, the only change in statistical significance is with

5. There were many significant results among the location dummies.

the labor variable in the biomass transfer regression, which now becomes significantly positively related to frequent use.

There is a positive link between wealth and the uptake of the technologies, in contrast to the findings in the pilot villages. This finding may reflect the extra attention given to reaching the disadvantaged groups within the pilot villages. It could also be partly attributable to different measures of wealth, because the surveys were not identical within and outside the pilot villages. Also, although farm size and labor constraints were apparent in reducing the uptake of improved fallows and biomass transfer in the pilot villages, such constraints did not emerge in the nonpilot areas. There is a marginally positive effect of labor on biomass transfer in the nonpilot areas, but the impact of farm size is almost nil. This issue requires further investigation.

Impact on Livelihoods

Evidence from Qualitative Case Study Syntheses

It is generally observed that, from the farmers' point of view, the various SFR technologies adopted have increased farm yields, raised household incomes, and improved food security and the ability to mitigate vulnerable situations. Among the 44 cases studied, 21 had tried the agroforestry technologies and 18 observed an increase in maize yield. Of the 18, 12 provided a quantitative estimate of the yield increase that was 100 percent or more for the season following the fallow. Generally, people appreciate the range of technologies that have become available over time. The choice to use them is shaped by certain incentives and disincentives. The biggest incentive is the incomes deriving from the sale of seed, increase in yields, reduction in the "hunger period," the medicinal value derived from some of the shrubs, and general improved welfare due to raised farm incomes. The various case study accounts, however, also suggest that actual impact depends on the circumstances under which these SFR technologies are adopted.

Generally, the SFR technologies adopted have given some members of the community an amount of social capital, especially in terms of their being seen as successful farmers and people who attract visitors from "far away." Indeed, some of these visits have been so eventful that several families have named their children after these personalities. However, the decision to adopt or not to adopt SFR technologies as a livelihood strategy has created jealousy and discord, some at the level of the family unit. In one case, both husband and wife now pursue different farming practices just because they would like to be different and even be seen to be pursuing different styles. In this case, it was the man that came into contact with the new SFR technologies, and because he had been a "drunkard" and held in low esteem within his community, the wife was not convinced that his farm practices would be anything to emulate. In at least one instance, then, the introduction of these SFR technologies has resulted in status inversion.

The full potential of some of the SFR technologies is realized on only a few farms. The qualitative studies show that, where some of the larger impacts have occurred, the successful households had above-average human capital resources or more diverse livelihood strategies on which to build. Some farmers were not yet able to benefit from SFR to a significant extent because they were too old or poor to undertake the complementary investments (such as the purchase of improved maize seed) to realize good yields. So these agroforestry technologies appear to have mixed implications for reducing poverty. On the one hand, their use by poor households is a positive sign. On the other hand, the success of these SFR technologies in generating significant improvements in welfare depends on the household's ability to manage the complexities and opportunities stemming from the introduction of SFR. To summarize, the impacts of SFR are noticeable in terms of yield increases, but for most households, these have not been large enough to translate into significant welfare improvements. Even in those cases with large increases in yields, the additional income was used by some men to take a second wife or to enter into the commodity market, taking away the control that women had over subsistence production.[6]

Nevertheless, there are those households that have succeeded. The question is: why? Generally, adoption is intertwined with ongoing social processes, and the success of SFR technologies then depends on the entire social framework. Who benefits and why can only be understood within the context in which these technologies are disseminated and implemented. Some of the key points of differentiation include people's resource base, the type of livelihood strategies that they choose to pursue, the nature of vulnerabilities facing them, the likelihood that these risks can be easily mitigated, and the gender and power relations governing their social system.

The various case studies suggest that social networks are crucial to one's ability to derive benefits from SFR technologies. For instance, some of the farmers only got to know about SFR from friends and neighbors who were already enjoying the benefits (see the next section). Besides being able to transfer the knowledge and skill required, such association was testimony to the potential benefits and a driving force behind the decision to take up SFR technologies. In cases where people may be unable to have direct contact with formal disseminators, there is still the possibility of acquiring information from other farmers.

Households that have diversified their sources of income cope better with some of the demands of implementing SFR technologies. Those households

---

6. This finding is consistent with other research showing how technology impacts are "gendered." For example, Carney (1992, 1993), in a study of technology change among Gambian subsistence farmers, shows how policies promoting a shift to irrigated rice and vegetables in wetlands initially increased women's production and household earnings. However, they also caused men to enter these activities, leading to arguments over work and income distribution. Eventually, many women lost control of the income they previously had.

with little land or that are unable to cope with unpredictable labor demands (for example, through substituting household labor with hired labor) found it difficult to participate in SFR. The various SFR technologies have managed to improve availability of cash incomes. However, the sharing of these resources depends on the social relations in each household.

Whether these technologies have been successful in targeting the poor depends on who is classified as poor and whether they are actually controlling and able to access the resources that are required for implementing these practices profitably. Generally, both biomass transfer and cultivation of improved fallows do best among smallholder farmers, most of whom engage in subsistence production and are most often poor. But this category of rural dwellers is vulnerable to many things beyond their control, such as rainfall patterns and poor infrastructure that inhibit the strengthening and use of people's resources (development of skills through education, and marketing of crops). And, even in the event that they do have a surplus for sale, they are faced with obstacles that include lack of markets and market information, poor and noncompetitive prices, and relatively little negotiating power at the economic and political levels. Thus smallholder farmers face multiple constraints, and the effects of any single type of intervention, such as technology, will be seriously limited unless accompanied by other interventions to overcome other constraints.

Furthermore, the possibility that the SFR technologies will succeed in involving the most destitute households becomes remote with a reduction in the farmers' resource base. Even though the physical location of the project could be appropriate, the requirements of the technologies may not always accommodate farmers that are very poor. Most of the inputs required, including labor, are not necessarily available among the poor in amounts adequate to make the adoption sustainable.

Evidence of Impact from Quantitative Assessments

EFFECT OF SFR ON MAIZE PRODUCTIVITY. In this section, we present the results of several analyses linking the use of SFR and crop yields. This section is distinct from the econometric analyses in the following sections because unlike other indicators, no baseline yield estimates had been collected from farmers' fields.[7] We include data from our surveys of farmers and researcher-designed / farmer-managed trials.

Within the pilot villages, two analyses were made. First, we present findings from farmer-managed trials of improved fallows and biomass transfer within

7. Because ICRAF scientists had been conducting rigorously measured experiments with farmers, the need for a baseline among randomly selected farmers was not appreciated. In practice, establishing a baseline is tricky, because farmers would select a portion of a large maize plot on which to try a fallow. Because there is within-plot heterogeneity, it would have required yield measurements at very small scales.

the pilot villages. These trials are the most reliable in terms of assessing the biophysical effect of SFR because while farmers managed them, they kept non-SFR management options similar between their plots and technicians measured yields accurately with scales. The improved fallow trials involved about 70 farmers, and yields from control and treatment plots were carefully measured by technicians for four consecutive seasons. The control was the planting of maize with no nutrient inputs in every season.[8] The improved fallow trials involved one season under trees with two or three seasons of maize (so that the opportunity cost of land is included in the calculation). The yield effects were noticeable—in a four-season rotation (one season fallow, three cropping), the total production in the improved fallow system was 5.1 tons, or 1.71 per season compared against 4.4 tons and 1.1 per season under the continuous maize system (four seasons of cropping). Another three-season rotation found total maize of 4.5 tons (or 2.25 per season) in the improved fallow system compared against 4.2 tons (or 1.39 tons per season) in the continuous cropping system. Both of these differences are statistically significant at the 95 percent level (see De Wolf and Rommelse 2000; Rommelse 2001), amounting to seasonal increases of 55–62 percent. The differences between improved fallow and natural fallow systems are even more pronounced (94 percent higher).

The economic analyses are also positive. The per hectare net present value for the three-year system using Crotalaria was US$351 compared to US$242 for the no-input control. Similarly, the four-year rotation using Tephrosia gave a net percent value of US$588 compared to US$405 for the control. Both these results are statistically significant at the 95 percent level (Rommelse 2001). These translate to discounted seasonal per hectare net gains of between US$36 and US$61 per hectare, both being 45 percent more than the base case of no inputs. The same set of trials also assessed the returns to labor from fallowing systems that were found to be around US$2.17 per day, 33 percent higher than from no-input, continuous maize production. These results are largely congruent with the information generated from the qualitative case studies reported above. Although the magnitudes are not large in the sense of lifting people out of poverty, it should be noted that maize yields have been flat in Africa for decades despite adoption of high-yield varieties, including in Kenya. Moreover, our case study respondents unanimously felt that the trend in their maize yields prior to the study period had been downward. Finally, economic returns are not the only consideration. Where farmers are growing maize primarily for food rather than

8. There is no single established practice. Only 20 percent of the 1,600 farmers in the pilot area used fertilizer. Many farmers (70 percent) use manure, but principally on kales (60 percent of farmers), napier (41 percent of farmers) and other cash crops (18 percent of farmers). About 17 percent of farmers indicated that they used no nutrient inputs at the beginning of the dissemination of SFR. Trial data on fertilizer effects are available but not for manure.

cash, they may be reluctant to spend heavily on monetary inputs for maize, making SFR an attractive option.

Returns to biomass transfer on maize in researcher-managed trials conducted during the late 1990s were significantly positive, but farmer-managed trials were inadequately designed, testing combinations of large doses of biomass with large doses of rock phosphate. To find more profitable opportunities, farmers have directed soil nutrient inputs to higher-value crops rather than to maize. Farmer-managed biomass transfer trials with kales and tomatoes have shown that similar increases in yields are obtained on these crops. Because they fetch much higher prices, returns to land are much higher than on maize. For example, returns to biomass transfer on vegetable production were high, with returns to land reaching as high as eight times that with no nutrient inputs (resulting in average net gains of between US$600 and US$1,000 per hectare).

A second analysis is a production function estimation involving many of the same households covered in the household welfare impact analysis below, but for the 2003 long rains production season, the year after the main study was completed. The data on yields and inputs are from farmer estimates; although many of the variables are reasonably reliable, there are problems with the biomass transfer and improved fallow variables. There was a mismatch between farmers with agroforestry in this survey and those from the annual monitoring exercise, and only 10 biomass transfer observations were obtained among 150 plots. Moreover, the average size of an improved fallow plot in this estimation is 0.14 hectare, which is significantly larger than the average size fallow measured by enumerators (reported earlier as 0.04 hectare). Thus it is unclear whether the measured yields actually correspond uniquely to the effect of the fallow. For all these reasons, the reliability of these results is questionable. Nonetheless, we felt it important to conduct the tests. For biomass transfer, we found a positive and significant result on yields ($p = .04$). However, the coefficient value was extraordinarily high, beyond reasonable effects based on nutrient content of the biomass and thus is difficult to explain. For improved fallows, the coefficient was not statistically significant. As noted earlier, it is unclear whether the fallow plot and maize yields were properly matched up through the survey. Also, agroforestry researchers had observed that several farmers were testing improved fallows on their most degraded land, and we were not able to control for this possibility. Trying to assess yield effects of SFR is fraught with many challenges.[9]

In the nonpilot villages, we had less credible means for assessing the effects of SFR on yields and thus relied on farmer estimates and recall. Notably,

9. By comparison, fertilizer quantity had a positive and significant effect on yield, but the effect of manure quantity was found to be insignificant.

fertilizer, improved fallows, and biomass transfer all led to positive yield changes in almost all cases. Fertilizer was claimed to have a positive effect in 93 percent of cases and improved fallows and biomass transfer in 88 percent. Farmers claimed that the impacts were more than double that of the no-input case, but these observations could not be used in a formal analysis, as the residual effects in subsequent years were not made clear. Farmers in the nonpilot villages were asked to provide data on the area under maize in 2001 and 1997. One hypothesis is that if SFR can improve yields, it may catalyze shifts into other higher-value crops. The data show that 67 percent of farms did not change area under maize, 9 percent had decreased, and 24 percent had increased the area. ANOVA and correlation tests did not find any relationship among change in maize area and the use of biomass transfer, improved fallows, or combined use of SFR practices.

EFFECT OF SFR ON HOUSEHOLD LEVEL WELFARE INDICATORS. In the following three sections, econometric models are used to assess the effect of biomass transfer and improved fallow systems on changes in asset values, changes in nonfood expenditure, and changes in food and nutrition indicators. All these variables are tested on our pilot village sample, so the rest of this section relates solely to the pilot villages and that sample of households ($n = 103$).

The testing of the effect of the use of improved fallows and biomass transfer is not straightforward because they are also endogenous variables. So two-stage methods must be employed, in which adoption of agroforestry is explained in the first stage and the predicted values used in the second-stage impact regression. One requirement for this analysis is the identification of variables that may affect adoption intensity but not impact. This is not easy to do from a theoretical point of view, because adoption and impact on welfare are very closely related. However, we were able to find six variables whose relationship with adoption is much stronger than with impact (for example, the jobs and social positions held by their fathers). The full rationale for this exercise is in Place et al. (2005). In addition, the measurement of adoption of the agroforestry systems must be addressed. Here we are interested in how different degrees of use may affect changes in asset holdings and other welfare indicators. This aspect is best measured by a continuous variable that can capture intensity of adoption over space and time. For improved fallows, the sizes were verified by enumerators in annual monitoring exercises. The sizes of plots on which biomass transfer was used were not measured.[10] Hence, for the pilot villages, our intensity variables are the sum of total area under improved fallow and the sum

10. They were not measured by enumerators because they could not be observed in the field (the biomass is incorporated into the soil and is not visible), unlike most fallows that were in the field during the annual monitoring.

of the number of seasons for which biomass transfer was used. Both of these variables include only those seasons relevant to the study period (six seasons).

The intensity variables are continuous variables, but they exhibit a non-normal distribution in that there are large concentrations at the value of zero—which reflects all the households who never used the technology. We therefore have two approaches we can use in the two-stage procedure. The first is to proceed and run both first- and second-stage regressions using ordinary least squares. This method leads to biased coefficient estimates in the first stage, owing to the nature of the dependent variable, but the standard errors for the predicted values in the second-stage regression are unbiased. It is computationally the easier method, as it is simply a two-stage least squares procedure. The second is to run tobit models in the first stage, which gives unbiased estimates in the first-stage regression. The predicted values from the tobit model can then be used in the second stage. However, in this case, the standard errors are biased and a technique such as bootstrapping (see Place et al. 2005) needs to be employed to be able to give reasonable estimates of unbiased standard errors for the coefficients. Both methods were undertaken and tests were made for the strength of the first-stage regressions, for over-identification, and specification errors, and these are discussed in detail in Place et al. (2005). To summarize, the instruments selected for the most part seem to do reasonably well in terms of correlating with the actual value and reducing the correlation to the error terms (statistical tests raised concern in one case, that of the predicted improved fallow variable in the expenditure equations; see Place et al. 2005).

Three models are run for each indicator. A two-stage least squares model is run with such household variables as age, sex, education, and ethnicity of household head used as explanatory variables.[11] The household variables reflect pre-adoption values (in reality, few variables changed over time anyway). A second model is a difference equation and tests for the link between agroforestry use and the welfare indicators, factoring out household structural factors. The third model is a two-stage approach in which tobit models are used in the first stage to generate predicted values of the intensity of agroforestry use. In the second stage, bootstrapping techniques are used to improve the estimates of the standard errors of the coefficients. In the text, we present tables showing only the second stage of the two-stage least squares results. The other models gave virtually identical results.

EFFECT OF SFR ON ASSETS. If the yield impacts from SFR investments lead to sustainable increases in livelihoods, then one would expect to observe

11. We did attempt to include the base-period welfare indicator as a regressor, but, due to our inability to overcome possible measurement error and regression to the mean, these results from the models cannot be properly interpreted.

some degree of asset accumulation. The qualitative research found that this was indeed occurring for some households, but not all. Patterns were difficult to detect with a small sample, but it was evident that because of rampant poverty, households were hard placed to convert any gains from increased yields into tangible assets. The few that were able to increase assets reported gains in livestock and housing. Several quantitative analyses were undertaken to confirm whether these mixed results hold across larger populations.

Before discussing the links between SFR and assets, it is important to understand the context of assets and their change during the study period. Looking at the actual values, livestock composes 70–80 percent of the value of all liquid assets. The mean total liquid wealth held by households was US$408 in the current year in the nonpilot villages and US$236 in the pilot villages, whereas that of livestock was US$302 and US$178, respectively. A large number of households suffered through disinvestment in both livestock assets and total assets over the period. This result is remarkably consistent in both sites, with percentages incurring disinvestment ranging tightly between 47 and 54 percent. In general, households with higher initial wealth fared poorly compared to the less wealthy. Some of the more wealthy households saw their livestock holdings collapse, through sales to meet obligations (such as funerals) and disease (especially for poultry).

Two-stage methods are used to first predict the use of SFR and then to measure the effect of the predicted SFR variables on assets (see the results in Appendix Table 5A.5). Neither agroforestry use variable is significantly related to the change in assets. In fact, the only significant variable is farm size, where it is found that asset-holding positions changed in more positive directions where farm sizes were smaller. This result suggests that non-land assets are not highly correlated with land assets, a reflection of market imperfections in land relative to other assets. The general lack of significance among other variables indicates the existence of complex relationships that are not easily captured by more structural household variables.

EFFECT OF SFR ON EXPENDITURE. Expenditures were collected for the pilot village subsample of 103 households both in 1999–2000 and in 2002. The April 2000 survey matches exactly the period of the 2002 resurvey, and thus we report on and examine only the expenditures reported at these two visits. Expenditures were collected on a three-month recall, and therefore we exclude all food expenditures from this analysis (they are too difficult to estimate over three months; food consumption is handled separately below).

We analyzed changes in nonfood expenditures per household and per capita. For the latter, we divided expenditures by the number of household members. Mean nonfood expenditures in 2000 were US$97, whereas the median was US$60, indicating that there are relatively wealthy households bringing up the mean. The mean level of nonfood expenditures rose slightly to

US$104 over the period, and the median behaved similarly over time. Per capita nonfood expenditures, however, were flat over time, with a mean and median of US$16 and US$10, respectively. Despite this sign of stability, many households (44–48 percent) experienced a setback in welfare, as measured by nonfood expenditures.

Turning to the econometric analysis, the two agroforestry variables have the opposite sign (see Appendix Table 5A.6), though the positive effect of biomass transfer is not significant. The coefficient estimates for the fallow variables are negative and significant at around the 5 percent level (although in the tobit model the significance level is somewhat reduced after bootstrapping). There is no logical causal link between use of improved fallows and reduction in expenditures, so the result suggests the improved fallow variable may be capturing effects of omitted variables. For example, it could be that improved fallows are being tested by households with relatively less access to remunerative livelihood strategies. None of the other included variables were significant.

EFFECT OF SFR ON CONSUMPTION. Food consumption and nutritional measures were based on 24-hour-recall surveys of households (three visits in 2000 and two in 2002) during a relatively hungry period before the long-rain harvest. Household-level indicators of intake and nutrition were calculated based on age requirements of all consuming household members. Nutritional indicators were taken from Food and Agricultural Organization of the United Nations (FAO) and U.S. Department of Agriculture (USDA) sources, depending on which more accurately reflected the specific type of food consumed (for example, cooked kales).

The average household scored well in terms of energy, carbohydrates, iron, riboflavin, and niacin in both years. An analysis of baseline data revealed that maize accounts for 75 percent of total energy. There is some diminished sufficiency in folic acid in 2002 and low levels of protein sufficiency are reported in 2002. But even for those nutritional indicators that appear favorable in the aggregate, often many households are unable to meet the recommended needs. For instance, in 2002, 42 percent of households had consumed less than the recommended minimum requirement of energy, 53 percent were deficient in folic acid, and 73 percent were deficient in protein. A general decline in nutritional status occurred over the two-year period—in fact, none of the variables exhibits improvement over time.

Econometric analyses focused on those nutrition variables that exhibited significant change over time: energy, protein, iron, and folic acid. Neither of the agroforestry adoption variables was found to be significantly related to changes in food intake and nutritional status (Appendix Table 5A.7). In fact, the only significant variable in each regression was gender of the household head: female heads are associated with positive change (or less negative change) in each of the three indicators. Therefore the dynamics of food intake and nutritional status are very complex processes. They are not easily pinned down to initial characteristics of households and are likely to be related to myriad decisions and livelihood changes that took place during the period.

Dissemination of SFR Technologies: Comparing Methods, Experiences, and Impacts

Although much of the study focuses on technology adoption and impacts, the dissemination processes themselves were also studied. This focus was important because dissemination approaches used by organizations in western Kenya are intended not only to disseminate technology but to strengthen human and social capital such that farmers can continue the dissemination process inside the village and ultimately in other villages. In addition, dissemination methods and experiences affect these organizations' ability to reach the poor and women—in other words, the process of dissemination can have as much impact on adoption as the nature of the technology itself. It is thus important to understand the different approaches used by different organizations in western Kenya, people's perceptions of the implementation in practice, and the effectiveness in achieving the objectives identified above. The dissemination of technologies needs to be understood as part of a social process, the dynamics of which shape how people read technology. In one group discussion with women, a participant mentioned that "agroforestry is for Mzungu [foreigners or white people]." The women referred to the situation that had occurred 8 years earlier, when an ICRAF project led by a white person started with agroforestry trials in her village. ICRAF negotiated with the school for a piece of land, hired labor to dig and plant trees, and so on. Asking her to explain what she meant, the participant said, "we women are not able to negotiate for a piece of land and we do not have the money to hire labor. You see this is why it is for Mzungu." This and other case materials show that technology and the style of its introduction and dissemination are not neutral; they hinge on the nature and transformation of social relations.

As Table 5.1 shows, a range of organizations are disseminating SFR technology using different approaches. All share certain characteristics but differ in other ways. They all enter villages with the assistance of local administrators and seek to determine local problems and solutions through broad meetings or local groups. They then work with groups to facilitate the dissemination of new locally adapted technologies in a sustainable manner. These may be existing community groups (such as women's, youth, church, or self-help groups, or those based on clans or SFR practices) or new groups formed for this purpose. In some approaches, the long-term goal is for these groups to disseminate technology to other villages. Some also use umbrella structures formed of representatives from groups across different villages, to provide support structures and link to external organizations. All approaches use a variety of teaching methods, involving field days and demonstrations, observation, use of schools, and others (see below), though they have different emphases. For example, the approaches of Training of Resource Persons in Agriculture for Community Extension (TRACE; used by CARE) and Participatory Learning and Action Research (PLAR; used by KARI-Kakamega) emphasize substantial training of lead farmers who are then to disseminate knowledge to others (the PLAR approach selected farmers to include a range of farmer types). The village approach (used by ICRAF) worked both with groups and individuals, through a more formalized representative committee. The Ministry of Agriculture and Rural Development (MoARD) extension service's "catchment" approach was also implemented through a committee but did not interact much with existing groups. Figure 5.3 provides an example of the complex relations of institutions involved with SFR information exchange through the eyes of a group of poor male farmers in Bukhalalire.

FIGURE 5.3 Village map of institutions involved with SFR information exchange

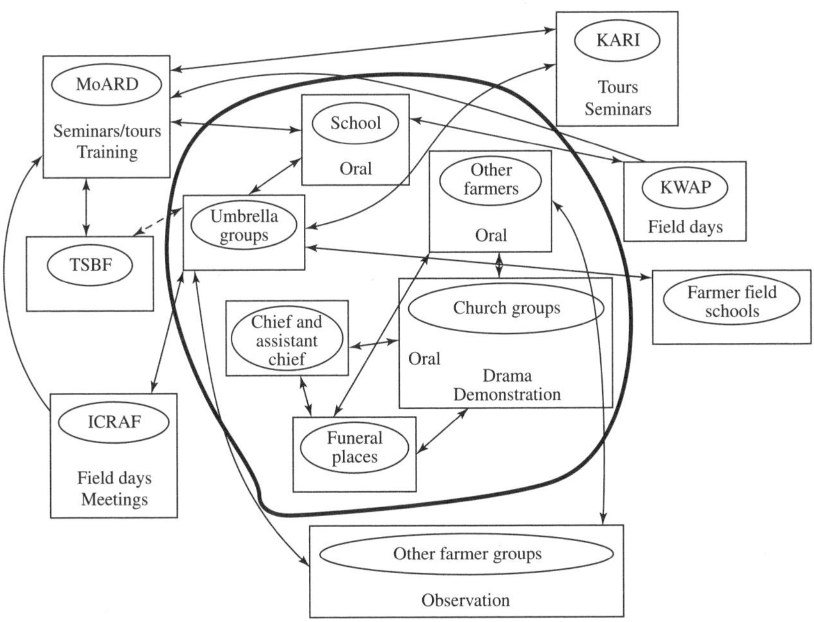

NOTES: Bukhalalire focus group, poor men. ICRAF, World Agroforestry Centre; KARI, Kenya Agricultural Research Institute; KWAP, Kenya Woodfuel Agroforestry Programme; MoARD, Ministry of Agriculture and Rural Development; TSBF, Tropical Soils Biology and Fertility Programme.

Evaluation of External Disseminating Organizations

Most villages in the study had one or more governmental organizations or NGOs working with SFR at some point. In each of the six focus group villages, there were multiple interventions that we touch on where villagers mentioned them. However, an effort is made to focus on the approach and institution that were particularly influential. Comparing organizations across communities is an imperfect measure because not all organizations were equally active in each village. However, the rough uniformity in evaluations and the fact that the most active organizations usually score higher suggests a basic degree of satisfaction. Aggregating across all focus groups, the four most active organizations (ICRAF, MoARD, CARE, and KARI) score approximately equally in the PRA exercises that evaluate them according to their usefulness and importance. Only ICRAF scores slightly higher. The Kenya Woodfuel Agroforestry Programme (KWAP) was also very popular in the one village where it worked. ICRAF was also favored for the time and effort spent with the communities and farmers. One of the main criticisms of external organizations is that they do not spend enough time with farmers.

There are still few differences when groups are disaggregated by gender (a result that holds for the qualitative and quantitative analyses). In the aggregate, CARE, KWAP, and KARI score higher among non-poor residents, whereas ICRAF and MoARD score higher among poor residents. It is not clear why; however, because poor farmers seem to have had less positive experiences with groups, they might appreciate ICRAF's and MoARD's greater use of direct farmer visits. Overall, the assessment of disseminating organizations is positive. The main problems raised were insufficient staff, insufficient time given to farmers, and brevity of intervention. As noted in a focus group of non-poor women, "what limits full implementation is that they are usually left before standing on their feet." Insufficient monitoring was also a problem.

Farmers' Assessments of Teaching Methods

Each dissemination approach uses a combination of teaching methods. These respond to criticisms of earlier dissemination approaches that were found to be overly top-down, insufficiently aware of local conditions, and favoring better-off farmers. The first group of methods involves forms of training organized by external organizations, such as demonstrations, field days, tours, exchange visits, and farm visits. A second set involves different types of meetings, formal or informal, that target specific individuals or are open to the public and that discuss future plans, resolve issues, monitor progress, or identify needs. Finally, there is observation of others' fields and conversations.

All three forms of teaching were popular, and people prefer a mix of them. Although they varied greatly by village, informal means, such as learning through observation, are highly rated, even though they did not involve external resources or organized activities. According to the focus group of non-poor women in one village, "few people learn from formal ways . . . but many do so informally through observation on other farmers' farms or orally from other farmers."[12] Poor men rank meetings very low, consistent with their comments that meetings are often dominated by elite men. Non-poor men's preference for conversation may represent their ability to rely on their social ties for information, or education and experience may be at play, where wealthier farmers may be more easily able to digest verbal descriptions and convert them to achievable plans. But synthesizing all the results shows that very few differences in opinions on methods emerge based on gender and wealth status. Significantly, people value the formal methods a great deal. Some specifically said that they would prefer more visits in their homes—the more traditional approach. This bias reinforces the key challenge for dissemination—how to balance the need for engagement with individual farmers with the need to reach a large number of them.

For the most part, the picture painted is one of information flowing mainly from disseminator to farmer, and less in the other direction. However, some degree of farmer input was solicited in all approaches, and groups from at least three of the six villages mentioned this specifically. For example, in one study village, poor women said that "in the initial stages, when contact is strong, our ideas are usually taken into consideration." In another study village, both poor men and women noted that ICRAF asked for their input in developing training manuals. Logistical issues raised were problems with meeting times interfering with funeral and market days, dissemination staff arriving late, overly long meetings, and the use of Swahili instead of the local language.

Local Institutions and Dissemination

In each of the dissemination approaches, the external organization introduces technologies and conducts training. However, these organizations cannot reach all farmers effectively, and a range of local institutions can be used to further the process. Furthermore, one objective of these approaches is to build capacity within the villages, including human and social capital, so that residents can continue carrying out dissemination activities with other farmers, and eventually in other villages. There are several means by which dissemination takes place using local institutions. These include *barazas* (community-wide meetings called by the local administrator); informal learning from other farmers; schools, where training is given to schoolchildren who are expected to teach

12. These findings support other studies that have found social networks to be very important to the diffusion of other types of innovations in the region. For example, informal women's groups were found to have facilitated adoption of birth control where cultural values and beliefs discouraged adoption and where other programs had failed (Rodgers et al. 2001; Behrman, Kohler, and Cotts Watkins 2002).

their parents; and "local leaders," referring to administrators, chiefs, and others. People also learn from the "contact farmer." The contact farmer experiments using the technology and adapting it to local biophysical conditions. Finally, there are what we call "SFR groups"—village committees, catchment committees, and umbrella groups, or sometimes women's groups, church groups, or the like. Because of the importance of the groups to each approach, and the implications they have for human and social capital, they are given considerable attention here.

Aggregating all villages, there is not a great deal of variation in ratings of the importance/usefulness of these internal providers. *Barazas* rank consistently highest, possibly because no one is excluded from them. However, although it is very good for imparting information, it does not facilitate exchange of information among participants, nor is it regular enough to provide follow-up support for a new innovation. There are also few significant differences in the ranking of the local institutions based on ethnic groups, wealth status, or gender, suggesting that these have worked fairly and that there are no inherent cultural biases. Although it was hypothesized that Luhya farmers might be less comfortable in groups, this suggestion did not emerge from the data. Women's, church, and SFR groups got roughly equal evaluations from poor and non-poor women, implying group-based approaches are working well for poor women.

Social Capital, Social Relationships, and the Experience of SFR Groups

As discussed above, each dissemination approach relies on local groups for disseminating the technology across a wide group of farmers and for ensuring sustainability. These groups were scored as relatively important sources of information, and in some cases were said to be working well, as noted by the poor women's focus group in one village: "committee members participated very much in organizing and mobilizing farmers." Yet they have also experienced many problems. In most cases, the groups were said to have provided benefits to their members. But in most villages there were reports that they had performed poorly with respect to providing information to other farmers. One problem is the lack of participation in the groups, either because of self-exclusion or exclusion by group members. Low levels of participation directly in the groups would not be as large a problem if the groups were conducting dissemination activities with other farmers, as envisioned. However, this effort has also been insufficient. Five of the six villages reported one or both of these problems, suggesting that these problems are not specific to Luo or Luhya villages. In some cases group members serve as models for other farmers to observe, as in one study village where they were said to "envision commitment and hard work as ways to spread technology, so that other farmers can observe the technologies as practiced by the committee members." Although this approach is helpful, the members are intended to engage in more active dissemination. Another of the study villages, where KWAP used an "umbrella group" approach, presents a very different picture from the other villages, however. All four focus groups described training and dissemination that the groups were carrying out, inside and outside the catchment. Women were particularly vocal about the group's activities. Although it is not possible to know whether this positive assessment is the result of the umbrella group approach or a more socially cohesive community, it is probably a combination of both factors.

In general, women tended to be more positive in their evaluation of groups than did men, particularly about women's groups.[13] Poor and non-poor women alike said that domination by men in the groups reduces women's participation and learning, reiterating the importance of having separate groups for men and women. In one study village, some women, church, and welfare groups were also agriculture groups that contributed to funerals, which can strengthen social capital as well as addressing people's priorities in the context of widespread AIDS. Existing groups that incorporated dissemination tended to be more active and sustainable than new groups formed for this purpose.

The study began with four hypotheses concerning social capital and social relationships:

1. Dissemination through local groups will enhance a community's social capital—the social networks, relationships, and organizations that facilitate access to resources;
2. Interventions might also introduce new social divisions;
3. Interventions, through skills training and participation in groups, can increase the confidence of farmers, leading farmers to make more demands of the groups and groups to make demands on external institutions; and
4. Interventions could reinforce existing power relationships.

All four hypotheses were found to be true to varying extents. In five of the six villages selected for the dissemination research, at least some groups said that SFR interventions and extension activities had brought their community closer together. However, local groups also introduced social tensions and politics. One or more of the following issues were reported in all villages except one: uneven distribution of resources, discord over extra attention that some farmers received from external organizations, failure of extension staff to visit farmers, the ability of some to amass wealth through the process, conflicts over resources, rivalry among leadership, and mismanagement of funds. In most of the villages, it was recognized that the interventions led to competition and conflict in some ways and to cooperation and cohesion in others.

13. This trend is also supported in the studies referred to earlier (Rodgers et al. 2001; Behrman, Kohler, and Cotts Watkins 2002) on the importance of women's informal networks.

Power relationships were reinforced by the exclusion of some people from groups. In five of the six study villages, elites or better-off farmers were said to dominate the groups to some extent, and in some cases the intervention strengthened their position. It is very difficult to avoid domination by local elites, especially in new groups formed for managing new resources. Also, the propensity to seek community leadership positions often hinges on the socio-economic status of an individual. In some cases, poor participants saw the process as less equitable than did non-poor participants, though even the latter sometimes acknowledged uneven capturing of benefits. Sometimes poor farmers acquired some power through the process, however. In one village, poor men said that farmers made demands on the committee when it was active and the committee, in turn, made demands for extension services on MoARD. With respect to the third hypothesis, some individuals reported that the interventions increased people's confidence, and in at least two villages, there were explicit reports of people making successful demands for changes to the external disseminators. Of the nine focus groups where confidence came up, six were women's and five were in poor groups.

The contact farmers (referred to earlier as the "agents") are mainly seen as the point of contact for outside organizations. Focus groups in three of the six villages indicated problems with the contact-farmer method (though in one, only poor men were critical). Although many groups were positive about the method, the findings provide important insight into unintended social consequences of dissemination methods: First, the contact farmer was seen as unfairly receiving too much attention from external organizations, as illustrated by comments from poor men's focus groups in two of the villages (for example, "the wealthy and educated who are frequently visited and make others feel left out and different from the preferred farmers," and "model farmers gained more prestige and control over other farmers as they trained them"). Second, contact farmers in two of the villages were not seen to have shared information with other farmers. Although it was possible to observe his fields, others did not copy what they observed because they did not believe that what was being done was for their own benefit. Some participants perceived that contact farmers were selected by the external institutions, although according to the organizations, they are to be selected by the villagers. Nevertheless, contact farmers are important for testing technologies and practices and adapting them to local conditions before they are disseminated to other farmers. There may thus be a period in which considerable contact between the contact farmer and the external organization takes place, before many other farmers are brought into the process. However, technological interventions involve social processes: social context affects adoption outcomes, and interventions affect social relationships. It may be necessary to bring the wider community into the process at an earlier stage to make sure people understand the role of the contact farmer and approve of the choice.

School Programs

In each village under study there was evidence that dissemination messages introduced in schools reached school children. Students had made efforts to train their parents, albeit weakly in some instances. In one of the study villages, women said that children made vegetable gardens, planted trees, and trained their parents. Children are said to still be practicing what they learned and have earned income through the activities. In some cases, parents "learn through observation in the school compounds." However, this means has been weakening over time. In three villages, focus groups noted that trees from school woodlots were a source of income and building materials for the schools.

The major challenge identified with the schools approach was that, in some villages, students hardly convinced their parents to adopt technologies. Additionally, formal agriculture lessons in the schools were perceived as so theoretical that students had nothing tangible to disseminate. Furthermore, the approach has shortcomings in that not all parents in the farming community have children in school. Moreover, in an African rural setting where children are considered ignorant and have no established forums for discussion with their parents and other adults in the community, the flow of information from students to their parents and communities is restricted.

Knowledge Acquisition

Although focus group participants have varying opinions of disseminating organizations and their methods, a good measure for assessing the performance of these organizations is the amount of knowledge people gained through the dissemination efforts. Seven SFR technologies were mentioned as introduced across all six villages: *Tithonia,* farmyard manure, compost, commercial fertilizer, rock phosphate, improved fallow, and terraces. Crop residue was mentioned in all but one village.

Focus group participants used "ladders" to show the amount of knowledge on the technology they had before and since the intervention, with zero meaning no knowledge and 100 being full knowledge (see Figure 5.4 for an example from a group of poor women in Sauri). For most groups, the starting point was zero. The most surprising finding about the amount of knowledge gained is its uniformly high levels. Although the range is from 33.3 to 76 percent, most are clustered around the mean of 49 percent. The two lowest-scoring technologies are rock phosphate and commercial fertilizers, probably reflecting that they are more expensive to obtain and thus fewer are using them.

For the two technologies of primary interest to this study, *Tithonia* and improved fallows, knowledge gain was 47 and 44 percent, respectively, consistent with the average for most technologies. The highest gains were claimed in the village that also reported the highest levels of satisfaction with the dissemination process (including success with group-based methods, using KWAP's um-

FIGURE 5.4 Knowledge acquisition: Sauri poor women's group

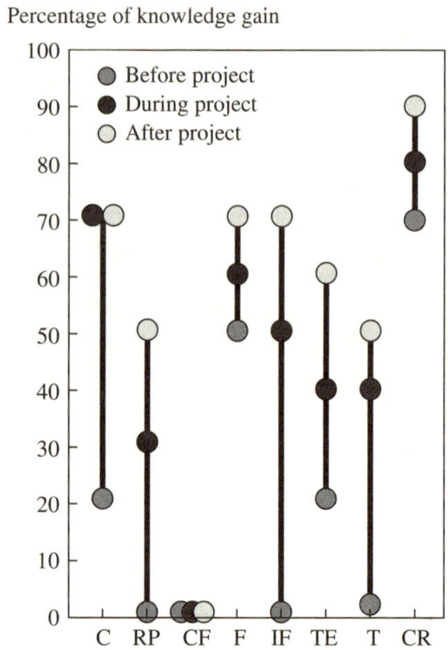

NOTES: Includes participants with at least four years of primary education. C, compost; CF, commercial fertilizers; CR, crop residue; F, farmyard manure; IF, improved fallows; RP, rock phosphate; T, *Tithonia;* TE, terraces.

brella group approach), suggesting a possible correlation. Participants in this village generally claimed to have learned more than the average on most technologies, implying that the dissemination method may be more important than the nature of the technology, at least with respect to learning about it. Quantitative analysis found that agroforestry knowledge acquisition was linked to direct contact with ICRAF, NGOs, or community-based organizations, but not to direct contact with extension or other farmers.

According to participants, those with more education generally learn more about the technologies than do those with less. Nonetheless, the difference is less dramatic than one might expect, indicating that disseminators are reaching vulnerable groups. There are not particularly large differences by gender and wealth. For improved fallows, poor participants reported learning more than did non-poor participants (the reverse was true for commercial fertilizers). There was no difference for *Tithonia.* Men on average claimed they gained more knowledge on *Tithonia* than did women, at 55 versus 43 percent.

Sustainability

Sustainability has two main dimensions here: financial and institutional. Project activity costs are an important challenge to sustainability after external initiating organizations phase out. In some of the villages, farmers demonstrated their willingness to share costs under certain conditions. A group of poor men in one village said that "farmers are ready to work . . . on a cost-sharing basis if only the organization is ready to stay in the village and tell farmers what will be benefited and steps to follow whenever problems arise." In another village poor women said that farmers also "provide plots and labor and take the risk associated with experimentation on the farm." Poor women in another village saw cost-sharing in a negative light. They explained that "a mere mention of the word money, paying for something, is enough to send some members of social/farmer groups packing." However, the major problem with sharing costs relates to the high levels of poverty in the project areas.

In terms of institutional sustainability, local administrators came across as important, because of their influential positions and ability to convene *barazas.* However, their involvement was ad hoc, with no training, which limits their effectiveness. Some committees and groups set up or adopted for dissemination work by external organizations continued to exist after the latter left, though in some villages groups had collapsed. Poor management, especially of finances, had also kept some groups and committees weak and ineffective in dissemination, and led to collapse in some cases. However, financial mismanagement was not a pervasive problem. Still, future projects should focus more on leadership and management training, to provide safe grounding for project activities and approaches after external organizations phase out. Farmers emphasized the need for external organizations to monitor what happens in the village after they leave, which would help to identify problems and encourage resolutions where possible.

Reflections on Methods

Reflections on the Sustainable Livelihoods Framework

As guided by the wider study of the impact of agricultural research on poverty, the sustainable livelihoods (SL) framework was used to identify the key research questions, provide a cross-disciplinary conceptual framework and language that helped to facilitate an integrated research design, and ensure inclusion of many key issues and relationships. However, the research team was conversant in a number of paradigms and methods that included similar concepts, and it is not clear that the SL framework was necessary to achieve the results we did. The team introduced other concepts not included or explicit in the framework as needed. Because the SL framework is not specific with regard to major direct relationships among variables, most variables are related to one another, resulting in the formulation of similar, overlapping, or duplicated research questions.

In the implementation of the research, the SL framework was implicit but not explicitly discussed. This observation is less a valuation of the framework than a reflection that the research team had agreed on the importance of a comprehensive and diversified research approach.

Integration of Qualitative and Quantitative Methods

As much as possible, qualitative checklists and survey questionnaires were formulated to provide insights into common issues and questions. The integration was useful for understanding different types of information—quantitative results led to identification of general patterns of adoption and impacts, and qualitative results helped to explain processes behind adoption choices, information flows, and impacts. Quantitative results are better at showing general trends, given the large numbers, but qualitative data were essential for uncovering issues related to culture, normative frameworks, and social dynamics. Integrating the methods was useful for investigating poverty, adoption, and dissemination issues. It worked less well with impact analysis, mainly because the quantitative research had a baseline as a guide but was limited to a small set of indicators, whereas the qualitative research did not benefit from a baseline but was broader in its scope. Although there is a great deal of complementary and supporting information, true integration requires the researchers to sit together and compare and contrast results. Limitations on time and funding, and the timing of the last survey, resulted in the team not being able to spend sufficient time to jointly analyze the research results. The research is therefore too compartmentalized into results from different methods.

Another issue was the usefulness of generating numbers from PRA exercises. These exercises were quite valuable in understanding relative assessments of different institutions, methods, and knowledge acquisition within villages. However, they were less useful and difficult to compare and analyze across villages, because of the different meanings attached to numbers, and because different categories were identified across villages. The categories could have been standardized, but doing so would have undermined the participatory nature of the exercise, stifling generation of local categories. Still, there was enough comparability across categories to allow for some meaningful comparison, if in places in broad strokes.[14]

Dialogue among Different Institutions and Stakeholders

Significantly, stakeholder meetings helped to plan and review the research. Having issues raised by residents of the region added a degree of objectivity and relevance to the research design that might not have been achieved had the design been driven by ICRAF and its partners alone. It also increased the practical relevance of the results and ensured an audience for the results. Some of

14. For more on this topic see Place, Adato, and Hebinck 2007.

the same stakeholders acted as sources of information for the project as well and therefore provided for some triangulation of results.

Arm's-Length Data Collection

The qualitative and quantitative fieldwork was undertaken almost exclusively by persons not attached to the intervention, as the intention was that field researchers be viewed by villagers as completely detached from ICRAF. It is not clear that this method was entirely successful, as ICRAF is well known in the area and much of the enquiries were related to agroforestry. The participation of sociologists from external institutions ensured independent critical analysis and provided ICRAF with potential collaborators for the future. The disadvantage is that capacity for doing such research was not built within ICRAF.

Viewing Poverty from Multiple Perspectives

Poverty is a slippery concept. Yet the task was to see what impact SFR technologies had on the poor. Rather than devising a single qualitative or quantitative classification, the team was open to alternative views and ways of comparing poverty levels across households. This approach best fit with the reality that households are vulnerable to poverty in different ways and engaged in a range of survival strategies. Certain types of poverty indicators may move together, while others may not. For example, we found that expenditure and consumption changes over time were quite similar, but asset portfolio changes behaved somewhat differently.

Highlights of Empirical Findings

Although there is no doubt that poverty is pervasive in western Kenya, distinguishing the poor from non-poor is not straightforward. People often do not accept being labeled as poor. When pressed, people will admit that poverty implies the lack of certain basic needs. The study used a variety of methods to assess poverty levels, including quantitative measures from surveys, enumerator ratings, and farmer self-assessments. These produce different outcomes, so that which households will be classified as very poor will depend on the criteria used.

Welfare or livelihood outcomes worsened for many households. In general welfare indicators deteriorated during the period of study. This observation holds true for assets, expenditures, and food consumption. Households with relatively high welfare indicators in the initial period suffered the greatest losses, due partly to the large number of adverse shocks affecting households and the cultural obligations felt by all community members (for example, the wealthier households contribute animals for slaughter at funerals).

SFR technology interventions imply assumptions about the role of agriculture in people's livelihoods that have varying validity. The role of agriculture in people's livelihoods is determined by economic circumstances, culture, normative frameworks, and social identities. The assumption that poverty can be reduced through farming is not necessarily reflected in the investments in livelihood activities made by people in thc region. Their decisions are embedded in their economic circumstances (including assets and institutional environment), cultural and normative frameworks, and social identities. For example, people who perceive themselves to be farmers are more likely to adopt SFR or agrotechnologies than are those who in essence look down at agriculture as a way of life. Decisions about agricultural investments are also shaped by structural phenomena, such as the squeeze on agriculture that does not guarantee adequate return to human and physical capital investments. In western Kenya, farmers are very aware of this squeeze in making livelihood decisions. Whereas researchers may evaluate agroforestry in terms of its role in generating agricultural production, rural people assess its ability to contribute to the variety of objectives they pursue.

Households do see the value of SFR—and there have been many effects on human capital. Both the qualitative and quantitative research found significant knowledge acquisition taking place, not only for agroforestry methods, but for general soil management and farming practices. People valued this information and have often put it into practice.

The poor are using SFR strategies at the same rates as the non-poor. Use and adoption rates are not outstanding but they are encouraging, with about 20 percent of all farmers using the technologies on a regular basis (a similar percentage among the poor) and a sizable percentage of farmers newly testing. Thus unlike some agricultural technologies historically, SFR is not biased toward people controlling and managing resources above a certain threshold.

At this relatively early stage, adoption takes place on small areas of the farm. Although an encouraging number of households are using or testing the SFR practices, the size of plots on which they are applied remains small. It is not yet known whether this fraction is indeed a ceiling or whether it is a consequence of the early stage of dissemination.

Farmers favor being exposed to multiple dissemination opportunities and methods. The dissemination analysis found that farmers appreciated some aspects of different disseminating organizations and the many different methods tried. They particularly appreciated direct contact and field observation methods. However, information flows were not guaranteed, individuals may not be able to make scheduled meeting times, and different methods benefit some social groups more than others. Farmers thus favor being able to access information through a variety of channels.

Social status and social relationships within villages affect the outcomes of different dissemination methods. New technologies can also reinforce or transform these relationships. Although characteristics of SFR affect whether people adopt, aspects of the dissemination process also influence adoption. The dissemination analysis found that the main feature of most dissemination approaches—group-based methods—can strengthen human and social capital, and farmers of different social status have benefited from them. However, this analysis also found that group-based approaches may also disadvantage farmers of lower social status and women who are less likely to participate in or dominate groups. But women's groups have worked well for women. Furthermore, the dissemination analysis and case studies found that the use of adaptive research farmers generated new social tensions, due to the amount of attention received by individuals from outsiders. These findings reinforce the conclusion that use of a variety of methods is best and point to the importance of understanding local social dynamics in designing dissemination interventions. It is important for organizations involved in dissemination to be cautious in dealing with those who are the first to participate, to learn more about the roles of different individuals and groups within a community, to be vigilant over the course of an intervention, and to visit the field often. Certain processes and outcomes cannot be controlled, but can be improved with engagement.

Sustainability of dissemination structures and processes is possible but tenuous. Sustainability has proved to be possible, but challenging, because of problems encountered by groups, limited capacity of local administration, social dynamics within villages, and limited cost-sharing ability. Monitoring would help to pick up these problems so that resolutions can be sought where possible.

SFR does significantly raise crop yields compared to the no-input system that is commonly used by the poor in the region. Respondents in the case studies and formal surveys consistently report very significant increases in yields from the use of SFR practices. This result is consistent with farmer-managed trial data.

SFR on its own cannot bring about a turn in poverty reduction. This conclusion is drawn from the body of impact assessment work. Even though SFR is being used by a number of poor households and having an impact on yields, it has not had a measurable impact at the household level. The impact is limited by the small percentage of land under SFR and the weak rural economy, which is not conducive to investment and development. Thus technological innovations alone are likely to have a limited short-term impact. They will be constrained without other interventions to overcome other impediments. Poverty alleviation interventions must be multidimensional.

Conclusion

Pathways out of poverty are varied and highly uncertain. Identifying clear strategies through agriculture is equally difficult due to low prices, variable climate, and the high costs of profitable investments. Small landholdings in turn limit the amount of diversification that households are willing to undertake. For widespread poverty alleviation to take place, many components of the rural economy need to be functioning well. Even if progress is made, the study found that households can easily slip back into poverty. Therefore, in addition to generating production and income, there is need for insurance through investment in risk-buffering assets and an expansion of rural based, nonagricultural economic activities.

Within agriculture, poor households can take initial steps by building on the crops or enterprises that they already have. The strategy under consideration in this study was a relatively safe one of increasing yields of the basic staples of maize and beans. What is the future for agroforestry? The soil fertility systems being disseminated are useful options for farmers, and these options are being tried by many with no prior record of investment in soils. There are clear limitations to the use of improved fallows and biomass transfer, however. Small farm sizes limit the extent to which niches can be found to produce the green manures.[15] The technologies are therefore best perceived as feasible and viable components of farm-level integrated soil fertility management strategies. Consequently, dissemination strategies should encompass a range of management practices for addressing the problem of poor soil fertility.

15. At this early stage of dissemination in eastern and southern Africa, the improved fallow system appears to be extremely beneficial in places like Zambia, where farm sizes are about 3 hectares and phosphorus is not a major limiting soil nutrient. In western Kenya, which has very small farms and more widespread phosphorus deficiency, the effects of improved fallows on yield and household welfare are less noticeable. One could ask why we did not focus on the Zambia study instead of Kenya. The reason is that the case studies were selected on the basis of existing poverty indicator baselines, which we had for western Kenya but not for Zambia. Furthermore, the extent of impact was not predetermined before the study but rather was the focus of the study.

Appendix: Econometric Regression Analyses

APPENDIX TABLE 5A.1 Multinomial logit results for household factors related to adoption of improved fallows in the pilot villages, 1997–2001

	Outcome		
Variable	Used early and dropped	Used recently only	Used throughout period
Constant	–3.0833**	–2.7064**	–2.5034**
	(.0000)	(.0000)	(.0000)
Focal village	.6555**	–.1494	.8041**
	(.0006)	(.4238)	(.0000)
Luo household	1.3505**	.2413	.9998**
	(.0000)	(.2714)	(.0000)
Number of adults	.2685**	.1331**	.0944**
	(.0000)	(.0189)	(.0214)
Female head—husband away	.6750**	.4922	.0461
	(.0318)	(.1336)	(.8414)
Female head—no husband	.1070	.3812	.0262
	(.6892)	(.1480)	(.9150)
Male head—polygamous or single	.6628**	–.3149	.1717
	(.0136)	(.4238)	(.4238)
Secondary education	–.8548**	–.2650	.2335
(1 = yes, 0 = no)	(.0246)	(.4840)	(.3682)
Upper primary education	–.2314	–.1058	.1763
(1 = yes, 0 = no)	(.4008)	(.7589)	(.4231)
Lower primary education	–.2194	.2804	–.0686
(1 = yes, 0 = no)	(.4377)	(0.94)	(.7642)
Age	–.0168**	–.0055	–.0059
	(.0358)	(.5389)	(.3174)
Owned land area	.1417**	.0846	.2306**
	(.0246)	(.2302)	(.0000)
Wealth index	.0418	.1270	.0395
	(.5828)	(.1216)	(.4840)
Percentage of cases observed	9.1	7.6	22.0

NOTES: Omitted outcome is the group of farmers never trying improved fallows; the *p*-values are in parentheses; $n = 1{,}583$. ** indicates statistical significance at least at the 5 percent level.

APPENDIX TABLE 5A.2 Multinomial logit results for household factors related to adoption of biomass transfer in the pilot villages, 1997–2001

Variable	Outcome: Used early and dropped	Used recently only	Used throughout period
Constant	−1.765**	−1.9317**	−3.6500**
	(.0002)	(.0000)	(.0000)
Focal village	−.1868	.4200**	.7082**
	(.2714)	(.0070)	(.0000)
Luo household	.1926	1.0225**	1.9524**
	(.3174)	(.0000)	(.0000)
Number of adults	.1019**	.1660**	.2045**
	(.0456)	(.0004)	(.0000)
Female head—husband away	−.0801	−.3833	−.1384
	(.7642)	(.1936)	(.6892)
Female head—no husband	−.0854	−.1599	.0303
	(.7644)	(.4840)	(.9204)
Male head—polygamous or single	.3162	.0911	−.0365
	(.2302)	(.6892)	(.9220)
Secondary education	−.3323	.0778	.7820**
	(.3174)	(.7890)	(.0094)
Upper primary education	−.5478**	.0254	.5783**
	(.0456)	(.9204)	(.0214)
Lower primary education	−.2762	.0638	−.1561
	(.2714)	(.7895)	(.5486)
Age	−.0130*	−.0218**	−.0041
	(.0892)	(.0010)	(.5486)
Owned land area	.0770	.0693	.1352**
	(.1336)	(.1616)	(.0026)
Wealth index	.2596**	.1679**	−.0172
	(.0004)	(.0070)	(.7889)
Percent of cases observed	10.4	14.6	15.0

NOTES: Omitted outcome is the group of farmers never trying biomass transfer; the *p*-values are in parentheses; $n = 1{,}583$. ** indicates statistical significance at the 5 percent level or less; * indicates statistical significance between the 5 and 10 percent levels.

APPENDIX TABLE 5A.3 Multinomial logit results for adoption of improved fallows in nonpilot villages

Variable[a]	Technology use	
	Infrequent	Frequent
Constant	–4.34396**	–2.81731*
	(.024)	(.094)
Luhya	2.24805*	.07469
	(.077)	(.940)
Female-headed household	–.77250	.60553
	(.148)	(.142)
Polygamous male-headed household	.37230	.68091
	(.389)	(.123)
Primary education of head	–.49873	.34139
	(.338)	(.531)
Secondary or greater of head	–.67662	.70998
	(.280)	(.239)
Age of household head	–.01375	.00307
	(.331)	(.813)
Number of household members	.04306	.06291
	(.578)	(.401)
Farm size	.00359	.03117
	(.947)	(.494)
Wealth—logarithm of assets	.09571	.11995
	(.549)	(.424)
Wealth—farmer-generated index of wealth indicators	.10673*	.14969**
	(.057)	(.004)
Wealth—enumerator-generated middle wealth level	.92191**	.20254
	(.018)	(.564)
Wealth—enumerator-generated high wealth level	1.3722**	.26358
	(.024)	(.666)

NOTES: The three alternative wealth specifications are tested in separate models. Explanatory variables reported are for the wealth—logarithm of assets specification. Where the results of the nonwealth variables change across specification, it is noted in the text. The two reported columns are to be compared to the omitted outcome of never having used the technology. The p-values are in parentheses; $n = 361$. ** indicates statistical significance at the 5 percent level or less; * indicates statistical significance between the 5 and 10 percent levels.

[a] Eight location variables not reported.

APPENDIX TABLE 5A.4 Multinomial logit results for adoption of biomass transfer in nonpilot villages

Variable[a]	Technology use	
	Infrequent	Frequent
Constant	–4.53487**	–4.5567**
	(.006)	(.006)
Luhya	.59624	.16505
	(.564)	(.861)
Female-headed household	–.23176	–.07907
	(.555)	(.845)
Polygamous male-headed household	.01958	–.40696
	(.961)	(.413)
Primary education of head	.50588	.34027
	(.290)	(.492)
Secondary or greater of head	.24172	.28078
	(.669)	(.625)
Age of household head	–.00776	.00975
	(.527)	(.447)
Number of household members	.04704	.10073
	(.491)	(.179)
Farm size	–.00333	.01840
	(.949)	(.696)
Wealth—logarithm of assets	.18310	.26398*
	(.190)	(.071)
Wealth—farmer-generated index of wealth indicators	.15621**	.15984**
	(.001)	(.002)
Wealth—enumerator-generated middle wealth level	.55798*	.34610
	(.085)	(.307)
Wealth—enumerator generated high wealth level	.74071	.10013
	(.183)	(.877)

NOTES: The three alternative wealth specifications are tested in separate models. Explanatory variables reported are for the wealth—logarithm of assets specification. Where the results of the non-wealth variables change across specification, it is noted in the text. The two reported columns are to be compared to the omitted outcome of never having used the technology. The *p*-values are in parentheses; $n = 361$. ** indicates statistical significance at the 5 percent level or less; * indicates statistical significance between the 5 and 10 percent levels.

[a] Eight location variables not reported.

APPENDIX TABLE 5A.5 Econometric results from second-stage regression of agroforestry on changes in value of nonfixed assets in the pilot villages

Variable	Two-stage least squares	
	Coefficient estimate	Significance level
Predicted area under fallow	4.773	.346
Predicted area under biomass transfer	1,076.104	.708
Luo ethnic group	572.121	.852
Female-headed household	4,277.755	.269
Household head obtained primary education	2,461.805	.470
Household head obtained secondary education	–3,058.495	.464
Household head age	–48.551	.626
Household size	925.546	.125
Farm size	–1,681.671**	.035
Constant	–6,047.048	.442
R^2	.062	
Probability of *F*	.283	

NOTES: ** indicates statistical significance at the 5 percent level or less; $n = 97$.

APPENDIX TABLE 5A.6 Econometric results from second-stage regressions of agroforestry on changes in nonfood expenditures and per capita nonfood expenditures in the pilot villages

Variable	Changes in nonfood expenditures per household	Changes in nonfood expenditures per capita
Predicted improved fallow area	–9.973*	–1.691**
	(.066)	(.046)
Predicted number of seasons with biomass transfer	5,101.192	736.170
	(.156)	(.189)
Luo ethnic group	–4,875.978	–735.134
	(.115)	(.128)
Female-headed household	3,701.186	501.621
	(.353)	(.420)
Household head obtained primary education	–192.820	–110.445
	(.955)	(.837)
Household head obtained secondary education	2,216.158	446.506
	(.604)	(.504)
Household head age	96.070	13.369
	(.362)	(.417)
Household size	–208.409	–14.103
	(.732)	(.882)
Farm size	–16.756	32.896
	(.983)	(.795)
Constant	–3,431.242	–617.102
	(.689)	(.645)
R^2	.00	.00
Probability of *F*	.665	.636

NOTES: The *p*-values are in parentheses. ** indicates statistical significance at the 5 percent level or less; * indicates statistical significance between the 5 and 10 percent levels; $n = 102$.

APPENDIX TABLE 5A.7 Econometric results from second-stage regression of agroforestry use on nutritional measurements

Variable	Energy	Protein	Iron
Predicted improved fallow area	.0054	–.0208	–.0711
	(.831)	(.547)	(.394)
Predicted number of seasons with biomass transfer	23.3291	23.8093	79.8342
	(.169)	(.302)	(.152)
Luo ethnic group	–13.3120	–10.4403	–13.3354
	(.360)	(.598)	(.779)
Female-headed household	43.1884**	47.5732*	130.6258**
	(.023)	(.066)	(.036)
Household head obtained primary education	18.2851	27.0920	45.5115
	(.260)	(.222)	(.393)
Household head obtained secondary education	5.3631	21.0868	34.7518
	(.790)	(.443)	(.600)
Household head age	.3742	.6219	.4415
	(.452)	(.359)	(.787)
Household size	.9970	2.3064	12.0369
	(.728)	(.556)	(.204)
Farm size	–3.6886	–1.3343	–3.0956
	(.336)	(.798)	(.805)
Constant	–74.4262*	–114.5048**	–287.2588**
	(.068)	(.040)	(.033)
R^2	.00	.03	.00
Probability of *F*	.601	.793	.583

NOTES: The *p*-values are in parentheses; ** indicates statistical significance at 5 percent level or less; * indicates statistical significance between the 5 and 10 percent levels; $n = 102$.

Acknowledgments

The authors thank Wesley Ongadi and Pamella Opiyo for carrying out the case study fieldwork; Maggie Lwayo and Edward Ontita for assistance with fieldwork supervision and data analysis; Mary Nyasimi for supervising the focus groups and Chris O'Leary for help with focus group data analysis; and Collins Otieno-Obonyo, Aggrey Otieno, and Qureish Noordin for their input into research planning and logistical assistance. We also thank Anthony Bebbington, Jere Behrman, and Robert Chambers, who served as external advisors to the project, and, along with Lawrence Haddad and Ruth Meinzen-Dick, provided valuable comments on earlier drafts of this work. Interested readers are encouraged to obtain the full project report (Place et al. 2005), as it further documents the findings described here and provides additional results.

References

Behrman, J. R., H.-P. Kohler, and S. Cotts Watkins. 2002. Social networks and changes in contraceptive use over time: Evidence from a longitudinal study in rural Kenya. *Demography* 39 (4): 713–738.

Carney, J. 1992. Peasant women and economic transformation in The Gambia. *Development and Change* 23 (2): 67–90.

———. 1993. Converting the wetlands, engendering the environment: The intersection of gender with agrarian change in The Gambia. *Economic Geography* 69 (4): 329–348.

De Wolf, J., and R. Rommelse. 2000. *Improved fallow technology in western Kenya: Potential and reception by farmers.* Nairobi: International Centre for Research in Agroforestry. Mimeo.

Mango, N., and P. Hebinck. 2004. Cultural repertoires and socio-technological regimes: Maize in Luoland, west Kenya. In *Seeds of transition. Essays on novelty production, niches and regimes in agriculture,* H. Wiskerke and J. D. van der Ploeg, eds. Assen, The Netherlands: Royal Van Gorcum.

Place, F., M. Adato, and P. Hebinck. 2007. Understanding rural poverty and investment in agriculture: An assessment of integrated quantitative and qualitative research in western Kenya. *World Development* 35 (2): 312–325.

Place, F., M. Adato, P. Hebinck, and M. Omosa. 2005. *The impact of agroforestry-based soil fertility replenishment practices on the poor in western Kenya.* IFPRI Research Report 141. Washington, D.C.: International Food Policy Research Institute.

Rodgers, J. L., H. P. Kohler, J. R. Behrman, and S. C. Watkins. 2001. The density of social networks and fertility decisions: Evidence from South Nyanza District, Kenya. *Demography* 38 (1): 43–58.

Rommelse, R. 2001. *Economic analyses of on-farm biomass transfer and improved fallow trials in western Kenya.* Natural Resources Problems, Priorities and Policies Programme Working Paper 2001-2. Nairobi: International Centre for Research in Agroforestry.

6 Assessing the Impact of High-Yield Varieties of Maize in Resettlement Areas of Zimbabwe

MICHAEL F. C. BOURDILLON,
PAUL HEBINCK, AND JOHN HODDINOTT
WITH BILL KINSEY, JOHN MARONDO,
NETSAYI MUDEGE, AND TRUDY OWENS

High-yield varieties of maize have been widely adopted in Zimbabwe. Although germplasm from the Consultative Group on International Agricultural Research (CGIAR) system has been used in the development of hybrid maize varieties since Zimbabwe's independence, research and dissemination activities involved several organizations in both the public and private sectors. Further, even though adoption of earlier hybrids was widespread—in 1985, more than 85 percent of smallholder maize area was planted with hybrid maize and production doubled over the period 1979–85—rural poverty and child malnutrition remain endemic. Some observers argue that the gains from these hybrids have been concentrated in a few agroclimatic areas and that there has been little impact on child nutritional status. This argument has implications for policy debates not only about raising nutritional status within Zimbabwe but also for the CGIAR system, given its mandate to link improvements in agricultural technology to better nutrition.

This chapter examines the diffusion of hybrid maize varieties in selected resettlement areas of rural Zimbabwe and their impact on incomes, assets, and —indirectly—child nutritional status, paying particular attention to varieties made widely available to farmers in the mid- to late 1990s. Using mixed research methods, we address three questions:

1. What factors affected the diffusion of new maize hybrids in the mid- to late 1990s?
2. How did the introduction of maize hybrids influence the development of asset bases, livelihood strategies, and livelihood outcomes?
3. What is the relationship among these asset bases, livelihood strategies, and nutrition outcomes?

This chapter summarizes our findings. It begins with a description of the research methods used and the localities where primary data were collected. It then provides an overview of the analysis of the questions above. The final section summarizes our results, discusses methodological issues, and comments on future directions for hybrid maize in Zimbabwe. The full report for this study (Bourdillon et al. 2002) provides considerably more detail on the issues and findings mentioned here.

Research Methods

This study used both quantitative and qualitative research methods. Our starting point was the existence of a unique, longitudinal survey, covering three resettlement schemes (Mupfurudzi, Sengezi, and Mutanda) in different agro-ecological zones of Zimbabwe.[1] The initial survey was conducted during 1983–84, and the sample households were re-interviewed in the first quarter of 1987 and annually, during January to April, between 1992 and 2000. These surveys contain extensive information, inter alia, on agricultural activities, non-farm activities, assets, and child nutritional status. Opportunistically, it was possible to include, on two occasions, questions on the adoption of hybrid maize.

Although these surveys were rich in quantitative data and had little sample attrition, there remained substantial information gaps. To redress these, in 2001 the researchers held a workshop in Harare, at which stakeholders were invited to identify, discuss, and prioritize research questions for this study. We took the sustainable livelihoods framework, added in the comments of stakeholders, and developed from these a research design matrix (see Bourdillon et al. 2002, Table II.1). We determined that qualitative field methods and analysis would clarify transforming structures and processes, allow us to understand more fully the vulnerability context, and obtain participant-defined characterizations of livelihood strategies and outcomes.

In implementing the qualitative fieldwork, we decided to mix approaches. The core method was a series of household-level case studies, supplemented by participant observation in villages found in two resettlement areas. The plan was that fieldworkers would be resident for six months in these areas and collect material for their case studies based on issues that came up in the stakeholder workshop. The case study work would be followed by focus group discussions in the selected villages, together with some focus group discussions in the third resettlement area to confirm findings of individual case studies, reconcile divergent findings, and allow a wider range of voices to be heard. Because of the deteriorating political situation, the fieldworkers were in each area for a total of four and a half months.

The first step was to choose villages for the case studies. In addition to logistical considerations, choice was based on information gleaned from the

1. Although there are a number of similar resettlement schemes in Zimbabwe, historically and structurally they differ from the majority of smallholder farming settlements. In the resettlement schemes, farmers came from diverse communities, and at least in the early stages received particular attention from government services.

quantitative surveys to ensure that we had villages with different histories of adoption of maize hybrids. We selected two villages, two each in Mupfurudzi and Sengezi. In each village, the quantitative data were used to profile all households in the longitudinal study, including demographic characteristics; maize varieties grown in the previous five years; time of adoption of new hybrids; whether the household also grew other cash crops; and wealth estimated by holdings of livestock, housing quality, and ownership of physical assets. Fieldworkers selected 30 respondents (14 in Mupfurudzi and 16 in Sengezi), guided by a summary of this quantitative information. We also included three additional households not in the quantitative study, two in which the heads were politically important and wished to be in the study, and one in which the head was using maize as part of a cattle-feeding project of interest to us.

A legitimate concern associated with the use of case studies is the degree to which they represent a broader population. There is a danger that the interactions between locally resident enumerators and respondents will lead to a selection of case studies, that, while interesting, are too idiosyncratic to be informative of broader tendencies. Given this concern, Table 6.1 indicates that in terms of two types of observable characteristics—demographic and rates of adoption—the case study households are, on average, broadly similar to other households in the same settlement scheme.

The case studies involved a series of visits of differing duration as well as participant-observation work. A fuller understanding of certain processes can be enhanced through direct observation—seeing the pathways through which information and seeds enter a community rather than relying on how these are reported; attending field demonstrations to see how these function, and so on. Further, a significant attraction of collecting case study narratives with both historical and current content is that they can be compared with the household survey data.

Political tensions curtailed the program. The focus group discussions were not conducted in Mupfurudzi for this reason. In Sengezi, before the fieldworker was withdrawn, he held discussions with 14 informal groups consisting of 3–10 people. Half of these groups consisted of youths. The groups covered topics identified in a midproject seminar (organized to discuss preliminary findings) as of interest for wider comment, principally critical events in the area that had affected the whole village, confusion surrounding advice from extension agents, and investment in communal capital.[2] In both study areas, the researchers attended various village meetings (including a village court session in Mupfurudzi), where information on various issues was disseminated, especially issues of governance within the communities. We also carried out a series of more fo-

2. Extension staff members were formerly employed by Agritex, the Department of Agricultural, Technical and Extension Services under the then Ministry of Lands, Agriculture and Rural Resettlement. At the time of our study, Agritex was being merged with the Department of Research and Specialist Services. The merged departments are now the Department of Research and Extension Services (AREX).

TABLE 6.1 Characteristics of households selected for case studies

	Mupfurudzi		Sengezi	
Settlement scheme	Case studies	Longitudinal data	Case studies	Longitudinal data
Characteristic				
Mean household size	9	9	7	7
Mean age of head (years)	62	56	57	58
Percentage of female-headed households	28	23	31	33
Percentage of households adopting new high-yield varieties in:				
1995–96	21	34	12	23
1997–98	64	65	38	30
1999–2000	100	90	100	71

cused key informant interviews, both in the field sites and in Harare, to tap information available from persons with specialist knowledge.

Hybrid Maize in Zimbabwe

The development of improved varieties of maize in Zimbabwe started in the early 1900s, when a Department of Agriculture was established to reorganize agricultural production through insights from agrarian sciences. In 1919, commercial farmers founded the Maize Breeders Association to promote selection and production of better seed, and scientific maize breeding started in 1933 (Mashingaidze 1994). The first hybrid maize varieties bred outside the United States were produced to fit the country's climate. Commercial farmers established the Seed Maize Association of Southern Rhodesia in 1940 to ensure the timely production and supply of high-quality seed. Experiments in the post-1945 period showed that these new hybrids provided significantly higher yields in both normal and drought years (Rattrary 1956). A second milestone was the release in 1960 of SR52, the world's first single-cross hybrid. By 1970, 98 percent of Zimbabwe's commercial maize area was planted in SR52.[3] In the late 1960s, attention shifted to breeding three-way-cross hybrids (crossing three cultivars instead of the usual two), such as R201 and R215, which showed good

3. Initially, breeders did not consider the needs of smallholder farmers in communal areas. In the mid-1950s, the Department of Agriculture started to develop maize varieties that would be suitable for areas with less rainfall, where most communal farmers lived. The program generated four improved open-pollinated varieties (OPVs) and one top-cross hybrid. However, only the hybrid variety was released and the breeding program was eventually discontinued.

adaptation to areas of unreliable rainfall and sandy soils (Masters 1994; see also World Bank 1991). In 1973 the Plant Breeders' Rights Act was passed to protect ownership of maize varieties. Subsequently, the Seed Maize Association established the country's first private research station in 1974, which tested experimental varieties that came out of public research programs.

After independence in 1980, the state-funded maize-breeding program was decimated by loss of experienced staff and severe funding reductions. Public-sector breeding efforts were boosted in 1985 with the arrival of the International Maize and Wheat Improvement Center (CIMMYT) in Harare, which introduced both expertise and germplasm. In 1983 the Zimbabwe Seed Maize Association and the Crop Seeds Association merged to form the Seed Cooperative Company of Zimbabwe (Seed Co), which initially worked in cooperation with the government. Changes in policy in 1995 cleared the way for increased foreign investment in Zimbabwe's maize seed industry. Although Seed Co began to face competition from large international seed companies, which invest more resoures in maize breeding than do the government and CIMMYT combined, at the time of this study, Seed Co was the most important player in Zimbabwe's maize seed industry.

The most dramatic change in the early postindependence period was in the pattern of usage of hybrids. Between 1950 and 1975, adoption was largely limited to commercial farmers. Subsequently, agricultural extension workers started to encourage the adoption of hybrids among communal farmers as a way to ensure national food self-sufficiency. These efforts were complemented by government investments in rural infrastructure. In the postindependence period, adoption of R201 and R215 ("first-generation hybrids") skyrocketed (see Rohrbach 1988), with dramatic increases in yield (Rukuni and Eicher 1994).

In the 1990s, Seed Co's maize breeding program paid relatively more attention to resistance to diseases of concern to commercial farmers and improved drought tolerance, rather than emphasizing increased yields. Seed Co subsequently produced a wide variety of hybrid maize seeds ("second-generation hybrids") with these improved traits, although the improvements may not be visible to all farmers. These are the SC40x, SC50x, and SC60x series of seeds (see Bourdillon et al. 2002, for a more detailed description). These new varieties were marketed to farmers in a number of ways, including sponsorship of field days and trial or demonstration units, advertisements in the print and electronic media, and the production and dissemination of seed manuals written in both Shona and English. Institutions that provide inputs or input loans (such as the Grain Marketing Board [GMB]) also played a part in the diffusion of new hybrid varieties, as did Agritex. As of the 2000/2001 crop year, Seed Co discontinued production and dissemination of R201 and R215.[4]

4. Recognizing that many popular hybrids were poorly adapted to the marginal production conditions faced by most communal farmers, Seed Services (Zimbabwe's sole seed certification

Study Areas

Background

The study areas are three resettlement locations: Mupfurudzi in Mashonaland Central (north of Harare), Sengezi in Mashonaland East (southeast of Harare), and Mutanda in Manicaland (southeast of Harare, but farther away than Sengezi). As the history of the people in these resettlement areas is somewhat unusual, we provide an introduction to their background.

Access to land has long been an issue of major economic and political importance in Zimbabwe. Anger at the gross disparities in landownership between blacks and whites was a rallying point during Zimbabwe's liberation struggle, and the new ZANU-PF government immediately began to deliver some of the promised equitable redistribution of land in postindependence Zimbabwe. As part of this commitment, households were resettled on farms previously occupied by white commercial farmers, in most cases in peripheral areas bordering on communal areas. Initially land was acquired for resettlement on a willing-seller, willing-buyer basis. The farmers in Sengezi, Mupfurudzi, and Mutanda are among the 56,000 families resettled by the government immediately after independence. The vast majority of farmers in Sengezi joined the resettlement scheme in 1980 because they had no lands in the villages where they lived, most of those in Mupfurudzi settled in 1981, while farmers in Mutanda moved into their villages in 1981 and 1982.

Criteria for selection into these schemes included being a refugee or other persons displaced by war, being unemployed, being a landless resident in a communal area, or having insufficient land to maintain themselves and their families (Kinsey 1982). In our sample areas, some 90 percent of households settled in the early 1980s had been adversely affected by the war for independence in some form or another. Before being resettled, most (66 percent) had been peasant farmers, with the remainder being landless laborers on commercial farms, refugees, and workers in the rural and urban informal sectors. At the time of settlement, the household heads were also supposed to be married or widowed, aged 18–55, and not in formal employment. Families selected for resettlement were assigned to these schemes, and the consolidated villages within them, largely on a random basis.

Families settled in these schemes were required to renounce any claim to land elsewhere in Zimbabwe. They were not given ownership of the land on which they were settled but instead were given permits covering residential and farm plots. Each household was allocated 5 hectares of arable land for cultiva-

---

authority) introduced Kalahari Early Pearl, an improved OPV from Botswana, and persuaded a number of local companies to multiply this seed for distribution. But the Ministry of Trade and Commerce prohibited the sale of OPV seed. Seed Co now produces OPVs for sale only in neighboring countries.

tion, with the remaining area in each resettlement site being devoted to communal grazing land. In return for this allocation of land, the Zimbabwean government expected male heads of households to rely exclusively on farming for their livelihoods. Until 1992, male household heads were not permitted to work elsewhere, nor could they migrate to cities, leaving their wives to work these plots. Although this restriction has been relaxed, with male heads being allowed to work off farm (provided that household farm production is judged satisfactory by local government officials), in this sample agriculture continues to account for at least 80 percent of household income in nondrought years.

Resettlement was intended to create a rural farming community that would move from subsistence to commercial production. The government worked to provide an enabling environment for sustainable economic growth in the resettlement areas. It provided appropriate infrastructure, such as roads, to ensure the successful marketing of produce: in Sengezi and Mupfurudzi (and to a lesser extent Mutanda), there are well-developed road networks. The government also provided staff housing, clinics, and schools in the resettlement centers to improve production and the quality of life. Initially, the government provided widespread access to agricultural extension services, with virtually all resettled farmers being visited by Agritex staff in the early 1980s. Loan facilities were made available to the farmers through the Agriculture Finance Corporation. By means of the seed packs given to resettled families in 1980, the government was a vehicle for the initial diffusion of the R215 and R201 hybrid maize varieties.

Institutions, People, and Structures

GOVERNANCE AND ACCESS TO POWER. In both Mupfurudzi and Sengezi, farmers remain largely loyal to ZANU PF, which is seen to have helped them in the past. The ruling party is a visible actor in the resettlement villages, and there is no tolerance of dissent. The party imposes strict rules that control the behavior and activities of other organizations and their personnel in the area. Agricultural extension officers, teachers, and nurses are supposed to support the party views.

All respondents maintained that there was no relationship between power and wealth, and in Sengezi, positions of authority relate to former participation in the liberation war rather than to wealth. Nevertheless, in Mupfurudzi wealth brings influence: at the *dare* (community court) the people who dominate the discussions or whose views merit serious consideration are the wealthy people. The most influential individuals in both villages were usually the rich people or the more successful farmers and, in the current scenario, those with political office. The top positions of political office at district level were held only by wealthier people. Another limited form of power is a relationship with traditional spirits or a reputation for powers of witchcraft.

INSTITUTIONS. A number of institutions operate in these areas, including Seed Co and Agritex. The seed from Seed Co is used in Mupfurudzi, Sengezi, and Mutanda. Its direct presence is most marked in Sengezi, where it sponsors a demonstration plot. In addition, maps and charts indicating the varieties suited to different areas were displayed in stores in Wedza, where farmers from Sengezi buy their seed. By contrast, these were not displayed in Mupfurudzi in 2000. Agricultural extension officers employed by Agritex perform several roles that include offering periodic courses (such as those leading to a Master Farmers' Certificate), holding field days prior to the planting period, and visiting farmers' fields. Seed Co works with Agritex, historically an important mechanism through which technical information is passed to farmers. Agritex staff played an important role in the dissemination of hybrid maize in the immediate postindependence period and continued to help with local trial plots of new varieties of seeds.

In these localities, farmers perceive that Agritex officials focus on the best, and therefore the wealthiest, farmers, although the official policy is that all farmers should receive help. Nevertheless, farmers growing cash crops still receive priority. We were struck by discrepancies between official policies propounded in Harare and the reported practices of officers in the field. The incorrect prediction of drought by Agritex in the 1999–2000 season and the current political climate have led to an increasingly mistrustful atmosphere. Some Mupfurudzi farmers view the phasing out of the older "more reliable" varieties and replacing them with the newer "less reliable" varieties as a conspiracy among Agritex officers and Seed Co to discredit the government. In Sengezi, Agritex officers are said to have only impractical, book knowledge about maize but are trusted for advice on the main cash crops.

In addition to Agritex and the seed companies, the GMB operates in all survey areas. A number of other organizations—Purity, Farmer's World, Cottco (the Cotton Company of Zimbabwe, Ltd.), Cargill, and Agribank (formerly the Agricultural Finance Corporation)—operate only in Mupfurudzi.

KINSHIP. Kinship has little relevance to the dissemination of technology. There were cases of information being passed through close kin, but even this transmission was very limited when the individuals involved did not live in the same household. Although there were many links by marriage in the communities, in-laws maintain social distance, and no one mentioned in-laws as a source of information about maize. Although intermarriage may lead to social cohesion, it did not increase the flow of information within the community. When people were asked why they had to go outside the community to obtain information, they pointed out that people are jealous, suspicious, and stingy with information. Such attitudes can be partly explained by tense relations among in-laws, as many of the households are related by marriage.

In Mupfurudzi, relationships that enabled the exchange of information were those based on *usahwira*—formal joking friendships. These are institutionalized friendships, and involve extensive exchange of services and gifts (Bourdillon 1976, 81–82). They are taken up by choice (rather than kinship, which is ascribed) and provide a system of support and security in times of need, which in other societies are often attributed to kinship. When asked to mention sources of information on new hybrids, people often mentioned joking relationships both outside and within the village rather than any kin relationships. In Sengezi, the communities were more heterogeneous in their origins and had fewer kinship links. Information was shared among people sharing the same totem and calling one another *sahwira* or interpreted as distant kin.[5]

GENDER. Women were resettled in these villages in their own right only in very small numbers, which, together with existing roles of men and women within households, has consequences for the control and use of resources. In the public sphere, men occupy most public positions. In the domestic sphere, men make most of the decisions, including those related to the disposition of the produce and distribution of proceeds within the households. This unequal distribution of power is reflected in many domestic disputes, which frequently call for the intervention of the village court: women are contesting inequality because they perform most agriculturally demanding tasks. Even in female-headed households, men, such as adult sons, are frequently expected to make most decisions concerning agriculture. In Mupfurudzi, men have better access to inputs and to formal markets. So merely counting women household heads tells little about gender politics.

CULTURE, MAGIC, AND RELIGION. Cultural beliefs are relevant to understanding causes of success and failure in agriculture. In the communities studied, the vast majority of people believe that magic can affect the performance of one's crops and that it is possible for certain people to steal the yields of others through magical means. People frequently attribute magical powers to those who achieve unusually high yields. Fears and suspicions of witchcraft were pervasive. In Mupfurudzi, they resulted in people being unwilling to show interest in the crops of others. Even to observe how others grow the crops is liable to arouse suspicions of witchcraft. In Sengezi, there is a widespread belief that implements or animals lent to other farmers could be returned bewitched. Belief in witchcraft can inhibit the spread of improved technology.

Vulnerability, Assets, and Livelihoods

Three factors particularly affect the vulnerability of farmers. One is erratic rainfall (see Kinsey 1999), something of considerable concern, given the reliance on rain-fed agriculture. A second is the HIV/AIDS epidemic. AIDS affects availability of labor and restricts sources of income. Farmers did not perceive different labor requirements to affect choice of maize varieties, but input requirements are relevant. A third factor is the very volatile economic situation in

5. The totem is a clan name, usually that of an animal that acquires ritual significance for members of the clan. See Bourdillon (1976, 37–38).

the country, particularly as it affects the costs of inputs and opportunities for nonagricultural incomes.

When farmers were asked why they perceived themselves to be relatively well-off or relatively poor, the key criterion was ownership or lack of cattle (26.3 percent and 27.1 percent, respectively; Kinsey 1999): cattle provide draft power, fertilizer supplement, security against drought or other disaster, and a source of cash for inputs or other needs. Other important criteria were having enough food (15.8 percent), having access to remittances of employed family members and to savings (14.4 percent), good farming skills (11.4 percent), having farm equipment (7.9 percent), and having access to good land (5.7 percent). Relative poverty was blamed on lack of savings and remittances (16.4 percent), lack of farm equipment (15.8 percent), lack of labor (7.9 percent), lack of social support (6.8 percent), lack of food (4.5 percent), and lack of land or good land (4.5 percent).

Crops and livestock are the dominant sources of livelihoods for these households. Crop income is the single most important source of household income, accounting (in nondrought years) for 73 percent or more of total household income and 70 percent of income for households defined as poor. This high percentage partly reflects the legal restrictions that limit nonagricultural work by household heads but also is indicative of the relatively large amount of land that resettlement bestowed on these households, relative to communal area farmers. The exception to this general pattern is drought years, such as 1994–95, when other sources of income—most notably government transfers and selling off livestock—play an important role.

Growing enough food makes more money (whether from agriculture or remittances) available for farming and equipment. Farmers in Sengezi ranked maize as the most important crop even when they did not use it for cash. As it was not widely considered a commercial crop, farmers were reluctant to spend heavily on monetary inputs for maize. However, its importance suggests that vulnerability would be decreased with increased drought resistance of the varieties.

Diffusion and Adoption of New Varieties of Hybrid Maize in the Late 1990s

We summarize our findings in three steps: describing the patterns of adoption of the new hybrids; exploring the role played by policies, institutions, and processes in the dissemination and adoption of these new hybrids; and assessing the determinants of adoption using econometric techniques.

Patterns of Adoption

Table 6.2 provides quantitative information on the adoption of hybrids made available in the 1990s (SC 40x, SC50x, and SC60x). The more rapid adoption of these varieties in Mupfurudzi reflects its better agroecological potential for

TABLE 6.2 Adoption of new varieties of hybrid maize by year and location

	Percentage of households adopting new varieties of hybrids			Percentage of maize acreage sown to new varieties of hybrid maize		
Year	Mupfurudzi	Sengezi	Mutanda	Mupfurudzi	Sengezi	Mutanda
1994–95	13.1	2.0	4.6	12.4	5.6	8.3
1995–96	34.1	23.2	13.8	22.6	15.2	20.6
1996–97	47.2	38.4	32.3	35.4	27.7	22.7
1997–98	64.9	30.3	33.8	63.4	41.2	36.6
1998–99	81.1	50.5	56.9	80.6	55.2	54.5
1999–2000	90.3	70.7	80.0	90.2	75.0	75.8

SOURCE: Survey data.

maize as well as the presence of a greater number of organizations supporting farmers. In Mupfurudzi the newer varieties gradually displace the older hybrids, with only small changes in total acreage under maize cultivation. In Sengezi and Mutanda adoption of new hybrids coincides with an increase in land devoted to maize cultivation. In Mutanda mean acreages planted in maize more than doubled. When looking at these data, it is also important to note that the older hybrids—R201 and R215—were gradually withdrawn from circulation in the late 1990s. Households continuing to use these older varieties are either drawing on stocks of seeds purchased several years earlier, obtaining these seeds from older inventories held by a merchant or trader, or using seeds from the previous harvest.

An interesting pattern in terms of the types of new varieties adopted is that in Mupfurudzi, initially SC50x type varieties dominated. Over time, however, SC40x and SC60x varieties became more popular. The SC40x varieties are attractive for farmers who are risk-averse, as these varieties have superior tolerance to heat and drought. SC40x varieties silk faster, making them less susceptible to midseason droughts (such as the one that occurred in 1994–95 in Mupfurudzi) and mature more quickly. By contrast, SC60x varieties silk and mature more slowly, but offer the prospect of significantly higher yields.

In Sengezi initially no single type of new hybrid dominated, but gradually the proportions of different new varieties evolved to a pattern similar to that found in Mupfurudzi. In Mutanda, by contrast, the drought-resistant SC40x varieties initially dominated, but over time these appear to be less favored compared to SC50x. By 1999–2000, patterns of adoption in Mutanda are comparable to those of Mupfurudzi.

Policies and Institutions in the Dissemination of New Hybrids

An attempt to reduce poverty on a sustainable basis requires the dissemination of information. In this section, therefore, we explore pathways of information and dissemination: how these new seeds enter into the study areas, how farmers learn about them, and how they receive information in the context of their prior knowledge of farming.

PHYSICAL AVAILABILITY. Availability of seeds plays a significant role in the adoption of certain varieties. Our interviews with households and local retailers revealed several points of interest. In 2001, only two people (both in Mupfurudzi) said they were always able to secure the kind of seed they wanted: most farmers have to settle at times for varieties they had initially not intended to cultivate. Second, local retailers do not play a dominant role in making seeds available to local farmers. Three farmers in Mupfurudzi bought their seeds from local retailers while the rest of the farmers in the sample went to nearby towns or, rarely, to Harare. In Sengezi, those who planted certified seed usually went to Wedza, about 30 minutes' travel by bus, rather than to the shops located in the area. Local retailers appeared passive in the selection of seed varieties they offer, although they do appear to take advice from Agritex officers, particularly in Mupfurudzi. One of the retailers claimed that there was a high demand for R201 and R215 from farmers, but he never received these varieties from Seed Co. He said that he simply accepted the seed supplied by the seed companies, which claimed to supply the seed most suited for the area (he was supplied with SC501 and SC403). He was oversupplied with the latter, which the farmers did not like at all. This claim was supported by remarks from several respondents: "We adopted the new seed varieties because our trusted variety R201 is no longer available. If it comes back from wherever it is, we will go back and grow it." "When we changed from R201, we planted SC501 because R215 was not there in the shops in Shamva and Bindura." Another said:

> When you go to the shops to buy maize, you just get whatever seed is available or you risk planting late or not at all. That is why this year I bought SC513 because that is what I found when I went to buy seeds at the shops. . . . Last year we bought seed from Chakonda (SC513) and Harare (R201). I never really looked for SC601 but I bought the seed I came across.

For those who receive loans, the time seeds became physically available also affected the seed variety they planted. On being asked why he had changed from the Seed Co variety he had planted the previous year to planting a Cargill variety, one household head maintained that he had received SC501 from the GMB later than the CG4141 he had planted. As a result of this delay on the part of the GMB, he had planted CG4141 and intended to plant the SC501 the next season. Two farmers had planted larger proportions of saved seed in their fields because they received seeds from the GMB late.

SOURCES OF INFORMATION. The government played a critical role—including providing seed packs, fertilizers, and technical support through Agritex—in resettlement and communal areas in the 1980s, leading to rapid adoption of R201 and R215. Adoption was further spurred by the radical increase in yields these varieties produced. A striking feature of the environment in the late 1990s and carly 2000s was the greatly diminished role of state institutions. Sources of information on new hybrids became more diffuse, with the media, neighbors, seed companies, and Agritex all playing a role. The current political climate also contributes to this state of affairs. In Mupfurudzi increased mistrust was noted between farmers and Agritex. In Sengezi information obtained from commercial farmers had to be kept secret or disguised.

One possibility that we had not considered when fielding the quantitative surveys was that youth and schools might also be sources of information. We used the opportunity to undertake qualitative fieldwork to pursue this dissemination route in Sengezi. Young people had a different outlook on knowledge from that of their elders. In group discussions, and contrary to the older farmers, young people trusted the knowledge from Agritex officers, whom they regarded as reliable because they were trained staff and had several years of experience in their field. In contrast, the older people trusted their own experiments and demonstration units. This difference in attitudes might be because the younger individuals had higher levels of formal education and could communicate with Agritex staff as peers. There was also a generation gap in the way the youth and their parents obtained new information. Parents mostly depended on observation, experimentation, and demonstration units, whereas youths mentioned radio, advertisements placed in buses, and booklets of seed companies as sources of information. Adults regarded their practical knowledge as superior: young people's knowledge was regarded as mostly theoretical. This difference might be due to the mobility of youths and their greater exposure to the world outside the villages.

This perceived superiority of practical knowledge over theory was the consensus in group discussions with both men and women. Trusting knowledge from the youth might be gendered: men claim a monopoly over farming knowledge and are unwilling to admit to their limitations. Culturally, the greater a person's age, the wiser he is supposed to become and the more people come to him for advice. This conventional attitude is being challenged by the learning of the youth. To remain the controllers of knowledge and maintain their positions, elders have to belittle knowledge of young people.

Farmer Agency in the Adoption of New Hybrids

EXPERIMENTATION. Another way to acquire knowledge about varieties of maize is through personal experimentation. Farmers indicated that they were eager to individually experiment with the different varieties. Before fully adopting the new varieties accessed though the market, they plant a larger proportion of the seed they are used to and a smaller proportion of the new seed. At harvest time they compare the yields, resistance to pests, and drought tolerance, and then decide whether to adopt the new varieties. In the first three years that the newer hybrids were available, roughly one out of three farmers practiced such experimentation.

However, farmers who depended on seeds given to them by relatives had no opportunity to experiment with the new seed while continuing with the old seed. Similarly those who obtained seeds from GMB loans grew the variety given and did not practice experimentation on small pieces of land. Two farmers in Mupfurudzi simply used the new hybrid varieties on all their land allocated for maize, accepting the recommendations of Agritex officers. Increased seed costs may also contribute to the likelihood of experimentation. For example, people found it cheaper to buy larger packs of one variety than smaller packs of different varieties.

ADAPTATION OF THE MAIZE PACKAGE. An intrinsic element of the adoption process is that farmers redesign technologies, such as hybrid maize. Intercropping, fertilizer use, maintenance of soil fertility, and the use of saved seed are examples of such redesigning in the study areas.

INTERCROPPING. Intercropping is widely practiced in the study areas, but appears to be contested. In Sengezi and Mupfurudzi, farmers intercropped maize with "peripheral" crops or those that are considered as women's crops, including cucumbers, pumpkins (pumpkin leaves are a preferred relish in the village), sweet cane (*magunde*—the stalks are eaten, like sugarcane), and/or cowpeas (*nyemba*). Farmers believe that these crops do not disturb or compete with maize. Farmers said they wanted the bean crop to benefit from the fertilizers they applied to the maize crops and that the maize is protecting their beans from wind, excessive rains, and sun. Some argued that intercropping saves labor and time.

However, ten Mupfurudzi farmers contested intercropping practices. They argue that cowpeas and beans climb on maize stalks, making the harvesting of maize difficult. These farmers also said intercropping results in unnecessary competition for soil nutrients among crops. Five farmers pointed out that Agritex officers had taught them that intercropping results in declining yields. Another five said their experience with intercropping before they were resettled taught them that the practice reduces yields. Interestingly, the acting chief of crops in Agritex encouraged intercropping, which he saw as beneficial for the crops and ideal in situations of land scarcity.

In Sengezi, farmers stated that the practice of intercropping influenced their choice of maize variety. All Sengezi farmers in the sample preferred to intercrop the R215 or R201 variety with beans. The farmers felt that these varieties can withstand the competition from other crops, as they do not require a great deal of fertilizer. The farmers felt that the resistance to drought of these varieties ensures that they can offer protection to bean crops from drought for a longer period. Commenting on the new varieties, farmers said they would intercrop with either SC501 or SC513 but they would increase the amount of fertilizers used.

FERTILIZER AND SOIL FERTILITY. The use of fertilizer is redesigned in two ways: applying less than recommended amounts or not applying at all. The patterns of fertilizer application differ from household to household for a variety of reasons. All farmers, however, complain of the cost of fertilizers, and many limit its use for this reason. For many farmers, the cost results in their preferring varieties of maize that they perceive to be less dependent on chemical fertilizer.

Three farmers in the qualitative sample pointed out that they applied different amounts of fertilizers to plants in the same field. Two farmers applied only top dressing on anthills and on areas on which they had spread cattle manure. Farmers' knowledge regarding the quality of their soils also determined the rate and pattern of fertilizer application. In Mupfurudzi, soil referred to as *shapa* (sandy soil) requires more fertilizer, and maize planted in such soil requires two applications of ammonium nitrate. Methods of fertilizer application also resulted in differential rates and patterns application. Nine farmers used cups found in fertilizer packs. These farmers mechanically applied equal amounts of fertilizer to their plants regardless of the differences in fertility on their soils. They were also of the view that different varieties of maize do not require different amounts of fertilizers. Differences in rates and patterns of fertilizer application in Mupfurudzi also resulted from the controversies surrounding the recommended amounts of fertilizers per acre. Some farmers maintained that Agritex had recommended three 50-kilogram bags of compound D and two 50-kilogram bags of ammonium nitrate, whereas others argued that it was three bags of compound D and one of ammonium nitrate per acre.

Despite its widespread, albeit redesigned, use, some farmers regarded chemical fertilizers as detrimental to the natural fertility of their soils. "Fertilizer spoils the soil" is commonly heard. However, others stated that fertilizer is not detrimental to their soil. These disagreements among farmers, and between some farmers and Agritex, point to the value of on-farm and participatory research as conduits of information as well as mechanisms for technology generation.

Farmers monitor declines in soil fertility through observation of the growth patterns of crops in their fields and through designing simple experiments. Twelve farmers in the qualitative sample regarded declines in yields, even with fertilizer applications, as signs of a soil that is losing its fertility. Some farmers interpret a decline in fertility as when maize crops develop yellowish or purple leaves and are thin and tall with small or no cobs. Others regarded the growth of witch weed in their field as a symptom of loss of soil fertility. One farmer in Sengezi said that the knowledge has been passed down to him from his forefathers. In addition to chemical fertilizers, farmers in the study area have adopted a variety of ways to retain soil fertility and enhance crop production, including applying cattle manure, spreading anthill soil, leaving land fallow, and practicing crop rotation.

USE OF SAVED SEED. A further adaptation is to save seed from one harvest for planting in the following year. Out of our 30 case studies, all but two households (both in Mupfurudzi) had used saved seeds at some point, with the majority (21 out of 28) using saved hybrid varieties (most of them old generation hybrids). Only four very poor households admitted to having used saved hybrid seed in the season prior to the study.

Saved seed (except for certain open-pollinated varieties) was never ranked positively compared to the newer hybrids. Saved seed was regarded as being prone to diseases, pests, and as having fewer cobs (which are smaller in grain size) and requiring a great deal of fertilizer to produce a good crop. Apart from Hickory King, the use of saved seed was regarded as a sign of poverty. Hickory King was not negatively viewed because the variety was not available in the shops and was well regarded on taste grounds, making it popular even among rich farmers.

Households used saved seed for a variety of reasons. Availability of money is one clear dimension to consider. The saved variety might mature early and be available for food before the hybrid varieties. Most people admitted to planting saved seed in periods of great distress or as a security precaution. Physical availability was another reason why people planted saved seeds. In Sengezi, where most people planted saved R201 and R215 seeds, they claimed that they did so because the seeds were no longer available in the shops. However, for Mupfurudzi the situation is different. Although R215 is no longer available in the shops, people do not plant saved R215 seeds. Instead, they buy new seeds and plant them alongside other saved open-pollinated varieties. It is only when they cannot afford to buy certified seed that they plant saved hybrid varieties.

Over the years people had developed skills in saving and using saved seed. This knowledge was mostly passed from parents to children, although people could also access this information through their social networks. When people were saving seed for the next season, they looked for certain characteristics, such as bigger maize cobs with large, well-matured grains, as this selection would ensure that the resultant crop would have these desirable qualities. There were many ways of preserving seed for the next season. Two household heads (both females) smoked seed as a preservation method, although such seed is susceptible to attack by rodents. Some households applied chemicals that guard seed from rodents as well as enabling them to save a large amount at the same time. People also adopted other less expensive methods when they did not have money to buy chemicals, including smearing Surf washing powder on the maize and using old eucalyptus leaves or tobacco, whose bitterness would stop the weevils from boring into the maize.

DECISIONMAKING, PERCEPTIONS, AND PREFERENCES. Processes of decisionmaking, household perceptions and preferences also affect the adoption of new maize varieties. In the majority of our case studies (16 out of 28) the selection of a maize variety was made by the household head or the person responsible for purchasing seeds, without consulting others. In the remaining 12 cases, household heads claim to consult other members on the selection of maize variety. The involvement of children in deciding which maize varieties to grow seems minimal and has much to do with the discussion on the relevance of book knowledge versus practical knowledge. In Mupfurudzi, women in all households were involved in decisions affecting the choices of such peripheral crops as groundnuts, roundnuts, rapoko, and open-pollinated varieties of maize that are regarded as women's crops. In female-headed households, it was not uncommon for sons to buy seeds without consulting their mothers.

Farmers' preferences for the different varieties of maize they adopt are shaped by a variety of factors. Most of the factors reflected the households' concerns with food security, obtaining a sufficient harvest in an uncertain environment. Taste and appearance, input requirements, marketing considerations, postharvest processing, and nutrition were also mentioned by respondents as affecting their choices.

The majority of farmers (78 percent of the case study households) prefer a maize variety that is drought resistant. This preference is not surprising in an environment characterized by substantial and unpredictable variations in rainfall. It also explains why farmers stated that they preferred the first-generation hybrid varieties (R201 and R215). The SC40x, SC50x, and SC60x series of seeds are still relatively new to these farmers, who are still acquiring knowledge regarding their pest resistance, yield potential, and fertilizer requirements. Farmers' perceptions regarding varieties with a high yield potential differ across and within households. Farmers did not agree on which variety has the longest cob among both the old generation and second-generation hybrid varieties. In part, this discordance may be because, unlike the varieties introduced in the early 1980s, newer varieties display less dramatic increases in yields. The practice of saving seed from the previous harvest can be explained by the farmers' trust in the older varieties in this regard. In group discussions, it was stated that the new hybrid varieties could not be successfully used as saved seed in comparison to the R201 or R215 old varieties. As a result, most people cultivated saved R201 varieties.

In considering the yield potential of a maize variety, farmers took cognizance of the extent to which different varieties resist pests, especially weevils. All farmers in Sengezi and Mupfurudzi deplored the SC401 variety for its lack of resistance to weevils both before and after harvest. Coupled with its poor resistance to conditions of excessive moisture, it was labeled a poor crop in terms of its yield potential. As a Mupfurudzi farmer explained, "401 is useless. No matter how much fertilizer you put [on it], the maize cob is small. When selling, it is very difficult to get a grade A when selling 401. We want R215 but we can't find it in the shop. R215 is a very good seed. Even if you plant late, you will get something, unlike these new varieties." This perception contrasts with that of Seed Co, which presents SC401 as early maturing and good for late planting, recommended for areas of high yield and to complement more drought-resistant varieties in areas of low yield.

Most respondents did not regard labor requirements as an important factor in the decision to adopt, because farmers weeded and cultivated their maize crops once only, regardless of variety. However, some pointed out that in cases of severe illness, such as AIDS, the concerned households had to adopt short-season varieties, such as R201, as they require less labor compared to other varieties of maize.

Taste and food qualities received some attention in both areas. In Mupfurudzi, men disregarded these for commercial crops: women, however, considered the taste of maize to be important when it is roasted or cooked as green mealies and for the quality of its cooked meal. Women grew open-pollinated varieties for their good taste. Although this gender dimension was clear in Mupfurudzi, in Sengezi there were no gender distinctions. Men also actively sought open-pollinated varieties and grew them both in their gardens and fields. The difference between the men in these two study areas is that in Mupfurudzi men consider maize as a cash crop; thus they were indifferent to the taste of maize. Sengezi farmers regard maize as a food crop, which may explain why both the men and women expressed concern about the tastes of various varieties.

Assets and Adoption

Using econometric techniques, we now consider the role that asset holdings played in the adoption of maize hybrids in the 1990s. The dependent variable is whether a household has adopted a new variety in a given crop year. Regressors are drawn from the sustainable livelihoods (SL) framework: human capital (age and education of household head, number of resident adults), physical/financial capital (value of livestock holdings), and natural capital (number of and distance to plots, land quality). Distance to the resettlement center (also called the "growth point") and whether the household had received at least one visit by an extension agent are included as measures of ease in obtaining information, as is sex of household head. We include village-level dummy variables to capture fixed village characteristics that might affect adoption and year dummies and lagged rainfall to capture differences in vulnerability context across years. As such, the model is a two-way (location and time) village-level fixed effects regression.[6] The model is estimated as a probit; standard errors are robust to heteroskedasticity and correlations in the disturbance terms within villages. Coefficients are reported in terms of marginal effects (Table 6.3).

6. We considered estimating a household-level probit fixed effects regression. However, doing so eliminates two very interesting classes of households: those who adopt in all survey years and those who never adopt. Given this rather serious limitation, we have chosen instead to estimate a village-level fixed-effects regression.

TABLE 6.3 Determinants of adoption of new hybrid maize varieties

Variable	Marginal effect
Age of head (years)	–0.001
	(0.58)
Grades of completed schooling, household head	0.016
	(3.09)**
Household head is female	0.0005
	(0.01)
Number of adults in household	0.014
	(2.43)**
Lagged number of oxen owned by household	–0.001
	(0.21)
Number of plots operated by household	0.028
	(1.07)
Distance to plots (minutes)	–0.001
	(0.78)
Soils are primarily loam	0.144
	(1.34)
Soils are primarily clay	0.200
	(1.55)
Soils are primarily sandy	0.163
	(1.70)*
Land is sloped	0.051
	(1.30)
Household receives at least one visit from extension agent	0.003
	(0.10)
Distance to local market (km) × 1995–96 dummy	–0.010
	(0.86)
Distance to local market (km) × 1996–97 dummy	–0.019
	(3.02)**
Distance to local market (km) × 1997–98 dummy	–0.012
	(1.42)
Distance to local market (km) × 1998–99 dummy	–0.016
	(2.00)**
Lagged rainfall (mm)	0.001
	(2.73)**

NOTES: Other regressors included but not reported are year and village dummies. Sample size is 1,710. Coefficients are reported in terms of their marginal effects. Standard errors are adjusted to account for cluster effects at the village level. Absolute value of asymptotic *t*-statistics in parentheses. * indicates statistical significance at the 10 percent level; ** indicates statistical significance at the 5 percent level.

We begin by noting that a number of covariates are not statistically significant, including female headship, age of head, and whether the household receives at least one visit from an extension agent. In light of our qualitative findings, these might seem surprising. For example, female heads are unlikely to attend group meetings and have considerably less interaction with formal markets. When we interacted sex of head with the number of adults in the household, we found that during the first years of dissemination, female-headed households had a lower likelihood of adoption, but as the number of adults increased, this negative effect diminished. This finding is consistent with the observation that headship is not the only route by which information enters the household. Similarly, age of head (or the logarithm of age, which we also tried) had no effect on adoption, but recall that young people are also a source of information about new varieties. Access to extension services was not associated with increased likelihood of adoption; again consistent with the results of the qualitative fieldwork described above.

Adoption increases with the schooling attainments of the household head and with the number of adults in the household, and decreases with poorer market access, as denoted by distance to the market center interacted with crop year. We have noted that an issue for many households is the physical availability of seeds; it may be that households find it easier to search for particular varieties either because they are less isolated or because there are more individuals who can undertake these searches. If we disaggregate the sample by settlement scheme (see Bourdillon et al. 2002), we find that distance to the resettlement center is more important in Mutanda and Sengezi, where penetration by outside institutions is less marked. We also find that in Mupfurudzi, wealthier households are more likely to adopt sooner than less wealthy households, but this difference does not hold true in the rest of the sample.

Hybrid Maize, Livelihood Outcomes, Asset Bases, and Child Nutrition

Over the past 20 years, living standards have improved considerably in these resettlement households. For example, housing quality has improved, households own more consumer durables, children are more likely to attend school and to obtain access to basic health care. Ascribing precise causes to these changes is, however, problematic. The process of resettlement and the initial diffusion of hybrid maize after 1980 occurred simultaneously. Furthermore, both processes were nearly universal. There is neither a "with and without" nor a "before and after" comparator group which could be used, as in the literature on project evaluation, to assess impact. In addition, the best farmers have a number of sources of income, which feed into one another. We have pointed to the importance of cattle for profitable farming. We also notice that those who successfully farm other crops also get good yields of maize. It is certainly likely that the early hybrids were important in allowing certain farmers to establish themselves and build up resources in capital and equipment. An adequate maize crop means that family subsistence is obtained at minimal cost, allowing other income to be spent on inputs for the next year's crops. Attribution of causality is made even more difficult by extreme weather fluctuations throughout the 1990s as well as the changes in input and output markets. These confounding factors do not deny the importance to farmers of developing new hybrids; they do deny any easy link between the development of hybrids and socioeconomic development.

Changes in Livelihoods

Farmers in Sengezi and Mupfurudzi perceived that positive changes have occurred in their livelihoods since resettlement. When prompted to elaborate, all farmers in Sengezi attributed changes in their livelihoods to adequate land from which they could grow different crops to ensure household food security. They all stated that the reliable yields they obtained from R201 and R215 had been important in improving their livelihoods. Only 5 of 14 farmers in Mupfurudzi attributed changes in their livelihoods to the availability of land and only seven attributed changes in their livelihoods to the cultivation of hybrid maize varieties. The remaining farmers attributed changes in their livelihoods to the cultivation of cash crops, such as tobacco and cotton, noting that producer prices for maize in the 1980s and early 1990s had limited returns from that crop.

As already noted, because the adoption of the first generation of high-yield varieties was so rapid, and occurred simultaneously with resettlement, it is not possible to quantitatively assess their impact here. Instead, we focus on the second generation of hybrids. Table 6.4 provides some descriptive statistics covering crop years 1994–95 to 1997–98. As a starting point, we look at four livelihood-related outcomes: gross crop incomes (calculated as physical quantities of output of all crops multiplied by their unit price, deflated by the consumer price index so that they are expressed in real [1992] Zimbabwe dollars),[7] maize production (in kilograms), productivity (maize production per cultivated hectare of maize), and acreage devoted to maize. Mean values for households who plant at least some portion of their land to these new varieties are given in ordinary typeface; mean values for households who did not adopt are given in italics.

In the first year, when relatively few households adopt, gross crop incomes are comparable for adopters and nonadopters, though maize production is considerably lower among adopting households. In subsequent years, gross crop incomes, maize production, and productivity are all higher for adopting households, with the magnitudes of these changes nontrivial. As time passes, acreage devoted to maize is higher for adopters, though this trend is largely a result of nonadopters reducing acreage. However, ascribing a causal relationship to these associations is tricky. A simple explanation could be that—apart from the drought

7. The 1992 exchange rate was Z$5.1 = US$1.

TABLE 6.4 Comparison of gross crop incomes, maize production, productivity, and acreage devoted to maize between adopters and nonadopters of second-generation high-yield varieties of maize

		Mean			
Crop year	Percentage of households adopting any second-generation variety	Gross crop incomes (real [1992] Z$)	Maize production (kg)	Productivity (kg/hectare)	Acreage devoted to maize
1994–95	9.2	1,822	791	336	5.22
		1,820	*2,058*	*835*	*4.99*
1995–96	28.2	7,456	5,743	2,420	4.83
		5,700	*4,992*	*2,111*	*4.79*
1996–97	42.4	5,117	3,606	1,475	4.94
		3,065	*2,024*	*874*	*4.27*
1997–98	50.9	5,479	4,267	1,694	4.80
		3,026	*2,229*	*985*	*4.11*

NOTE: Figures in italics are mean values for nonadopters.

year of 1994–95—wealthier farmers with better access to inputs are more likely to adopt, to obtain advice from extension agents, and thus more likely to have higher yields. But controlling for these observable characteristics may not be the whole story. Suppose that more capable farmers are more likely to adopt and that they are also likely to have higher crop incomes. As we can control at best imperfectly for such abilities, merely taking into account these observable characteristics may not be sufficient.

Given these concerns, we estimate the impact of the adoption of these second-generation hybrids as a "treatments regression" using Heckman's (1979) two-step consistent estimator. Define Y_{it} as the outcome of interest; $ADOPT_{it}$ as a dummy variable equaling one if household i adopts these new hybrids in period t; $\boldsymbol{X_{it}}$ as a vector of characteristics that also affect Y_{it}; $\boldsymbol{L_i}$ as a vector of location dummy variables; $\boldsymbol{T_t}$ is a vector of time dummy variables; β, γ', η', and κ' are parameters to be estimated; and e_{it} is the disturbance term. Outcome Y_{it} is determined by

$$Y_{it} = \beta \cdot ADOPT_{it} + \gamma' \boldsymbol{X_{it}} + \eta' \boldsymbol{L_i} + \kappa' \boldsymbol{T_t} + e_{it},$$

where $ADOPT_{it}$, the endogenous dummy variable, is assumed to reflect an unobservable latent variable $ADOPT_{it}^*$, which itself is determined by

$$ADOPT_{it}^* = \tau' \boldsymbol{w_{it}} + v_{it},$$

where $\boldsymbol{w_{it}}$ are covariates that affect adoption, τ are their associated parameters, v_{it} is a disturbance term, and the relationship between $ADOPT_{it}$ and $ADOPT_{it}^*$ is given by $ADOPT_{it} = 1$, if $ADOPT_{it}^* > 0$; $ADOPT_{it} = 0$, otherwise; and where e_{it} and v_{it} are bivariate normal. Among others, Maddala (1983, 120–122) shows that consistent estimates of β can be obtained by first estimating the determinants of treatment (here, adoption). From this probit, the hazard (λ) or inverse Mill's ratio is calculated and then inserted as an additional regressor. Denoting the parameter estimate for the inverse Mill's ratio as ω, this gives us

$$Y_{it} = \beta \cdot ADOPT_{it} + \gamma' \boldsymbol{X_{it}} + \eta' \boldsymbol{L_i} + \kappa \cdot \boldsymbol{T_t} + \omega \lambda_{it} + e_{it}.$$

Estimating this model requires that we identify covariates that plausibly affect adoption but do not directly affect the outcomes we consider. Our estimation of the determinants of adoption provides us with several ready candidates: lagged rainfall, soil type, and distances to plots and market centers. These covariates have already been shown to have a statistically significant impact on adoption. To see if they are uncorrelated with such outcomes as maize yields and crop income, we estimate reduced-form determinants of these outcomes. In these regressions, we exclude adoption, include these "instruments," and test to see whether these covariates are individually and jointly significant. None of these covariates has a statistically significant impact on outcomes, suggesting that they are suitable as instruments.

Table 6.5 reports the results of estimating these treatment regressions for four outcomes, gross crop incomes (expressed in real [1992] Zimbabwe dollars), maize output (in kilograms), maize output per hectare, and acreage devoted to maize. Our X_{it} vector is composed of the following variables: lagged number of oxen owned by the household (both to control for household wealth and to account for the ability of households that own oxen to better time the

TABLE 6.5 Select determinants of gross crop incomes, maize production, productivity, and acreage devoted to maize

Variable	Gross crop incomes	Maize production	Productivity	Acreage devoted to maize
Household grows new hybrids	2,474.4 (3.75)**	2,232.14 (2.48)**	993.26 (2.98)**	2.91 (7.58)**
Number of adults in household	78.72 (2.25)**	39.04 (1.13)	–0.82 (0.08)	0.09 (3.31)**
Grades of completed schooling, household head	–33.57 (0.90)	–30.95 (1.20)	–13.48 (1.09)	–0.02 (1.03)
Age of household head	–8.10 (1.10)	0.56 (0.08)	1.28 (0.45)	0.01 (0.60)
Household head is female	–736.63 (3.24)**	–352.52 (1.32)	–184.10 (2.36)**	–0.06 (0.29)
Number of plots operated by household	125.40 (2.00)**	103.47 (1.19)	71.78 (2.18)**	–0.09 (1.18)
Lagged number of oxen owned by household	270.94 (5.47)**	246.74 (5.78)**	102.49 (6.82)**	0.08 (3.19)**
Household receives at least one visit from extension agent	546.18 (3.51)**	626.96 (3.97)**	198.44 (3.11)**	0.30 (2.25)**
λ_{ij}	–1,086.50 (2.81)**	–1,080.63 (2.01)**	–460.69 (2.49)**	–1.54 (6.41)**
Wald test of independence of adoption and outcome	7.84**	3.80*	5.65**	31.69**

NOTES: Other regressors included but not reported are rainfall levels by year and settlement scheme, slope of land, and year and village dummies. Sample sizes are approximately 1,300. Standard errors are adjusted to account for cluster effects at the village level. Absolute value of asymptotic *t*-statistics in parentheses. * indicates statistical significance at the 10 percent level; ** indicates statistical significance at the 5 percent level. Wald test is distributed as a χ^2 test statistic with one degree of freedom.

plowing and planting of their fields),[8] household demographic characteristics (number of adults; education, age, and sex of the household head), characteristics of the land operated by the household (degree of slope, number of plots operated), whether the household received at least one visit from an extension worker, and measures of rainfall. Also included are village-level dummy variables ($\boldsymbol{L_i}$) and dummy variables for year of observation ($\boldsymbol{T_t}$). By doing so, we are effectively estimating a two-way (village and time) fixed-effects treatments regression. Abbreviated results are reported in Table 6.5; full results are available on request.

The key finding from Table 6.5 is that adoption of new hybrids is causally associated with higher crop incomes, maize production, maize yields, and acreage devoted to maize, and that this relationship is statistically significant. However, these are conservative estimates. Recall from Table 6.4 that the first year of adoption (1994–95) was a drought year when only a few farmers (9 percent) of the sample were experimenting with these new varieties. Inclusion of this year will lower the estimate of impact. To see if this is indeed the case, we re-estimated these models, excluding 1994–95. We retained the same set of controls so that these new results are also two-way (village and time) fixed-effects treatments regressions.

How large are these effects? Table 6.6 also shows the expected changes in outcomes, conditional on adopting the second-generation varieties, accounting for the endogeneity of adoption. Adoption of the new varieties increased maize productivity and maize production by about 20 percent. Households growing these new varieties slightly increase their acreage planted in maize. Also striking is the reduction in variability in outcomes, with the coefficients of variation for adoption markedly lower than that observed for nonadoption.

Development of Asset Bases

We also used our qualitative data to explore the relationships among hybrid maize, incomes, and the development of asset bases. Our interviews in Mupfurudzi indicate that households invest in a wide variety of assets. Although many said that maize was an important crop, not everyone linked it directly to the purchase of major assets. Income from maize was used to purchase solar panels, a water pump, a welding machine, consumer durables, improved housing, clothes, food, and livestock as well as payment for school fees and the hiring of labor. The largely successful cultivation of cash crops (tobacco and cotton) in Mup-

8. We experimented extensively with this representation of household wealth. For example, we tried the value of both contemporaneous and lagged livestock holdings. The former produces a parameter estimate on adoption that is virtually identical to the one we report here. Second, we reestimated the treatments regression dropping the livestock covariate. Doing so significantly increases the magnitude of the positive impacts that we find, indicating that controlling for livestock holdings generates a more conservative estimate of impact.

TABLE 6.6 Impact of dropping 1994–95 on the effect of adoption of new hybrids on gross crop incomes, maize production, productivity, and acreage devoted to maize

Variable	Gross crop incomes	Maize production	Productivity	Acreage devoted to maize
Estimated coefficients				
Household grows new hybrids	2474.4	2232.14	993.26	2.91
(includes 1994–95)	(3.75)**	(2.48)**	(2.98)**	(7.58)**
Household grows new hybrids	3081.66	3902.58	1493.42	3.15
(excludes 1994–95)	(3.76)**	(6.29)**	(4.26)**	(6.32)**
Predicted outcomes				
Predicted outcome, nonadoption	3744	2307	1109	4.33
Predicted outcome, adoption	4472	2747	1366	4.70
Percentage difference in predictions	19.4	19.1	23.2	8.5
Coefficient of variation, nonadoption	0.66	0.63	0.62	0.17
Coefficient of variation, adoption	0.56	0.53	0.50	0.16

NOTES: Model specifications as per Table 6.5. Absolute value of asymptotic t-statistics in parentheses. ** indicates statistical significance at the 5 percent level. Full results available from the authors on request.

furudzi may lead our respondents to downplay the importance of maize as a cash crop. In addition, three people pointed out that they sold maize later than other crops, so that by the time maize was sold they would have gotten most of the things they wanted from the sale of early cash crops: "The payments for maize usually come late so that we use the money from maize to buy more inputs. For example, we can buy cotton seeds or even maize seeds so that we can plant early. When the fertilizer loans come, the crops will already have germinated."

In Sengezi, 11 households purchased livestock (cattle) from returns obtained from the marketing of surplus maize and three used proceeds from maize to buy agricultural inputs. Sengezi farmers also invested their surplus maize into social relationships, giving maize to relatives residing in urban areas and sometimes to relatives in communal areas. Eleven out of 16 farmers in Sengezi invested their maize in such relationships. Farmers in both Sengezi and Mupfurudzi also invested in their children's education with money obtained from the sale of maize.

Given that the views expressed by our respondents suggest considerable heterogeneity in terms of the use of maize and other crop production for the accumulation of assets, it is worth exploring whether any general patterns can be discerned. We do so by focusing on two assets: livestock and agricultural tools. We choose these because our respondents associate possession of them with being relatively well off and their absence as a cause of being relatively poor. We estimate a very simple specification—a flexible accelerator model—in which investment is a function of total gross crop income (the value of production of all crops in a given year)[9] and existing stocks of capital. We disaggregate crop income into maize and other crop income to see whether increases in maize incomes have an effect on investment that differs from that obtained from increasing incomes from other crops. All values are expressed as logarithms so that the coefficients are also elasticities. Although the valuation of livestock is straightforward—the summation of the values reported by respondents—the construction of the value of agricultural tools, defined as tools and equipment used in crop production, is less obvious. For this reason, we describe it in detail here.

Agricultural tools include ox-plows, scotch carts, cultivator/harrows, ox-planters, water carts, cotton sprayers, wheelbarrows, tractors and tractor equipment, hoes, axes, spades, machetes, and slashers. Although information on possession of these items is readily available, their valuation is difficult. As part of the surveys, households were asked about what items they owned, when they were obtained, how much they paid for it, and how much it would cost today. Answers to these questions revealed two problems. First, some households did not remember, and therefore report, what they had paid for the item. Second, the prices of virtually identical items were highly variable among households, perhaps because of faulty recall or differences in knowledge regarding current prices. Rather than allowing the price of capital goods to vary across households, we impose a uniform price. Specifically, the median purchase prices of items both acquired and reported for the crop year 1995–96 were used as a base. These were then deflated using the consumer price index to derive prices for other survey years. As an example, the median reported purchase price for an ox-plow in 1995–96 was Z$775. When we compare the deflated values of this figure with the median reported buying prices in previous years, we find that the correspondence between these is reasonably close.

Note that the validity of the approach relies on assumptions regarding the deflator and the treatment of depreciation. Specifically, the prices of agricultural capital goods are deflated using a consumer price index. Implicitly, this operation assumes that changes in the prices of the former broadly mimic the latter. Assets typically decline in value as a result of wear and tear. One could argue that this reduction in value must be reflected in the capital stock measure. In other words, depreciation should be deducted from gross investment to calculate the increase in capital that is relevant to explaining increases in output. The implication of this view is that each $1 of depreciation reduces output by

9. We also estimated these models using net crop income and obtained similar results.

as much as each $1 of gross investment increases output. However, we could also argue, following Scott (1991), that a machine that produces the same quantity of output ceteris paribus, year in and year out, cannot be said to have experienced any depreciation. Scott argues that capital only depreciates when it becomes obsolete. For this reason, he argues that capital stock should be measured in gross terms, not net of depreciation.

This argument reflects the situation facing households in the sample. Because of the nature of asset ownership within the sample, it would lead to a gross underestimate of the contribution of capital to growth if conventional systems of growth accounting were used. Many households own and still use equipment handed down from previous generations. For example, in 1982–83 moe than 50 percent of households owned and used an ox-plow that was more than 10 years old (11 percent of households owned an ox-plow more than 30 years old). To measure the effect of capital in these circumstances, the appropriate measure of capital stock is a gross, not a net, figure.

Two sets of results are reported in Table 6.7. The top panel reports the results of estimating a two-way fixed-effects regression in which both household fixed effects and year dummies are controlled for. Denoting I as investment, K as our measures of assets, y as income, $\boldsymbol{X}$ as a vector of other covariates, such as rainfall and the year dummies, and ε the disturbance term, and where mean values are written in italics, we estimate

$$I_{ivt} - (I_{iv}) = \beta_K(\mathrm{K}_{ivt-1} - K_{iv}) + \beta_{\mathbf{y}}(\mathbf{y}_{\mathbf{ivt}} - \mathbf{y}_{\mathbf{iv}}) + \beta_{\boldsymbol{X}}'(\boldsymbol{X}_{\boldsymbol{ivt}} - \boldsymbol{X}_{\boldsymbol{iv}}) + (\varepsilon_{ivt} - \varepsilon_{ivt}).$$

A problem with this approach is that assets are a function of prior investments, which may result in correlation between lagged assets and the disturbance term. Arellano and Bond (1991) show that this can be resolved by estimating in first differences using generalized methods of moments (GMM). As a check on these fixed effects estimates, we also estimate

$$I_{ivt} - (I_{ivt-1}) = \beta_I(I_{ivt-1} - I_{ivt-2}) + \beta_y(y_{ivt} - y_{ivt-1}) + \beta_{\boldsymbol{X}}'(\mathbf{X}_{\mathbf{vt}} - \mathbf{X}_{\mathbf{vt-1}}) + (\varepsilon_{ivt} - \varepsilon_{ivt-1}).$$

We estimate this quantity using the one-step estimator proposed by Arellano and Bond. The lagged first difference in investment is treated as endogenous, as is the logarithm of income. All current and prior first differences and levels of exogenous variables serve as instruments. The Arellano-Bond one-step estimates are also corrected for heteroskedasticity. Results of this estimation are reported in the bottom panel of Table 6.7.

The top panel of Table 6.7 shows that a 10 percent increase in crop incomes is associated with a 1 percent increase in holdings of tools and a 4.8 percent in-

TABLE 6.7 Determinants of investment in agricultural tools and livestock

	Tools			Livestock		
Variable	(1)	(2)	(3)	(1)	(2)	(3)
Pooled results across all crop years controlling for household fixed effects						
Logarithm of crop income	0.010 (3.12)**			0.048 (3.85)**		
Logarithm of maize income		0.008 (2.25)**			0.017 (1.45)	
Logarithm of nonmaize income			0.009 (2.80)**			0.033 (2.71)**
F statistic, household fixed effects	3.09**	2.42**	3.12**	3.73**	3.40**	3.63**
Sample size	2,281	1,754	2,155	2,102	1,639	1,996
Pooled results across all crop years estimated using Arellano-Bond one-step estimator						
Logarithm of crop income	0.033 (2.89)**			0.076 (1.88)*		
Logarithm of maize income		0.032 (2.94)**			0.074 (1.81)*	
Logarithm of nonmaize income			0.047 (3.08)**			0.119 (2.32)**
p-value, Sargan over-identification test	0.24	0.83	0.06	0.095	0.06	0.05
p-value, Arellano-Bond test for AR(2) in first differences	0.39	0.61	0.83	0.44	0.06	0.64
Sample size	1,767	1,621	1,552	1,620	1,493	1,446

NOTES: Figures in parentheses are absolute values of *t*-statistics. * indicates statistical significance at the 10 percent level; ** indicates statistical significance at the 5 percent level.

crease in livestock holdings. The bottom panel of Table 6.7 shows slightly larger effects, a 3.3 percent increase in holdings of tools and a 7.6 percent increase in livestock holdings.[10] Put another way, these resettlement households save between 6 and 11 percent of any increase in crop incomes in the form of increased holdings of tools and livestock, with the majority of that savings taking the form

10. The Sargan test tends to not reject the overidentification restrictions in the tools regressions but does so for livestock. However, Arellano and Bond (1991, 291) provide simulation results

of livestock. Results found in columns (2) and (3) show that increases in maize incomes have the same effect on investment in tools and livestock as increases in nonmaize crop incomes. As suggested by our qualitative fieldwork, maize does not play a special role in wealth accumulation.

Resource Sharing, Conflict, and Differentiation within Households

To this point, our discussion of maize, livelihoods, and assets treats the household as a unitary entity. However, resource sharing and distribution is a highly contested issue within these households. Furthermore, people often gave deceptive answers when probed on household resource allocation (other household members and members of the local community frequently disagreed with the answers given by household heads). Hence, our discussion draws on a variety of sources and individuals as well as our own observations.

Men control land, the major means of production, as they were the ones who were resettled as plot holders (except in one case in Sengezi, where a female plot holder determined the cropping patterns of her sons, to whom she had given subdivided plots). Consequently, they determine cropping patterns and decide whether a plot is to be extended or subdivided and if so, by how much. On the issue of division of labor, most respondents agreed that men should plow fields and do the work that needs great physical strength, whereas the planting of seeds was regarded as women's work. Although a few household heads agreed that weeding and fertilizer application were women's work, a larger number claimed that they helped one another to weed, apply fertilizers, and harvest. However, in about half of these cases, such statements were contested by other family members and members of the local community, who maintained that the male household heads left most of the weeding and fertilizer application to their wives and children.

Most people agreed that when still in the field maize belonged to the men, but as soon as it was put in the granary it belonged to the women. When asked about the distribution of proceeds from the sale of crops, eight respondents maintained that after selling their crops they did not give one another individual shares. They emphasized the household as a single unit. Money from the sale of crops was family money, which would be used for buying food and clothes and paying for children's school fees. Presenting the household as a unified entity was misleading. Some women (both those whose households were in the sample and some outside of the sample) were not satisfied with the way their husbands used "family money." In some cases, men were also not happy

---

showing that this test statistic has a tendency to over-reject in the presence of heteroskedasticity. Note that in five of six specifications, we find no evidence of second-order serial correlation. This finding is important in that these estimates would be inconsistent if such correlations existed (see Arellano and Bond 1991, 278 and especially 281).

with the way their wives used money. This dissatisfaction resulted in people taking one another to court or in consulting others for mediation. Some of the women (not in the sample) went on to sell all the "family's maize" to *madhaiza* (informal traders) as a way of getting back at their husbands. Such conflicts were usually because of lack of transparency in the use of proceeds from agriculture—certain members of the family were regarded as using resources for their own selfish ends.

Family money was usually the household head's money, who in most cases was a man. We came across cases where other family members were disillusioned because they had very little or no control over family resources. One respondent maintained that all the money they get from selling crops was family money, yet the money was kept in her husband's bank account and she did not have a bank account in her name. The husband was always buying cattle with "family money" (cattle are traditionally regarded as male property). The wife was afraid that if her husband should die, she would be left with nothing because her husband's relatives would take all the cattle away. This issue led to conflicts between this woman and her husband. Thus in some cases resource conflicts are not about the husband's squandering of money but about his making investment decisions that are not gender sensitive. The "family money" becomes the household head's money depending on the kind of things the money is used for.

Resource sharing was also problematic in polygynous unions, especially where the husband insisted on having a single granary and combining labor in the field. In such a union, the wives did not have the freedom to use proceeds from agriculture in any way they wanted as they usually had to reach a consensus with the co-wives. Usually families with polygynous marriages were so large that women could not trade any maize with *madhaiza,* because all the maize had to be retained for feeding the huge family.

Hybrid Maize, Assets, and the Nutritional Status of Children

Finally, we turn our attention to the links among hybrid maize, assets, and the nutritional status of children. Consistent with both the literature on the determinants of nutritional status—as well as the observations of our respondents noted above—we focus attention here on the links between assets and child height.

Drawing on multiple rounds of the quantitative survey, we take as our dependent variable the growth rate of children initially aged 12–24 months. Using five rounds of these data—from 1993 to 1997, we are able to construct growth rates for four cohorts of these young children. These rates give us our dependent variable—growth in stature of children aged 12–24 months at the time of first observation—measured in centimeters per year. However, multivariate regression analysis finds no significant impact of assets on child growth once we take into account child characteristics (initial height, sex, age at first observation, duration of observation, the product of age and duration of observation), maternal height, age, schooling, relationship to the household head, holding characteristics (soil type and acres of landholdings that are sloped or steeply sloped), and village characteristics. We repeated these regressions for older children and again found no association.[11]

However, recall from our discussion of the context of these resettlement areas that the threat of drought is a very important component of their vulnerability context. We have also noted that households report that livestock are an important mechanism for coping with drought and that a drought occurred in the middle of this period (1994–95). So an interesting question to ask is whether livestock protects these children's health in the aftermath of drought. Below we replicate two key tables from Hoddinott and Kinsey (2001). Specifically, we stratified the sample by livestock holdings as measured in 1995. Recall that these were measured just prior to the realization that 1994–95 would be a drought year for these households. Table 6.8 presents the results of dividing the sample of children initially aged 12–24 months into two groups: those residing in households below and above the median value of predrought livestock holdings. Drought only affects the growth of children residing in poorer households (those below the median). The coefficient for the "drought cohort" is not significantly different from zero for children living in households with predrought livestock holdings above the median.

Does this protective effect really matter? This growth slowdown is unlikely to be important if it is only transitory. To investigate, we examine the determinants of heights of children aged 60–72 months. Because stature by three years of age is highly correlated with attained body size at adulthood, these heights are good predictors of likely completed heights. The dependent variable is expressed as the child's height for age *z* score. Table 6.9 reports the results of estimating a maternal fixed-effects model.

Children aged 60–72 months, measured in early 1999, are the children who were initially aged 12–24 months in the year after the 1994–95 drought. These children have *z* scores about six tenths of a standard deviation below that of comparable children measured in nondrought years. The right-hand column in Table 6.9 interacts the 1999-year dummy with a variable that indicates whether the household's predrought livestock holdings were below or above the sample median. Children from wealthier households appear to have suffered no long-term effects from this drought. Children from poorer households, by contrast, appear to have experienced a growth slowdown that has persisted to age 60–72 months.[12] Linking this observation to our earlier results, we have already seen

---

11. A much more detailed description of these findings, together with a conceptual model, is found in Hoddinott and Kinsey (2001).

12. A caveat: we tested to see whether the difference in coefficients between the interaction terms was statistically significant. Although there is a difference of more than half a *z*-score, it is only significant with a relatively high *p*-value, .25. We do note, however, that given that we have a relatively small sample and we are using maternal fixed effects, we have low statistical power to detect such differences.

TABLE 6.8 Livestock, child growth, and drought, children aged 12–24 months

Variable	Child resides in household with predrought livestock holdings: Below the median	Above the median
Child in drought cohort (cm)	–2.202	–1.281
	(1.795)*	(0.844)
Height (cm)	–0.323	–0.284
	(3.062)**	(2.174)**
Mother's education (years of completed schooling)	0.843	–0.542
	(1.748)*	(1.090)
Mother's education squared	–0.075	0.014
	(1.713)*	(0.353)

NOTES: Dependent variable is annual (12-month) growth rate in child height. Asymptotic t-statistics based on Huber-White standard errors in parentheses. * indicates statistical significance at the 10 percent level; ** indicates statistical significance at the 5 percent level.

TABLE 6.9 Maternal fixed effects estimates of determinants of child height for age z-score, children aged 60–72 months

Variable	Specification (1)	Specification (2)
Child was initially aged 12–24 months during years 1995–96	–0.602	
	(1.927)*	
Child was initially aged 12–24 months during years 1995–96 and predrought livestock holdings were below sample median		–0.907
		(2.201)**
Child was initially aged 12–24 months during years 1995–96 and predrought livestock holdings were above sample median		–0.379
		(1.028)
F statistic on joint significance of maternal dummies	2.60**	2.47**

NOTES: Dependent variable is child height for age z-score. Other regressors included but not reported are child sex, mother's age, logarithm of the value of livestock, and dummies for years 1993–1997. The omitted year dummy is 1998, representing the cohort directly preceding the drought cohort. Sample size is 265. There are 124 different mothers. Absolute value of asymptotic t-statistics in parentheses. * indicates statistical significance at the 10 percent level; ** indicates statistical significance at the 5 percent level.

that to the extent that the adoption and use of hybrid maize increases crop incomes, higher crop incomes are associated with investment in livestock. Holdings of livestock are not directly associated with improved child nutritional status—as measured by growth in stature. However, they play an important protective role in the aftermath of droughts.[13]

Summary and Conclusions

In this section we summarize our findings, provide an assessment on methodology, and comment on future directions in the development of hybrid maize in Zimbabwe. Although detailed findings are in some cases specific to the cases we studied or to resettlement schemes generally, in what follows we draw conclusions that have broad application to smallholder farming and research.

Summary of Findings

Zimbabwe's "green revolution" was characterized by the widespread adoption of hybrid maize varieties (R201 and R215) and significant increases in yields. The diffusion of newer varieties that replace these has occurred more slowly and has had a more modest impact. Several factors account for this deceleration.

One factor is the changing role of the private and public sectors. In the early 1980s, the government was heavily involved in the dissemination of hybrid maize as well as the development of supporting institutions, such as credit and marketing. Government's current role is much reduced and one that increasingly focuses on "better farmers." Private sector institutions that have entered the maize sector operate mainly in areas of high agricultural potential. Consequently, adoption partly reflects choice but also the (sometimes) limited physical availability of varieties. A further factor is the nature of the technology being introduced. Although R201 and R215 were initially bred to meet the needs of commercial farmers, they have characteristics (high-yielding and drought resistant) that made them attractive to smallholders. Newer varieties

13. A reviewer encouraged us to construct a simulation of combined effects to see, in the end, how much adoption of new maize varieties contributes to protecting children's health from droughts, noting that because the causal chain is long, the result was likely to be quite small. We considered this idea carefully but ultimately decided not to undertake this simulation. Although the initial components of such a simulation are straightforward (adoption increases crop incomes, which in turn increase assets, which has a feedback effect via the relationship between assets and income generation), the final component is not straightforward. We can only quantify the protective relationship between child health and assets in relatively general terms (whether the child is in a household above or below the median for livestock holdings). Simulation results would require us to model the distribution of assets, with particular attention to those households around the median. Depending on how we did this, we could plausibly claim a big or a small effect, but it is not clear that such results would be meaningful.

are bred to meet the evolving needs of commercial farmers, but these new needs—most notably improved disease resistance—are not shared by the farmers in our survey and are not associated with significantly higher yields where the use of fertilizers is limited.

These two factors point to the limitations of relying on the private sector for expanding the options for smallholders. Current conditions in Zimbabwe suggest that smallholder farmers value drought resistant low-input varieties (such as open-pollinated varieties [OPVs]); it is unclear—quite apart from legal restrictions on the development and dissemination of these varieties—whether private firms are best placed to respond to this demand. If private suppliers do not find it profitable to service smallholders, it seems that intervention by government and related institutions is desirable, and perhaps necessary, to provide the services or to subsidize the private sector in supplying them.

In addition, information is disseminated via multiple channels and in a fragmentary fashion in an environment where tolerance of dissent is limited, the behavior of neighbors is viewed suspiciously, and some actors involved in dissemination (such as extension workers) are increasingly viewed with mistrust. The presumption that farmers learn from one another is less applicable in such circumstances.

Our case studies indicate links between the production of maize in excess of subsistence needs, the accumulation of such assets as livestock and tools, payment of school fees, and the acquisition of such inputs as fertilizer and labor for the subsequent cropping season. These observations coincide with the views of farmers who see high-yielding varieties of maize as an influential factor in raising livelihood above the level of poverty that prevailed when they first moved into the area.

However, new varieties appear to have increased incomes only modestly. Not only is this the view from farmers themselves, it is reflected in our multivariate work. When we control for farmer characteristics and the endogeneity of adoption, use of these new varieties increases crop incomes by about 20 percent. Additionally, many respondents convey the view that there is nothing special about maize production, which is confirmed in our multivariate analysis. A 10 percent increase in maize income is associated with an increase in livestock holdings ranging from 6 to 11 percent. However, it also shows that income from maize and nonmaize crop production has approximately equal effects on the accumulation of assets.

That said, these modest impacts result in an improved ability to deal with vulnerability. Hybrids do raise productivity in maize production. Higher income from maize, and from other crops, leads to investment in livestock. And livestock holdings are an important means through which child health is protected when drought occurs. Protection of child health is taken as an indicator of protection from impoverishment.

Assessment of Methodology

THE SL FRAMEWORK. We used the SL framework and found it provided a useful checklist of issues to be researched. It also provided a useful base for conversations across disciplines. It did not, however, serve as a behavioral model. We do not believe it fundamentally affected our research and analysis, because our team would have been able to communicate well whatever framework we adopted. We found that the framework offered a model that could not always accommodate nuances of particular situations, and many topics appear in a variety of places in the framework, a feature that could pose a problem of repetition for less experienced fieldworkers.

USE OF DETAILED CASE STUDIES. Although a method involving six months of fieldwork to cover few households poses problems for replication and generalization, the depth of the understanding gained compared to a more rapid assessment approach is substantial and thus should be considered as a potential method for future impact assessment work.

We found detailed case studies helpful for several reasons. Repeated visits to homesteads led to trust and a willingness to talk about issues on which people had initially been silent, such as witchcraft and politics. Repeated visits also enabled us to verify data and hear the perceptions of the different household members on such topics as the sharing of resources. We also noted that a person might give different answers, depending on who else was present during an interview. Observation enabled us to verify data and access information that people did not report, such as on intercropping and the use of open-pollinated varieties.

Case studies gave different information than did group discussions. In the group discussions in Sengezi, people did not mention planting Hickory King, which was a very popular variety among the villagers. That the book knowledge of youth is unreliable was the consensus view in groups, whereas there were several dissenters in private. However, we heard more on the community's critical events in group discussions when private discussions focused only on personal crises. The groups also gave the fieldworker an opportunity to talk with youths, who in most cases were excluded by adults from taking part in the interviews at their homes.

In the time available, some questions remained too personal or too sensitive to warrant a forthright answer. In Sengezi, people remained reserved on issues related to magic, levels of education, and ranking other farmers, and women did not want to answer questions related to AIDS and illness. Ideally such studies should cover a full agricultural cycle. The timing of the fieldwork was dictated by constraints of a larger international study (this volume) and not coordinated with the agricultural season in Zimbabwe. Although research in the off-season meant that farmers had more time to talk, it limited our observation of agricultural activities. We were not able to check on discrepancies in the answers of men and women about making decisions. We could not observe the use of fertilizers, the type of seeds planted, or the division of labor along gender and age lines. We could not observe how stated intentions and ideals related to practice. The issue of gender also posed problems for the researchers in the field. In Sengezi, the male researcher was restricted by husbands in his access to women's views. In Mupfurudzi, some male respondents became less forthcoming when their sexual advances were spurned, a problem that was largely overcome by time.

A further problematic issue was the provision of short-term rewards to participant households who gave time and attention to our study. Although these were very small in terms of costs of the study, they were large enough in local terms to create jealousy and conflict.

INTEGRATING QUALITATIVE AND QUANTITATIVE TECHNIQUES. An attractive feature of our approach was our ability to iteratively integrate the qualitative and quantitative analyses. A good example of the benefits of this integration is our analysis of aspects of gender and technology adoption. Our qualitative work indicated that women do not have access to many of the channels through which information on new hybrids is diffused. But our quantitative data showed no difference between male- and female-headed households. These seemingly contradictory results were reconciled by further qualitative work that indicated that other adult males, such as youth, provided an alternative conduit for information on new hybrids. Reliance on only one approach would have been unsatisfactory here.

An integrated approach to this topic, together with the use of the SL framework, allowed us to develop a rich understanding of processes of adoption and their impact. In particular, our approach helped to reveal several variables relevant to knowledge as well as access to resources and benefits, creating differentials between households and between categories of persons even within households. Yet it is costly to obtain such understandings, particularly in terms of time. One of the dangers of the SL framework is that it can be associated with the discourse of rapid appraisal methodologies. However useful rapid techniques may be in terms of economy and time, and in eliciting noncontroversial answers to specific questions, a detailed understanding of what is happening in agricultural communities cannot be obtained in a hurry.

Future Directions in the Development of Hybrid Maize in Zimbabwe

The current maize landscape in Zimbabwe is significantly different from the pre-independence and postindependence period up to the mid 1990s. These differences are such that the present setting can be labeled as a *third* stage in the production and adoption of (hybrid) maize. This stage is fuelled by a drastic change in the breeding and commercialization of hybrids by seed companies.

Historically, Seed Co, the dominant player in the provision of maize seeds in Zimbabwe, served a variety of clients, including large-scale commercial farmers as well as smallholders in resettlement and communal areas. In the pre-independence period, Seed Co largely responded to the needs of commercial farmers. Seed Co released the successful varieties R201 and R215, which were much favored by resettled and communal farmers because of their high yields and drought tolerance. In the late 1980s and early 1990s, when Seed Co reinvigorated its plant breeding efforts, greater attention was placed on developing disease tolerance, an important concern for commercial farmers. Seed Co's decision to terminate the production of these "old" hybrids in the late 1990s was taken for both agronomic (the need for varieties that were more disease resistant) and technical reasons (problems associated with continuing to attempt to produce older hybrids). But as maize breeders reminded us on several occasions, it is impossible to breed all desirable characteristics into any one variety. As a result, in the case of newer hybrids, there is not necessarily any dramatic increase in yield, nor are such characteristics as taste, number and size of ears, and flintiness taken into account. So it is not surprising that these new hybrids receive such mixed reviews. Although Seed Co sees itself as assisting these smallholders, the farmers themselves see their options, in terms of seeds available, being substantially reduced.

One of the relevant options provided by the current institutional environment in the region is the work of CIMMYT in the field of OPVs. Many are early maturing, resistant to maize streak virus, suitable for green maize production, and have flinty grain types. Thus they are well suited to smallholder and resettlement farmers. Furthermore, they require fewer inputs, an important consideration in an environment where reliability and cost of input supply are matters of increasing concern.[14] Breeding, and above all, the multiplication and marketing of OPVs do not come without problems. Significantly, OPV breeding is based on genetic variation, through cross-pollination and recycling. Maintaining genetic variation and accessing genetic material is hence crucial. One of the problems for Zimbabwe may be that maize is a recent crop, which implies that naturally present variation is somewhat limited. The OPV program is meant to supplement or improve the lack of genetic variation in the region and thus provide more options for local people's livelihoods.

To conclude, we summarize answers to our three research questions:

1. Key factors affecting the diffusion of new maize hybrids in the 1990s included:
   - A diminishing role of government and increasing role of the private sector, leading to a caution against relying too heavily on the private sector;

14. Some new developments have emerged since the completion of our fieldwork. The quantitative fieldwork for this study was undertaken in the mid- to late 1990s up to 2000, the qualitative fieldwork was conducted in 2001, and most of our initial analysis was completed in 2002. While our survey work was in the field, Dr. Marianne Banziger of CIMMYT was developing new varieties with several qualities considered desirable by our respondents. The first of these, called "Grace" or "Zm521," is now available. Zm521 is an OPV variety, requires less fertilizer, and is drought resistant (CIMMYT 2001; Muza, Waddington, and Banziger 2001).

- A less dramatic increase in yield offered by the newer varieties, and a neglect of the needs and preferences of smallholders;
- Information that is disseminated in a fragmentary fashion, and in a climate of mistrust; and
- Smallholders who are often more interested in lowering the costs of inputs than maximizing yield, a preference that should influence the development of seeds.

2. A correlation exists between maize production and the accumulation of assets, although the accumulation cannot be attributed to maize alone.
3. An improved asset base helps households to deal with such shocks as drought and thus reduces the effects of impoverishment.

References

Arellano, M., and S. Bond. 1991. Some tests of specification for panel data: Monte Carlo evidence and an application to employment equations. *Review of Economic Studies* 58 (2): 277–297.

Bourdillon, M. F. C. 1976. *Shona peoples.* Gweru: Mambo Press.

Bourdillon, M. F. C., P. Hebinck, J. Hoddinott, B. Kinsey, J. Marondo, N. Mudege, and T. Owens. 2002. *Impact of agricultural research on poverty reduction, case study 3: Assessing the impact of HYV maize in resettlement areas of Zimbabwe.* Washington, D.C.: International Food Policy Research Institute. Photocopy.

CIMMYT (International Maize and Wheat Improvement Center). 2001. Why seed matters: Hunger grows in a land of hybrids. Harare: CIMMYT. Press release.

Heckman, J. 1979. Sample selection bias as a specification error. *Econometrica* 47 (1): 153–161.

Hoddinott, J., and B. Kinsey. 2001. Child growth in the time of drought. *Oxford Bulletin of Economics and Statistics* 63 (4): 409–436.

Kinsey, B. 1982. Forever gained: Resettlement and land policy in the context of national development in Zimbabwe. *Africa* 52 (1): 92–113.

———. 1999. *Determinants of rural household incomes and their impact on poverty and food security in Zimbabwe.* Report submitted to SDAR/IG5 FAO on the Zimbabwe Rural Household Dynamics Project. Economics and Social Institute, Free University, Amsterdam.

Maddala, G. S. 1983. *Limited-dependent and qualitative variables in econometrics.* Cambridge: Cambridge University Press.

Mashingaidze, K. 1994. Maize research and development. In *Zimbabwe's agricultural revolution,* M. Rukuni and C. Eicher, eds. Harare: University of Zimbabwe Press.

Masters, W. 1994. *Government and agriculture in Zimbabwe.* Westport, Conn.: Praeger.

Muza, L., S. Waddington, and M. Banzinger. 2001. Preliminary results on the response of "Nitrogen use efficient" OPV and hybrid maize to N fertilizer on smallholder fields in Zimbabwe. In *Integrated Approaches to Higher Maize Productivity in the New Millennium: Proceedings of the Seventh Eastern and Southern Africa Regional Maize Conference,* 11–15 February 2001, D. K. Friesen and A. F. E. Palmer, eds. Nairobi, Kenya: CIMMYT and KARI.

Rattrary, A. 1956. Maize breeding. In *Maize production in Southern Rhodesia: Proceedings of the Second Annual Conference of the Professional Officers of the Department of Research and Specialist Services,* Gwebi Agricultural College, 19–20 March. Salisbury, Southern Rhodesia: Government Printer.

Rohrbach, D. 1988. The growth of smallholder maize production in Zimbabwe (1979–85): Implications for food security. In *Southern Africa: Food security policy options,* M. Rukuni and R. Bernstein, eds. Harare: University of Zimbabwe Press.

Rukuni, M., and C. Eicher, eds. 1994. *Zimbabwe's agricultural revolution.* Harare: University of Zimbabwe Press.

Scott, M. 1991. *A new view of economic growth.* Oxford: Clarendon.

World Bank. 1991. *Zimbabwe: Agriculture sector memorandum,* Vol. II: *Mai report.* Southern Africa Department Report 9429-Zim. Washington, D.C.: World Bank.

7 Improved Maize Germplasm, Creolization, and Poverty: The Case of Tuxpeño-Derived Material in Mexico

MAURICIO R. BELLON, MICHELLE ADATO, JAVIER BECERRIL, AND DUBRAVKA MINDEK

Improved maize varieties have been available in Mexico for more than 40 years, but diffusion of these varieties has been limited. Despite repeated government campaigns to encourage the use of improved seed, today only about one-fourth of the total maize area in the country is planted in improved varieties; most of this area is located in the commercial production zones of central and northwestern Mexico (Morris and Lopez-Pereira 1999). The relatively low rate of diffusion may provide a misleading impression, however, of the true impacts of improved germplasm on the welfare of rural households. A growing body of evidence suggests that many small-scale subsistence-oriented farmers have taken up improved varieties and planted them alongside local varieties. Through exposing improved varieties to their conditions and management, continually selecting seed of these varieties for replanting, and in some cases promoting their hybridization with landraces, either by design or accident, farmers produce what they recognize as "creolized" varieties (*variedades acriolladas*) (Bellon and Risopoulos 2001).[1]

Conventional germplasm impact studies usually focus on areas planted in improved varieties. To date, few attempts have been made to document the use of creolized varieties. The lack of studies in this area constitutes a major gap: if creolization is ignored, the benefits generated by formal plant breeding programs may be significantly underestimated. This study attempts to document how poor farmers in lowland tropical Mexico use improved maize germplasm both directly (by adopting improved varieties) and indirectly (by creating creolized varieties). In addition, the study attempts to determine how the use of improved germplasm contributes to the well-being of poor small-scale farmers. Our key hypothesis is that poor farmers benefit from improved germplasm through creolization. Although improved varieties provide desirable traits not found in landraces, they may also lack traits found in the latter. As a distin-

1. Wood and Lenné (1997) use the term "rustication" to describe the process through which materials produced by formal plant breeding programs change in the hands of farmers.

guishing feature of landraces is their local adaptation, choosing between them and improved varieties presents trade-offs for farmers. Creolization lessens these trade-offs by adapting improved varieties to local conditions most relevant to these farmers. Creolized varieties provide a combination of traits not supplied by landraces, while entailing fewer trade-offs than do improved varieties. Creolization thus provides farmers with new options, as they deliberately modify an improved technology generated by the formal research system to suit local circumstances and needs.

The study involves three separate but related activities: (1) measuring and explaining diffusion, local adaptation, and use of improved maize germplasm; (2) understanding how adoption choices are linked to livelihood strategies and vulnerability context of rural households; and (3) assessing the impacts of adoption on the welfare of rural households. The specific focus of the study is the Tuxpeño germplasm complex. Tuxpeño is one of approximately 250 maize landraces found in the New World. This maize race has been subjected to intensive breeding efforts, first by the Rockefeller Foundation and the Mexican Ministry of Agriculture and later by their successors, Centro Internacional de Mejoramiento de Maíz y Trigo (the International Maize and Wheat Improvement Center [CIMMYT]) and Instituto Nacional de Investigaciones Forestales Agrícolas y Pecuarias (INIFAP), respectively. Tuxpeño germplasm has been used by these and other institutions, including private companies, to breed both hybrids and improved open-pollinated varieties (OPVs).[2] This study was carried out in two regions: the coast of Oaxaca and the Frailesca in Chiapas. Oaxaca and Chiapas are two of the poorest states in Mexico.

The chapter is divided into six sections. The first section describes the methods used in the study, in particular, how the study was designed and conducted. It is followed by a section describing the two study areas, especially how they contrast in terms of development, degree of commercialization, and maize production. Then we describe the different types of maize germplasm studied and present a history of their diffusion, including the origins of the seed used and its management. We show that farmers plant many different types of maize germplasm, of various origins and management histories that affect their current choices. The penultimate section presents results of adoption of different germplasm types. It shows that adaptation, management intensity, cultural factors, risk, and integration into the regional and national economies play a key

2. There are two types of improved maize varieties: hybrids and open-pollinated varieties (OPVs). A hybrid can be defined as the combination of two inbred lines—exhibiting hybrid vigor—whereas improved OPVs are populations that have been subject to selection by breeders. If seed from a hybrid is replanted it will not be as productive as the original seed. Therefore hybrid seed has to be purchased every season to maintain high productivity. However, seed from an OPV can be replanted without major drops in yield—usually up to three years. Hence OPV seed needs to be purchased once every three years.

role in the adoption process, although the process differs for each germplasm type. We then present the impacts of the various germplasm types on farmers' well-being. The effects are defined and analyzed in terms of the extent to which different types of germplasm supply farmers with traits they consider important and the trade-offs they entail. We show that farmers in both study areas—male and female alike—value multiple traits in their maize, that the different maize types provide these traits in varying degrees, which in turn translates into trade-offs among these maize types. There is no "perfect" maize type; nevertheless, as hypothesized, creolized varieties present a compromise between improved varieties and landraces for certain traits. The final section presents our conclusions.

Methods

The overall design of the research of this mixed-methods study used the sustainable livelihoods framework as a guide, sharing a common conceptual framework with four other studies of the impact of agricultural research on poverty (this volume; see also Adato and Meinzen-Dick 2003). Twelve communities were selected in areas of medium, high, and very-high marginality, defined according to an index used by the Mexican government to target its poverty alleviation program PROGRESA (CONAPO-PROGRESA 2000). They also included communities with indigenous populations. Site selection further considered agroecological conditions, as the study focused on a tropical maize germplasm, and government programs to diffuse the seed of improved varieties.

The qualitative research began with two sets of focus group discussions. The first was on local perceptions of poverty, livelihood strategies, and vulnerability; the second was on perceptions of maize traits and how they respond to these conditions. These discussions were followed by household case studies conducted in 4 of the 12 communities. The case studies were carried out by anthropologists, who spent several months living in these four villages and interacting with farmers in the household, field, and market, and during other activities. Ten case studies were conducted per village, with households selected to roughly represent "extremely poor," "average poor," and "less poor" farmers. The quantitative research involved a representative sample survey of 325 farming households covering all 12 communities. Finally, the project included a collection of all maize types grown in the communities and an agronomic evaluation of maize samples.[3]

The Study Area

The 12 communities included in this study are located in two highly contrasting regions: the coast of Oaxaca and the Frailesca, Chiapas (see Figure 7.1). For

3. For a complete description of the methodology, see Bellon et al. (2005).

FIGURE 7.1 Map of the communities included in this study, Oaxaca and Chiapas

NOTE: Asterisks denote communities where in-depth anthropological case studies were carried out.

simplicity they are referred to in the rest of the chapter as Oaxaca and Chiapas. The communities were selected systematically to sample the range of marginality levels, levels of improved germplasm diffusion, and ethnic makeup of both regions. However, because we have adopted a case study approach, we consider that the results presented in this chapter are valid only for the 12 communities studied.

Communities in Chiapas have better access to government-provided services and infrastructure—even for similar marginality levels—than those in Oaxaca. Productive activities are more oriented to the market, and the region has received strong support from state and national governments, particularly for agricultural development. This region produces important maize surpluses that are exported to other parts of Mexico; however, agriculture is still dominated by small-scale farmers that produce both for the market and self-consumption. There is an important dairy industry, and farmers can add value to their maize by using it as animal feed. The use of inputs and improved seed has been promoted through several government projects throughout the years.

In contrast, Oaxaca has been more isolated and has not received much government support for agricultural development. The state imports substantial amounts of maize from other parts of Mexico and from abroad. Although the coast has a better climate for maize and agricultural production than do other regions of the state, it is not an important producer of this staple. Commercial agricultural activities are biased toward extensive cattle ranching and maize production toward home consumption. Development has been more related to tourism, particularly in the southern part of the study area, where there are such resorts as Puerto Escondido, Puerto Angel, and Bahías de Huatulco.

As an example of these differences, Table 7.1 compares key characteristics related to population, farming, maize production, income sources, and poverty between the two study areas. The population is much larger in Oaxaca than in Chiapas. However, the farming area is much smaller; hence, there is stronger population pressure on the land in Oaxaca. Farming households in Oaxaca produce maize mainly for self-consumption, with little use of hired labor and fertilizers, and most do not sell or have to purchase additional maize for consumption. Cattle ownership is not very high, although those who own cattle have a fair number of them. Many farmers work outside their farms as hired laborers, and other sources of income are not very common, with less than a fifth engaging in them. Farming households in Chiapas, however, mainly produce maize for the market, although they also consume some of what they produce, use a much higher level of hired labor and certainly of fertilizers (the average rate of use for those that use fertilizers is almost four times as high as in Oaxaca). Almost no farmer in Chiapas needs to purchase additional maize for consumption. Cattle ownership is lower than in Oaxaca, both in terms of number of households who own and the average number of heads owned. A much lower percentage of farmers work as farmhands compared to Oaxaca, although a similar proportion have other sources of income outside their own farm.

Poverty

Poverty is pervasive in both study areas, even in the more commercialized and developed Chiapas. Poverty rates were calculated with data on household consumption obtained from the survey. These data included both purchased and self-produced items, to which local prices for similar goods and services were imputed.[4] Two poverty lines were constructed: extreme poverty and poverty.[5]

4. Per capita expenditure was calculated and adjusted to adult equivalents with the weights used by Skoufias, Davis, and Behrman (1999). Furthermore, household expenditure in Oaxaca was adjusted to make it equivalent to purchasing power in Chiapas, because prices for similar goods were higher in the former than in the latter.

5. The poverty lines were developed according to the methodology of Guevara-Sanginés et al. (2000). The extreme poverty line was defined as the expenditure necessary to purchase the Coordinación General del Plan Nacional de Zonas Deprimidas y Grupos Marginados standard food basket plus 27 percent more for basic nonfood items (MX$415 per capita per month in 2001). The poverty line differed from the extreme poverty line in that the former increased the amount of nonfood items to 125 percent of the cost of the food basket (MX$754.82 per capita per month in 2001).

TABLE 7.1 Comparison of key characteristics of the two study areas

Variable	Oaxaca	Chiapas
Total farming population	21,471	10,507
Number of households	3,539	1,994
Households that speak only Spanish (%)	34.5	94.3
Maize production		
Production objective (households)		
Self-consumption exclusively (%)	82.6	2.9
Market and self-consumption (%)	17	95.4
Agricultural landholdings (average per household; ha)	1.9	5.3
Use of hired labor (%)	27.1	60.5
Use of chemical fertilizers (%)[a]	45.3	99.4
Average rate (kg nitrogen per ha)[a]	52.9	202.6
Households that sell maize (%)	27	98.8
Households that purchase maize (%)	64.5	8.4
Animal holdings		
Households that own cattle (%)	36.4	27.4
Number of cattle per household (average)	10.9	8.5
Income sources (%)		
Performance of off-farm labor	63.5	45.0
Performance of nonfarm farm labor	17.6	19.8
Temporal migration	16.7	14.1
Remittances	10.6	13.6
Poverty indicators		
Number of farming households		
Extremely poor	2,645	1,261
Poor	666	521
FTG poverty indices		
Headcount index	0.80	0.72
Poverty gap	0.34	0.27
Severity of poverty index	0.17	0.13

NOTE: Estimates are derived from the sample survey using expansion factors.

[a] These data are based on one plot per household.

Based on these lines, three groups were defined: the extremely poor (expenditure below the extreme poverty line), the poor (expenditure between the extreme poverty and the poverty line), and the non-poor (expenditure above the poverty line).

Most farming households are below the extreme poverty line in Oaxaca and Chiapas—74.7 and 63.2 percent, respectively. On a population basis, however, the rates of extreme poverty increased substantially. Table 7.1 presents the Foster, Greer, and Thorbecke (1984) poverty measures using the extreme poverty line as a reference. The headcount index, poverty gap, and severity of poverty index show that there are higher numbers of poor individuals in Oaxaca, with a larger poverty gap and more extreme poverty, than in Chiapas.

Poverty has multiple dimensions, and consumption is only one of them. The qualitative work provided insight into other dimensions. Local indicators of poverty and wealth fall into several categories: material resources, culture, beliefs, and behavior. Resources are given the greatest emphasis, with access to and uses of land being most significant; others include access to money, planting of other crops (for example, coffee), performance of activities other than agriculture, ownership of animals and implements, amount of family labor that can be mobilized, ability to speak Spanish, receipt of remittances, and type of off-farm labor.

Another aspect of local perceptions of poverty is cultural: indigenous roots indicate poverty. Indigenous people live on the margins of the community with little land or money; illiteracy and lack of Spanish fluency keep people in poverty by limiting their ability to find work outside the area. Finally, poverty is also related to beliefs, practices, and behavior. Wealthier families are said to represent the best morals and practices: they are hard workers and frugal. They are also described as snobbish, untrusting, and stingy. Families of average wealth are described as hard workers, although they are held back by a lack of access to some vital resource. The poorest are described as having great difficulties. They have no money and no one to help them. Additionally, they are sometimes perceived to hurt themselves by wasting money from government programs on vices and to be perpetrators of domestic violence. Women cannot provide good homes because they have to work. Similarly, their children cannot study for lack of money. In some communities, religious affiliation seems to matter; in particular, evangelical Protestants are said to be wealthy because they are hard working and do not drink.

Several factors related to agriculture and maize production make people vulnerable to falling into, or deeper into, poverty:

1. Population growth: increases in population provoke land pressure.
2. Resource pressures: cash for investment in agriculture is scarce, "tired," hilly, and eroded land requires extra investment.
3. The local economic system: restricted access to markets, lack of stable wage work, little education, illiteracy, and monolinguism hinder employment prospects; falling coffee prices and low maize prices depress incomes. The institutional environment surrounding maize markets in Chiapas reveals and exacerbates social differences. Warehouses require a minimum quantity and quality, which the poorest farmers cannot meet. They thus must sell to intermediaries, better known as "coyotes." Coyotes are less demanding about quality and quantity, but they pay considerably less than the warehouses. Coyotes are seen as at least a necessary evil—they pay cash

up front, pick up the maize, and do not charge for transportation. Coyotes also provide loans, and many farmers must go into debt to plant.

4. Shocks: rapid and severe climatic, human, and animal health changes; delayed or excessive rainfall; and strong winds adversely affect agriculture. Pests and diseases affect larger animals.
5. Seasonal changes: shortages of maize occur and seed and agricultural investments must be made precisely when money and food are most scarce. The poor must leave their fields to work elsewhere during the planting season, and as a consequence do not tend their fields, thus lowering yields. Colds and influenza make working more difficult. Finally, the religious festival season requires the poor to harvest maize too early, before the ears are ready, and to sell the grain before the price reaches its maximum.

Role of Maize in Farmers' Livelihoods

Households in the two study areas have diversified livelihoods: they grow several crops, keep different types of animals, and participate in diverse off-farm and nonfarm activities. Besides maize, crops include beans, squash, tree fruit, coffee, tomatoes, red peppers, sesame seed, hibiscus, groundnuts, and cacao. All households grow maize, which is an important component of farmers' livelihoods in both study areas. There are, however, differences between them. More than three-quarters of farmers in Oaxaca grow maize for home consumption exclusively, whereas in Chiapas almost all farmers grow maize for both home consumption and market. Few farmers in both study areas produce solely for the market. More than half of farmers in Oaxaca did not produce enough to meet their maize needs in the past five years. Only about a third of farmers frequently sell maize, and most sell less than half of their production. Maize is sold mainly to families in local communities and, to a much lesser extent, local traders. However, farmers are very commercialized in Chiapas. More than 90 percent produced surpluses in the past five years, and almost all sold more than half of what they produced. They sold mainly to the government, private business, and local traders, or a combination of them. Almost none sold to other families in the community.

The price of maize varies between study areas. Maize is much more expensive (60 percent on average) in Oaxaca than in Chiapas. There are also differences between the purchasing and selling prices within the two study areas. Although maize is more expensive in Oaxaca, there is almost no difference between the selling and purchasing price. However, there are important differences in Chiapas, where the purchase price of maize is about 30 percent higher than the selling price.[6] Hence, it is significantly cheaper for a household

6. The reasons for this difference are not clear. It may be that the local markets are thin and traders incur high storage costs as well as risks (not only the physical storage costs, but also the money tied to the product in the absence of efficient financial markets) and hence they demand a high premium.

in Chiapas to produce its own maize than to sell and buy it. This may explain, to a certain extent, why in a commercialized system such as the one in Chiapas, production for home consumption remains an important objective of maize production.

The qualitative work supports many of these findings. In Oaxaca, people were found to be growing maize mainly for personal consumption. The poorest farmers depend on it for their food security ("it is the only thing that we cannot do without in the house"). Growing it enables them to use their scant money on other items ("I plant so as not to buy, and when I go to trade and make a little money it is to buy something else, for example sugar, soap, and what we need at home"). Although people in Chiapas are mostly interested in selling their crops, maize cultivation assures basic subsistence and is particularly important for the poorer farmers. As one extremely poor farmer in Dolores explained: "it is necessary to take out the portion that is our food because there is no work and if we don't plant we will die of hunger." Still, for many, maize is most important as a source of money, although they use a portion of it for their annual consumption.

Maize also plays important social and cultural roles in people's lives, though these sometimes also have economic effects. For example, maize plays a role in the cargo system (known as *mayordomia*), a ritual cycle in which people sponsor parties honoring a saint's feast day. This practice especially affects the extremely poor, who have to sell maize early to help pay for holiday expenses. An early harvest causes people to lose significant income. A few case study informants in Oaxaca said that they planted maize because of traditions. They also interpret the material benefits of maize in light of established practices. One man held, "I cannot accept not planting, because ever since I was little this was the job of my father, so I can't keep from planting, because when there are tender ears, you go and harvest whenever you want and the amount you want, and if you go to buy it isn't the same."

Maize Diffusion and Adaptation

Maize Germplasm

Farmers in both study areas plant numerous maize varieties ranging from hybrids to landraces. A collection of different maize varieties was carried out and a total of 126 samples assembled. Each sample corresponds to one distinct type recognized by farmers in a community.[7]

---

7. The sampling strategy was to collect all the different maize types recognized in each community. The types or varieties were identified during the focus group discussions as well as from the survey, as the collection occurred afterward.

The survey also elicited extensive information on the varieties planted, their names, origin, history, and management. We classified the maize varieties identified in the survey into five categories: hybrids, recycled hybrids, improved OPVs, creolized varieties, and landraces (also referred to as *criollo*).[8] The classification is based on (1) the name provided by the farmer, (2) whether the farmer said that the seed came from a "bag," (3) the number of years seed was used, (4) information on its origin from the farmer and focus group discussions, and (5) classification by a maize taxonomist of a collection of maize samples from all communities in the study. Table 7.2 presents the specific criteria used for each category. Classifying the maize types elicited from farmers in the survey entails a certain degree of arbitrariness. As Morris, Risopoulos, and Beck (1999) note, the dynamic nature of maize makes classifying its varieties into distinct and well-defined categories difficult and somewhat arbitrary. However, a classification is useful as long as the criteria are workable, defensible, and consistent. The criteria defined above were applied systematically, and we are confident that the classification is meaningful. This classification is the basis for the adoption and impact analyses presented below.

Furthermore, the 126 collected maize samples were evaluated for agronomic characteristics at the INIFAP Cotaxtla Experimental Station, together with three commercially purchased improved OPVs and nine landrace accessions from the CIMMYT genebank as controls. Figure 7.2 presents 95 percent confidence intervals for the mean of three key agronomic characteristics: anthesis (male flowering, an indicator of growing cycle), plant height, and yield for these samples classified according to the criteria presented in Table 7.2, plus the two controls. There are statistically significant differences among the groups, as well as clear trends. These results support the idea that our classification is biologically meaningful.

The figure shows that our classification is able to capture measurable and statistically significant differences and trends in key plant characteristics. Average plant height and anthesis increase, while yield decreases, as one moves from the groups classified as hybrids and recycled hybrids to those classified as

8. There are two types of improved maize varieties: hybrids and OPVs. For simplicity a hybrid can be defined as the result of the combination of two inbred lines, whereas improved OPVs are populations that have been subject to selection by breeders for a very specific set of traits. A recycled hybrid refers to the product of replanting saved seed from a hybrid. Creolized varieties have been defined in the text. Landrace or *criollo* refers to one of the many maize populations that farmers have inherited from their ancestors who domesticated the crop, but that continue to evolve through natural and farmer selection. There is increasing evidence from population genetics studies that small-scale maize farmers in Mexico shape the diversity of traits of their maize populations (Pressoir and Berthaud 2004). These definitions are used by breeders and ourselves; however, farmers may give different meanings to these terms, as discovered during our research and discussed in this chapter.

TABLE 7.2 Criteria to classify varieties identified in survey into five categories

Category	Criteria
Hybrid	Name provided by farmer was of a known hybrid Seed came from a "bag" and first year of planting Focus group identified the name as being introduced to the community by government or commercial outlet Maize taxonomist indicated that sample with same name was of a hybrid or recycled hybrid
Recycled hybrid	As above, but farmer had planted the seed from the previous harvest up to four years
Open-pollinated variety	As above, but name provided by the farmer was from a known open-pollinated variety Seed had been planted for first time or recycled up to four years
Creolized	Any of the above, but farmer had recycled the seed for more than four and up to fifteen years
Landrace	Name provided by farmer was of a known maize race (for example, Zapalote, Tepecente, Olotillo) Seed did not have a specific name (*maiz blanco*) but had been planted for many years either by the farmer or by somebody else in the community Seed did not come from a bag Focus group identified the name as a local variety Maize taxonomist indicated that the sample with the same name was a landrace

FIGURE 7.2 Confidence intervals for the means of anthesis, plant height, and yield

a. Anthesis

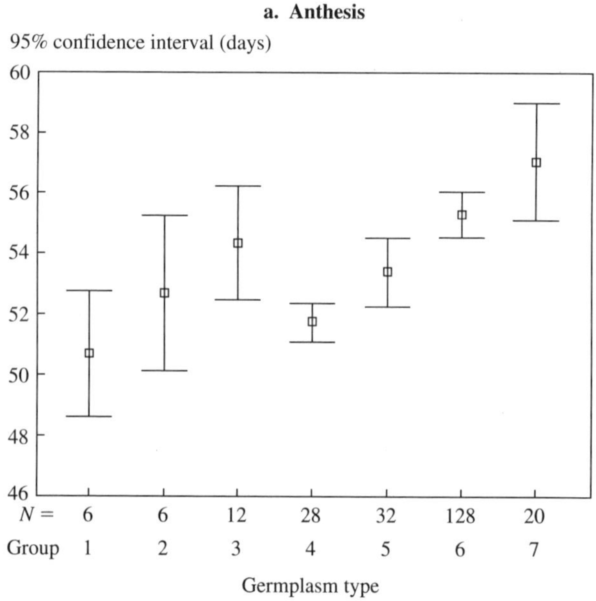

FIGURE 7.2 *Continued*

b. Plant height

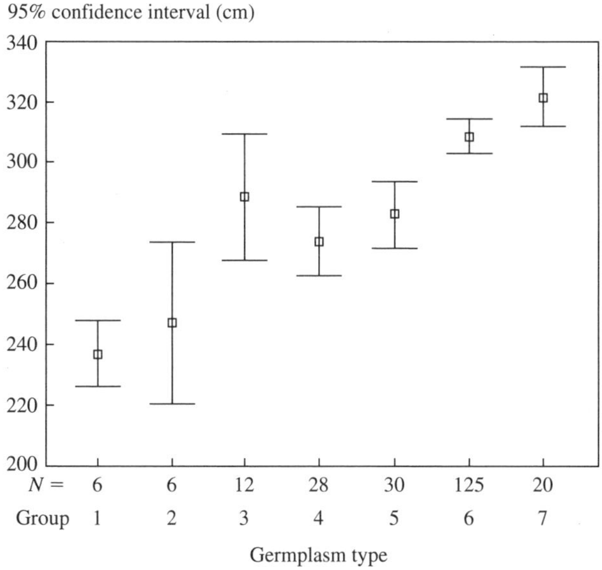

c. Yield

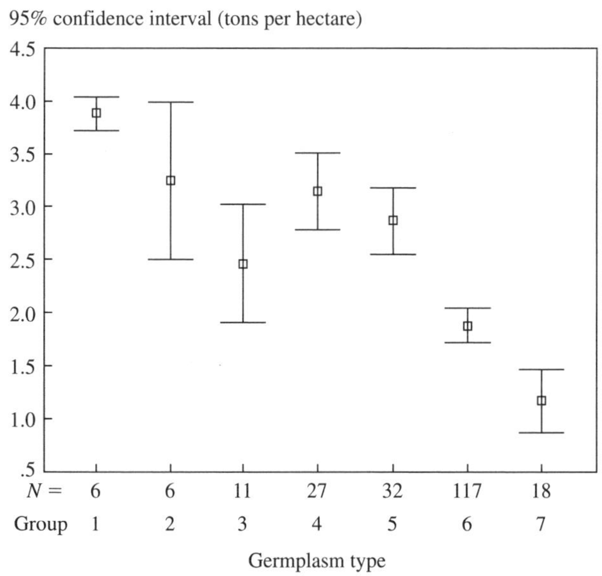

NOTES: Data are from maize samples collected in the study sites and classified according to criteria in Table 7.2. Group 1 refers to seed of a set of three commercial open-pollinated varieties (OPVs) purchased in a commercial outlet. Groups 2–5 refer to the seed of the 126 varieties collected in the field sites and classified according to criteria in Table 7.2 as: hybrids (group 2), recycled hybrids (group 3); OPVs (group 4); creolized varieties (group 5); and landraces (group 6). Group 7 refers to a set of nine representative accessions of landraces from the region obtained from the CIMMYT genebank. Groups 1 and 7 are the controls referred to in the text.

creolized varieties and landraces.[9] The control groups are at the extremes. The trends in the data are consistent with what one may expect with maize populations resulting from scientific breeding—they have shorter growing cycles and plant height, as well as higher yields, compared to populations that have been under long-term farmer management (landraces and creolized varieties).

A key finding of the household case studies is that local categories of seed types are not the same as the ones defined above. People generally classify seeds that do not come in a new package as *criollo,* regardless of whether they are formally defined as recycled, creolized, or landraces. In discussions of the case study findings, we use the local terms when referring to the perspectives of the informants. In Oaxaca, people distinguish between *criollo* and "variety" maize. The latter includes all those that come from agricultural secretariat programs. In Chiapas, people distinguish between *criollo*s and seeds from a bag. Among bagged seeds, they distinguish between those from the secretariat and hybrids or commercial seeds from veterinarians. In neither region do people distinguish between old or "original" *criollo* seeds (that is, landraces) and those that were "creolized" *(acriollado)* over time. Both types are called *criollos.*

Furthermore, people do not necessarily define varieties so much as describe them in terms of their advantages and disadvantages. Only in Oaxaca did some people refer to *criollo* as the original maize (that is, a landrace). They do have positive associations with these varieties: "it is good; it was the first one that began to help us." People in Oaxaca generally have better knowledge about the characteristics of each variety, probably related to a longer tradition of maize cultivation than exists in Chiapas.

Nonetheless, people have different levels of confidence in the various types of seed. Notably, farmers placed more trust in *criollo* seed (that is, the combined local category): "we consider it with confidence because we already know it, we have planted it before, and we have no doubts about it." Recycling (selecting seed from a previous harvest and replanting) is considered to be creating creolized seed. Most people regarded recycled or *acriollado* seeds to be *criollo* in a few years. Even in Chiapas some large-scale producers expressed their preference for *criollo* seed, even though they plant improved varieties. Key to classification as a *criollo* seed is that the seed has been "acclimated" to local soils and so is seen as adapted to these soils. According to one farmer in Chiapas: "at first it was like a hybrid and now, later, it is *criollo.* . . . It likes the soil. It acclimated." When asked whether this process was what makes a variety *criollo,* another farmer said, "yes, that is exactly what makes it *criollo.* After

9. The relatively large confidence intervals associated with the hybrid and recycled hybrid groups are consistent with what one may expect from this type of replanted germplasm, as there will be high rates of segregation (that is, plants resulting from replanting seed from a hybrid will be highly variable in their agromorphological characteristics).

some seasons it adapts and will produce any place. Because they planted it once and now it knows the land, and since the land is good [it produces]."

Sources of Seed

Farmers in both Oaxaca and Chiapas distinguished between maize kernels as grain and maize kernels as seed, although from a biological perspective they are the same. In the case of recycled seed, maize kernels used as seed are usually subject to a rigorous selection process. In farmer-to-farmer seed transactions, kernels for seed and grain show important price differentials. For example, landrace seed costs MX$3.88 per kilogram[10] and MX$3.51 per kilogram in Oaxaca and Chiapas, respectively, whereas landrace grain costs MX$2.41 and MX$1.82, respectively.

People's memories are not very precise regarding the particular history of introduction and adaptation of each variety. But they are aware that the old *criollo* varieties have been replaced by improved ones. One informant explained that the new seeds arrived "through institutions that came offering improved seeds, and the ones we had been planting for so long were left behind. . . . They said, 'Look, this seed is good.' And since our land no longer wanted to produce as much with the seeds we already knew, we were encouraged to try them." According to informants in the case studies, farmers can obtain seed in six manners: select it from their own harvest, obtain it through social networking (including buying and selling, and reciprocity from relatives, neighbors, and friends), buy it from the government through the *ejido*[11] commissary, buy it from *campesino* organizations, purchase it (at a greatly reduced price) through political campaigns, and buy it in veterinary clinics or seed stores. The previous harvest and social networks were the most common sources in both study areas and were prevalent among all social groups. Data from the survey support these findings. During the rainy season of 2001, most farmers in Oaxaca planted seed from the previous harvest (61.4 percent of seed lots).[12] In Chiapas this fraction was much lower but still significant (39 percent of seed lots). The rest of the seed was acquired from other farmers, the government, or stores. In Oaxaca the most common outside sources of seed are farmers' social networks—family, friends, and neighbors—followed by the government and the store. In Chiapas, however, the government is the main source, followed by social networks and stores. These patterns again illustrate the contrasting nature of maize

10. The average exchange rate for the period of the fieldwork was MX$9.25 per US$1.

11. *Ejidos* were part of the study communities in Chiapas and Oaxaca, but not all farmers in the sample were *ejido* members.

12. A seed lot is defined as "all kernels of a specific type of maize selected by a farmer and sown during a cropping season to reproduce that particular maize type" (Louette, Charrier, and Berthaud 1997, 24).

production in the study areas. Social capital plays a key role in accessing seed in Oaxaca, whereas this role is much weaker in Chiapas.

Once inside a community, new seeds spread mainly by informal networks. Social networks are key to diffusion because they are trusted—more so than the government—and because people can observe the fields of others and obtain successful varieties by buying or trading for them. Occasionally, they receive them as a gift. This practice of observing the performance of varieties is widespread, especially among the poorest farmers in Oaxaca, who can tolerate less risk:

> Sometimes the maize is unknown and you don't trust to buy it. Rather, you go with your people because you see that the crop grows well and the ears are pretty. So you ask if they have some stored and you buy a bit for planting. With the seed from the stores, there is no confidence. . . . You have to see it growing in the fields of your neighbors who have grown that variety. If not, you don't buy it.

In the opinion of producers in Chiapas, good and guaranteed seeds are expensive, and they are sold by seed companies. Even if considered the best, they are too expensive for many. In fact, people defined poverty in part by what kind of seeds one uses: "Poor people around here are the ones who plant ordinary varieties."

The government has played an important role in supporting maize cultivation, especially for less well-off farmers, through programs to promote access to seed, credit for purchase of inputs, and technical support. Several programs have existed, but with many problems. This experience has influenced people's perceptions and attitudes about the reliability of government support and the quality of government seed. Significantly, experience with government seed and related programs have made people wary of using improved seed more generally.

The Agricultural Secretariat is the main government program and promoter of seeds in both study areas. The agency manages two important programs: Alianza para el Campo and the Programa de Apoyos al Campo (Procampo). The former provided, among other things, subsidized seed from both public and private sectors, known as the Kilo por Kilo program. Originally through this program farmers exchanged an amount of grain for improved seed equivalent to the quantity needed to plant one hectare or several hectares; later the program changed, providing improved seed for the price of grain but through cash payment. The latter program provided farmers with a cash subsidy for the area planted in certain crops, including maize, provided they were planting at least one of these crops during any of three agricultural seasons prior to August 1993 and they were registered for the program. Farmers can use Procampo money to purchase seed and agricultural inputs at their discretion. These programs, but especially Procampo, are distrusted, and many do not register all or any of their land, because people believe that the programs aim to take land from them. In both study areas, the *ejido* commissaries are the most important local institution that connects government programs and farmers. The Kilo por Kilo seed is channeled through the *ejido* commissaries, which become an important source of seed. The majority of improved seeds used by *ejido* producers come through inexpensive technological packages. These packages have been, and remain, the principal source of improved seeds for our case study households, although they are not the most popular. The quality of the seed often is poor. An agronomist working in the region explained that government seeds are poor because municipal governments limit themselves to providing cheap seeds that are poorly adapted to local soils.

In addition, in both study areas seeds arrive at the wrong time—when it is too late to plant: "There is no faith in the government now, because they don't come through with what they promise. . . . The support comes so late that nothing can be done." In communities in both study areas farmers expressed a strong need for government technical assistance but also a lack of faith in the motives and reliability of government:

> We need someone to come and study the soils to know about the pests. . . . We need to know what kind of insecticide to spray. . . . Some people have died because they got some on them. . . . Well, later some technicians from the secretariat came by, but they came . . . when the time that we needed them has already passed. I think they come only so that we can get to know them, but when we need their knowledge, we never see them.

Politics also enters into seed distribution. In one community in Chiapas people explained the influence of political parties: "The commissary gives the seed to his group of people and sells what is left over to the townspeople. He calls his people very secretly and writes their names on the list." Seed is also politicized through political campaigns, when improved seeds are introduced in communities and given as gifts or at low prices to supporters. Complaints about politics are also heard regarding agricultural support services more broadly. In Chiapas the poorest people complain that support is mainly given to the people close to authorities. Another problem attributed to politics (though it may also relate to economics) is the frequent criticism that government programs stay in regional centers, with little help reaching small towns.

In Chiapas some expensive but subsidized seeds can be obtained by belonging to regional *campesino* groups, which also channel such government support as subsidized fertilizers, credit, and soil analysis. To belong to such a group can be difficult and expensive, however, and impossible for poor farmers. But it said to be worth the effort for those who are able to join.

Seed Management and Flow

Recycled seed from one's own harvest or from that of other farmers is the most important source of seed, even in the more commercialized Chiapas. Beyond its value as a source of seed, seed recycling has important genetic consequences for the maize varieties that farmers plant. Varieties change under farmer selection. By selecting the plants, and hence genes, that are carried from one generation to the next, farmers play an important role in shaping the genetic structure of their varieties.

The case studies revealed seed recycling to be a widespread practice in both study areas. When people are content with their harvests, they try to select and store seeds from it. Some farmers consider it embarrassing to "waste" seed from their fields. There is also the notion that "it is better to choose my own seed grain, the one I like" rather than buy unproven bag seed. Recycling is seen locally as one way in which *criollos* are created—through successive plantings, seeds are seen to adapt to local soils. Additionally, most consider seed too expensive to buy every year. In fact, we did not find a single person who bought all of his or her seed every year. Nonetheless farmers in Oaxaca were more likely to recycle selected seed than those in Chiapas, and they buy seed less frequently from government or informal networks. If the extreme poor plant maize from a bag, generally it means that the seed was free or cheap and that they obtained it through a government program.

Although all farmers recycle, poorer ones among the case study informants were more likely to do so than richer ones. However, some less-poor informants prefer to plant recycled improved seed that they obtain from the harvest of a neighbor who planted bag seed. Recycling provides access for the poor to improved varieties that they otherwise could not afford as original seed. The number of years that farmers recycle varies between the study areas: from four to five years among informants in the case study communities in Chiapas to longer in Oaxaca. After this process they do not distinguish the seed from those long in use. Farmers here consider that it is possible to creolize or adapt any seed and do not believe claims that replanting has negative consequences: "They tell us that the hybrids will not produce from one year to the next. But I think that this is a lie, because the seed companies are making money." Many farmers claim that getting recycled seed from neighbors is a way to improve their harvest.

When asked why farmers preferred to recycle, one explained: "Because we have always done it like this and, like I told you, we can't spend a lot on seed. Also, this way is safer because we have seen how the seed produces in the lands around here." However, farmers recognize that seeds degenerate over the years: "We change when the soil demands it, because sometimes the land just doesn't want the same seeds, because what happens sometimes is that the seed has degenerated." Because people observe other farmers' fields and see results, everybody buys and trades seed as well.

Farmers also shape the genetic structure of their germplasm by fostering gene flow among different varieties, something that has been documented in other parts of Mexico (Aguirre-Gómez 1999; Bellon and Berthaud 2001). In

Oaxaca farmers have mixed seed acquired from outside into 8.9 percent of their seed lots in the course of planting a seed lot; the same took place in 7.8 percent of seed lots in Chiapas. By mixing seed, we mean that a farmer added seed from a different variety or source to the seed lot, so that when planted, cross-pollination is highly likely. Further evidence of potential gene flow: farmers in Oaxaca said that in 2001 they gave seed to other farmers (exchange, sale, or the like) from 26.4 percent of their seed lots, while they received seed from other farmers for 29.7 percent of the seed lots they planted. These rates were much lower in Chiapas, as farmers only gave seed to other farmers from 7.8 percent of their seed lots and they received seed for 5.5 percent of their seed lots. Thus in Chiapas farmers seem to play a more limited but still significant role in shaping their germplasm.

The case studies collected information on and observed systems of maize planting, to learn how creolization may occur. Many *ejido* farmers divide their crop in several parcels, which are located on different slopes, and they plant each variety in different conditions. Most commonly, they will plant two varieties; however, some plant more. In Chiapas those who have the highest production of commercial maize maintain their lands separately and planted exclusively with only one kind, avoiding the contamination of the ears. However, we also found farmers who said that they planted more than one variety in the same plot, with little or no separation among them. This way of planting often presents a mixture of maize varieties that is not seen as a problem, as this maize is for household use, and the deformed or stained ears are fed to the animals. Farmers also stagger their planting of different varieties to serve various purposes and minimize the risk of loss.

Regarding the deliberate crossing of maize varieties, it appears that most farmers have limited knowledge about the process. Nevertheless farmers are crossing maize, intentionally or by accident. In Chiapas they know that a maize crop is always purest in the center of a plot, and that one finds mixed grains of different varieties on the borders. They are not very knowledgeable about the characteristics of different kinds of maize. They know that the maize can be changed or contaminated when seeds are mixed through improper handling. Even if the process of cross-fertilization is not understood completely, some farmers recognize it and do it on purpose. Some farmers in the case study households in Oaxaca explained ways in which they had crossed improved maize with *criollos* when they saw plants that they liked. In Oaxaca they tended to know more about this process than in Chiapas, possibly because Oaxaca has a longer tradition of maize cultivation, and because the offices of the agricultural secretariat are in Nopala, with eight technicians and demonstration plots. An extremely poor farmer from Nopala explained how: "A year ago, I planted the one we call '*tablita*' in one plot and in another together with another variety. But if I cross it now with 526 it produces half yellowish grains and the ear is a little bit narrower . . . but it became stronger. That is what we want—to cross a *criollo* with a variety to make it more resistant, so that it doesn't rot much." However, not everyone crosses maize varieties intentionally, nor do they know how to do it. Many just notice the cross because they observe a change in the color of the kernels or height of the plants as a consequence of having planted two varieties together: "We don't know why it happens, but it happens."

Adoption

Extent of Planting by Maize Type

The relative area planted and the proportion of farmers who plant each of the five types of maize germplasm varies between the study areas (Table 7.3). Landraces dominate in Oaxaca, followed by creolized varieties. The importance of creolized varieties is very similar across poverty groups. Few farmers planted improved germplasm, especially hybrids, and those who planted improved varieties did so in small areas. Furthermore, the use of landraces, even though they are dominant, is the lowest among these two groups. In contrast, the use of improved germplasm, and particularly hybrids, is dominant in Chiapas. All farmers, particularly those who are not extremely poor, plant improved maize types. All poverty groups also plant creolized varieties and landraces. Creolized varieties are the single most widely planted maize type in terms of relative area and proportion of farmers and are planted in roughly similar proportions by all poverty groups. In spite of the wide adoption of improved germplasm, landraces occupy more than a fifth of planted area and are planted by more than a fourth of farmers, particularly among the poor. The importance of landraces decreases with the poverty level. In both study areas (although at very different scales), there is a trend of increasing use of hybrids and improved germplasm with decreasing poverty and a downward trend for landraces. Creolized varieties seem, however, neutral to poverty level in both areas.

Factors Affecting Adoption

Adoption is a complex process, affected by many factors and circumstances. These factors can be grouped into five categories: (1) adaptation, (2) management intensity, (3) cultural values associated with maize consumption, (4) risk, and (5) participation in the regional or national economy. Adaptation refers to the performance of the germplasm in a particular agroecological environment. A well-adapted variety is one that performs well in a particular environment, whereas a poorly adapted one does not. Management intensity refers to the quantity and timing of inputs required by a variety for a good performance. A management-intensive variety requires a large amount of inputs and strict timing of planting, weeding, and fertilizer application to perform well; otherwise its performance is drastically reduced. A non-management-intensive variety is one that can withstand delays in these operations and responds to low

TABLE 7.3 Distribution of type of germplasm by area and number of farmers

	Extremely poor		Poor		Non-poor		Total	
	Area (ha)	Number of farmers	Area (ha)	Number of farmers	Area (ha)	Number of farmers	Area (ha)	Number of farmers
Coast of Oaxaca								
Total	3,012	2,645	833	666	321	228	4,165	3,539
Relative distribution (%)								
Hybrids	1.5	3.1	0.0	0.00	7.1	6.7	1.6	2.7
Recycled hybrids	2.0	3.1	8.5	8.7	12.2	13.3	4.1	4.8
Open-pollinated varieties	7.0	7.0	2.0	2.8	2.5	8.1	5.7	6.3
Creolized	14.3	10.4	12.8	15.4	24.2	20.0	14.8	12.0
Landraces	75.2	84.2	76.7	85.3	53.9	66.7	73.9	83.3
Frailesca, Chiapas								
Total	5,789	1,261	2,214	521	1,036	212	9,039	1,994
Relative distribution (%)								
Hybrids	19.8	30.9	22.2	31.1	63.3	54.8	25.3	33.5
Recycled hybrids	8.8	9.9	18.5	26.0	3.9	17.5	10.6	14.9
Open-pollinated varieties	20.0	33.1	12.8	22.8	4.3	10.5	16.4	28.0
Creolized	26.6	36.7	31.8	38.8	25.3	37.6	27.7	37.4
Landraces	24.9	32.6	14.8	10.3	3.1	11.1	19.9	24.5

quantities of inputs without a dramatic reduction in its performance. Cultural values associated with maize consumption are important because maize cultivation in Mexico is not just an economic activity but also has strong cultural values and preferences associated with it. Preferences for special characteristics, especially for culinary and ritual uses, are common among small-scale farmers, particularly indigenous farmers. Maize cultivation is a risky endeavor, particularly under the rain-fed conditions faced by these farmers. Risk is not only related to biotic and abiotic stresses associated with maize cultivation but also with the knowledge and understanding of the performance of different varieties when faced with those conditions. Participation in the regional or national economy provides farmers with opportunities to sell their surpluses, acquire seed of improved varieties, purchase inputs, and enjoy other income opportunities and access to cheaper consumer goods—including maize.

Actual adoption of a particular type of germplasm by a farming household depends on the interaction between the above-mentioned factors, the assets controlled by the household, and the conditions that it faces. To examine actual adoption, we included variables related to these factors in the survey[13] and included them in a regression framework to explain the area planted to the five different types of maize germplasm defined above. The variables included are farmer's age, household language, percentage of indigenous speakers in the community, household expenditure,[14] source of labor used in maize production, land quality, fragmentation (number of plots into which a farm is divided), access to extension services, participation in government programs, and distance to the main town. The regression models were based on an economic adoption model and its econometric estimation developed by Bellon and Taylor (1993).

13. Several reviewers questioned whether some of these variables were endogenous, for example, exclusive use of family labor, use of extension services, participation in government programs, and landholding fragmentation. For the first three variables, the information elicited referred to a period of five years prior to the year of the decision to plant a variety, so that these variables were fixed with respect to specific planting decisions. The fragmentation variable refers to the whole farm in the period previous to the actual planting, also fixed before the measurement. The only truly endogenous variable was expenditure, explained in the following footnote. Furthermore, there were questions on whether participation in government programs induced a systematic bias in the sample, as people may not have been eligible to participate. Based on our knowledge of these farmers, all qualified to participate in these programs, and if they did not participate, it was because they chose not to do so. Particularly for Procampo, farmers said that many did not participate because they were either absent or sick when the registration took place or because some were afraid that the government might take their land away or they might be taxed. These factors are not systematic, however. Participation in these programs was very high in any case.

14. To avoid endogeneity problems, for each region we estimated a regression of the logarithm of expenditure as a dependent variable against a set of explanatory variables associated with local perceptions of poverty and other measures of marginality thought not to affect adoption decisions. We then used the predicted values in the adoption regressions.

Bellon et al. (2005) describe in detail the model's application to these data, as well as full results. Here we present a summary of the results.[15]

Table 7.4 presents a short description of the variables as well as descriptive statistics for both regions, while Table 7.5 presents the hypothesized relationships among the independent and dependent variables (the latter being the area planted to each of the different maize types) and the rationale for the hypothesized relationship. These relationships are based on certain expectations about the performance of the different maize types. Improved varieties (particularly hybrids) are hypothesized as

- Having a limited local adaptation;
- Being suitable mainly to the best environments;
- Being management intensive, because they have been selected under optimal management conditions in research stations;
- Not having been selected for culturally important consumer traits, which are subjective and difficult to select for;
- Being riskier, particularly because they are less well known and understood, although they may actually not be riskier; and
- Being associated with good integration and interaction with the regional and national economies, as these economies would be the sources of the seed, information, and incentives to grow the variety.

Landraces, however, are hypothesized as:

- Having good and broad local adaptation, because they have evolved in the particular environments;
- Being nonmanagement intensive, because they have been selected under suboptimal management conditions (for example, late planting, high weed infestations, low inputs);
- Having been selected for culturally important traits, as they are the products of selection by farmers who value these traits; and
- Being less risky because they are well known and trusted, and even if they entail risk, farmers can evaluate these risks well.

Integration into the regional and national economies may not be important, as access to seed and knowledge are related to local social networks. Creolized varieties are expected to have traits that are somehow intermediate, but they are probably more similar to improved varieties than to landraces.

15. This report (www.cimmyt.org/english/docs/impacts/impmaize_05.pdf) presents in greater detail a description of the variables used, the rationale for the predictions, the quantitative results, and their in-depth interpretation.

TABLE 7.4 Description of variables used in adoption regressions for Oaxaca and Chiapas, Mexico

Variables	Definitions[a]	Oaxaca	Chiapas	Significance
Dependent variables				
Hybrids	Area planted to hybrids	0.025	1.144	****
Recycled hybrids and OPVs	Area planted to recycled hybrids and improved open-pollinated varieties (OPVs)	0.133	1.229	****
Improved varieties	Area planted to improved varieties (hybrids, recycled hybrids, and OPVs)	0.158	2.373	****
Creolized varieties	Area planted to creolized varieties	0.247	1.236	****
Landraces	Area planted to landraces	0.875	1.014	
Independent variables				
Total area	Total farm area[b]	2.138	5.365	****
Land type 1	Area under plow agriculture—best quality	0.298	0.475	
Land type 2	Area under plow agriculture—fair quality	0.304	0.565	
Land type 3	Area under pedregal[c] agriculture—best quality	0.808	1.600	****
Land type 4	Area under pedregal agriculture—fair quality	0.623	2.533	****
Land type 5	Area of worst-quality land	0.106	0.191	
Fragmentation	Number of plots owned by household	1.129	1.975	****
Age	Farmer's age (years)	50.01	48.66	

Household language	1 if household mother tongue is Spanish; 0 if otherwise	0.436	0.938	****
Percentage speaking indigenous language	Percentage of speakers of an indigenous language above 5 years of age in the locality	34.003	1.883	****
Expenditure	Predicted expenditure (pesos per capita per month)	377.75	388.94	
Family labor only	1 if household only used family labor to produce maize in the past 5 years; 0 if otherwise	0.718	0.401	++++
Extension	1 if household had contact with extension in the past 5 years; 0 if otherwise	0.123	0.160	
Kilo por Kilo	1 if household participated in the Kilo por Kilo government program in the past 5 years; 0 if otherwise	0.258	0.358	++
Procampo	1 if household participated in the Procampo government program in the past 5 years; 0 if otherwise	0.736	0.840	**
Distance	Distance to main town in travel minutes	74	78	

NOTES: **, **** indicate statistical significance at the 5 and .1 percent levels, respectively, for a two-tailed *t*-test. ++, ++++ indicate significance at the 5 and .1 percent levels, respectively, for a χ^2 test of independence.

[a] Means for continuous variables and fractions for dummy variables; all areas in hectares.

[b] These unweighted averages differ from the weighted averages in Table 7.1.

[c] In the pedregal system, land is not plowed and planting is done with a dibble stick.

TABLE 7.5 Hypothesized relationships between variables and adoption, and rationale

Variable	Rationale	Improved varieties	Creolized varieties	Landraces
Age	Indicator of risk attitudes. Older farmers more likely to be risk averse and have better knowledge of local landraces.	–	+	+
Household language indigenous	Indicator of cultural identity. Speakers of indigenous languages more likely to attach stronger values to maize consumption.	–	–	+
Household language exclusively Spanish	Indicator of the ability to interact with national economy. Spanish speakers better able to interact with regional and national economies and access new technologies.	+	+	
Percentage of speakers of an indigenous language in the community	Indicator of cultural identity. Even if indigenous languages not spoken by a household, there may still be strong attachment to maize culture in community.	–		+
Predicted household expenditure	Proxy for welfare and poverty. Poorer households more constrained to afford inputs and seeds and less likely to take risks.	+	–	–
Exclusive use of family labor	Indicator of constraint to mobilize labor to deal with labor-intensive varieties, particularly hired labor.	–	+	+
Land quality	Indicator of adaptation. Not all land equal.	+		–
Landholding fragmentation	Indicator of management intensity. It can make coordination and mobilization of labor more difficult.	–	+	+
Landholding fragmentation	Reduce risk by reducing chance of crop failure due to multiple production conditions.	+		
Extension	Indicator of links to regional or national economies. Access to new information and inputs.	+	+	
Participation in Kilo por Kilo program	Indicator of links to regional or national economies. Access to new seeds.	+	+	
Participation in Procampo program	Indicator of links to regional or national economies. Access to additional funds.	+		
Distance to nodal town	Indicator of links to regional or national economies. Determine costs of interacting with the outside.	–		+

NOTE: The signs indicate the hypothesized relationships between a variable and the area planted to a type of germplasm: + indicates a direct relationship, – indicates an inverse relationship, while a blank cell indicates that there is no a priori hypothesized relationship.

Table 7.6 summarizes the regression results for the different maize types and both study areas. In general, land quality is an important factor in the adoption of most maize types, both in Oaxaca and Chiapas, particularly for creolized varieties. Even if land quality per se is not important, the size of landholdings is, except for improved varieties in Oaxaca that are planted in a very limited area. Thus adaptation is an important consideration for adoption, although it affects the various germplasm types in different ways. Fragmentation is another factor that was significant for most types of maize germplasm, but it has two different and contrasting interpretations and hence merits some further explanation. A positive association with area planted with improved varieties in Oaxaca suggests that fragmentation is a risk-management option. In Chiapas the negative association for hybrids and positive association for creolized varieties and landraces suggest different degrees of management intensiveness of these germplasm types, consistent with the predictions in Table 7.5.

Language and culture also play a role in adoption, particularly for creolized varieties and landraces and particularly in Chiapas. The use of family labor (and conversely hired labor) is a factor in the adoption of certain types of maize germplasm, but not in all, indicating that certain types are considered more labor intensive than others. Expenditures, and hence welfare and poverty,[16] were not a significant factor in the adoption of any type of maize germplasm, except for creolized varieties in Oaxaca, where there was an inverse relationship between expenditure and adoption of creolized varieties, indicating that the poor tend to adopt these types of varieties. The lack of significance of expenditures indicates that there is not a direct relationship between the level of welfare of a household—at least in the narrow sense of expenditures—and its adoption decision, except in the case already described. Distance of the community to a major town was a factor that only influenced the adoption of certain types of germplasm in Chiapas, but not in Oaxaca. Government programs do not play a significant role for adoption decisions in Oaxaca, although they do for certain types of germplasm in Chiapas, particularly for creolized varieties. Directly they do not seem to have a negative impact on landraces in either region. However, in both study areas the different types of germplasm seem to compete with one another.[17] As will be shown in the next section, this result is not completely consistent with the farmers' perceptions of traits and trade-offs, which indicate that

16. Obviously expenditure is a partial indicator of welfare, as there are many other important dimensions of welfare that are not taken into account by this variable. It is, however, easy to measure and widely accepted, even with these limitations.

17. This observation is based on an analysis of the correlations between residuals of the regressions for the different types of maize germplasm. A statistically significant negative correlation indicates competition among germplasm types, whereas a positive correlation indicates that the types complement one another. The observed correlations were for the most part negative, indicating competition. For details, see Bellon et al. (2005).

TABLE 7.6 Regression results of the adoption models for different type of maize germplasm in Oaxaca and Chiapas

	Oaxaca			Chiapas			
	Improved varieties	Creolized varieties	Landraces	Hybrids	Recycled hybrids and open-pollinated varieties	Creolized varieties	Landraces
Age	0.002*						0.028**
Household language[a]		0.451***					
Percentage speaking indigenous language		0.004*				–0.083*	0.383****
Expenditure		–0.0005***					
Family labor only	–0.131***				–0.571*	0.326*	
Total area[b]				0.486****			0.093*
Land type 1 (*arado* system, best quality)[c]		0.120**	0.505****			0.247****	
Land type 2 (*arado* system, fair quality)		0.713****			0.310***		
Land type 3 (*pedregal* system, best quality)			0.333****				

Land type 4 (*pedregal* system, fair quality)		0.148***		0.093*		
Land type 5 (any system, worst quality)		0.073*			0.243**	
Fragmentation	0.194****		−0.725****		0.403****	0.524**
Extension						
Kilo por Kilo			0.707*		0.354*	
Procampo					0.421*	
Distance			−0.005**		0.003**	

NOTES: Only the estimated coefficients for the variables that were statistically significant at least at the 10 percent level are reported. *, **, ***, **** indicate statistical significance at the 10, 5, 1, and .1 percent levels, respectively. The sign indicates the nature of the relationship between the variable and the area planted to the maize type.

[a] This variable was not included in the Chiapas regressions, as there is almost no variation among households. Most are only Spanish-speaking.

[b] If total area is significant, it indicates that land quality did not have an effect on adoption; otherwise land quality has an effect on adoption and the specific land types that were significant are reported. Two regressions were estimated. In the first, all land qualities were combined into total area; in the second, land quality was disaggregated into five land types. If land quality effects are not important, both models should be equal; if they are not, land quality effects contribute to explaining adoption. Both models were compared statistically. If the models were statistically similar the results from the simpler one (land quality not disaggregated) are reported.

[c] Landholdings were classified into five land categories depending on the production system used and the quality assessed by the farmer (very good, good, poor). Production system refers to whether farmers plow the land (locally know as *arado*) or not (locally known as *pedregal*). The production system is correlated with the slope and stoniness of a plot; plots in *arado* being flatter and with low stoniness, whereas *pedregal* plots may be on steeper slopes or flatter but with high stoniness.

one type of germplasm may complement another by providing some traits that the other does not.

The hypothesized relationships between dependent and independent variables for the different types of maize germplasm were corroborated in most cases where there was a statistically significant association, except in the case of age and improved varieties in Oaxaca, where the relationship was the opposite. Thus older farmers may be less risk averse, or alternatively, improved varieties may be less risky than thought. Because the qualitative data stress the lack of confidence and trust of farmers for these varieties and the perception of riskiness associated with them, the first alternative is more plausible.

Case Study Findings on Factors Affecting Adoption

The case study findings complement the results presented above by examining the reasons given by people for why they adopt or do not. In Chiapas the most frequently cited factor that discourages adoption of improved varieties was expense: "It would be better to buy [seed from] the bags, since the yield they give is better. The only problem is that the seed is too expensive. Another problem with the bags is that you have to plant them very close together, with only one grain. It takes more work, more liquids, and fertilizers." Interestingly the reference to inputs was only made in Chiapas (though mainly by poorer farmers). This asymmetry is probably related to the more commercial orientation of farmers in Chiapas, who use a larger amount of purchased inputs. Also, the maize grain they produce has to comply with commercial standards; hence there may be a greater need to purchase improved seed, and thus price is a consideration. In Oaxaca, these considerations are not as important, given that farmers are subsistence oriented, use small amounts of purchased inputs, and use improved varieties in a very limited way. Curiously, a number of Oaxacan farmers said that price is not a determinant—they would find a way to buy it if they thought it was good quality.

The most commonly cited factor in Oaxaca explaining nonadoption was that new seeds are risky. The issue of risk comes up repeatedly in the studies; particularly among poorer farmers. However, even a less-poor farmer said, "I already decided that right now only the pure *criollo* is the safest; that way there is no risk." This concern over risk drives the most significant practice informing the decision to adopt or not—observation of good or bad yields and other characteristics in new maize varieties planted by family, friends, and neighbors. The issue of observing before doing surfaced repeatedly in the study. The majority of our informants in both Oaxaca and Chiapas said they prefer to observe how new seeds produce before trying them. The tolerance of risk followed a clear pattern: the poorer the farmer, the less willing to risk the harvest. A typical poor informant from Tiltepec, when asked if he would plant a seed an outside organization was actively promoting, said "No. Even if they would give it to me for free I would wait to see someone else's crop. Since I have my seeds,

I will plant my own. Why should I investigate? All I would do is expose myself to losses." Even where people have seen experimental plots planted by the agricultural secretariat that look promising, some individuals suspect that the technicians have added some secret substance and wonder whether the varieties will yield as well on their lands.

The relationship between soil type and maize varieties emerged in interviews with many case study informants, but in different ways in the two study regions. In Oaxaca, people expressed ideas about the correspondence between certain varieties of seed and certain land types, though they are not explicit in this regard. For example, in Oaxaca it is said, "the land chooses the seed," which they learn "by trial and error." No one in Oaxaca specifically mentioned soil type as a reason for planting a given variety. However, the ideas expressed above could imply a concern that a new, unknown variety might not fare well on their land, contributing to their risk aversion, and thus soil might be taken into account, if not directly acknowledged. In Chiapas, some farmers were more explicit on this point. For them, good and expensive seed is only justifiable if it is planted on good soils—flat bottomlands. For poor soils—those over-farmed, with considerable slope, where proper fertilization is not possible (it washes away)—it is most advisable to plant *criollo* seeds. Informants say that *criollo* maize does well on any type of soil, even those that are worn out and weed-covered. This opinion again reflects the notion that *criollos* have acclimated to the soils.

Another set of factors explaining adoption relates to access to different types of assets. Financial capital, needed to purchase seeds, fertilizers, and pesticides, was important to better-off farmers in Chiapas. Social and political capital influences access to seed—through informal social networks, particularly for poorer farmers, and through peasant organizations and political campaigns. Human capital helps as well—access to labor in good health and able to work the land, and access to knowledge—and matters to all farmers, both more and less poor. Access to natural capital—good quality land in sufficient quantities and the right type of soil—makes adoption worthwhile, although we explained above how different types of farmers regard the importance of soil.

Finally, the characteristics of the different varieties were an important set of factors explaining adoption. These factors and explanations are developed in more detail through the survey data, elaborated in the following section on impacts. Case study informants said the main characteristics they admire were the appearance of the ears, large kernels, flavor, resistance, and heartiness of *criollos.* The principal disadvantage that they find in these seeds is their height, which is so tall that the plants blow over. Farmers appreciate the weight, good yields, and lower height of improved varieties. However, they are not fond of the small ears, fragility, and propensity to rot and the kernels to crack. They also claim that these plants are so short that animals eat them. Additionally, improved seeds are expensive and require greater care and investment of time and inputs, and yields decline over the years. Finally, some farmers said that the improved varieties taste bland.

Impacts

Impacts are analyzed in this section based on the principle that households derive utility from the crop's traits or attributes, rather than from the crop itself (for example, see Adesina and Zinnah 1993; Barkley and Porter 1996; Smale, Bellon, and Aguirre 2001; Hintze, Renkow, and Sain 2003; Edmeades et al. 2004). Clearly, changes in the supply of valued traits have important welfare implications for farmers, beyond trade-offs between the level of expected yields and the variance (or variability) in yield performance (Edmeades et al. 2004). As many studies have shown, small-scale farmers who plant maize for subsistence —particularly those who also sell some of their production—value multiple traits in their crop. Usually no one variety can provide all of the valued traits, hence farmers continually face trade-offs in their variety choices (Bellon 1996; Smale, Bellon, and Aguirre 2001). To the extent that these trade-offs are reduced, farmers benefit because they can satisfy their preferences at less cost.[18]

A key hypothesis of this study is that farmers, particularly the poor, benefit from improved germplasm through creolization. Although improved varieties provide desirable traits or combinations of traits not found in landraces, they may lack other beneficial features found in the landraces. Hence choosing between one or the other presents trade-offs to farmers.

Creolized varieties can provide traits not supplied by landraces while entailing fewer trade-offs than improved varieties. To look at the impact of these varieties on farmers' well-being, one has to examine the importance that farmers attached to each trait and the supply of crop characteristics by different types of maize germplasm.

The survey included a section on farmers' evaluation of maize varieties. This evaluation was done for 19 crop traits or characteristics identified as significant in focus group discussions. The evaluation was made up of two parts. The first consisted of an assessment of the importance that farmers give to different crop characteristics. Male and female farmers rated each trait as very important, important, or not important in terms of their relevance for choosing a

18. Plant breeders recognize that there may be trade-offs among traits they select for in the crop breeding process, as traits can be negatively correlated (they also can be independent of one another or positively correlated). Plant breeders' objective usually is to improve for more than one trait simultaneously; however, the more traits included in what they select for, the slower the improvement will be for any given trait. Responding to multiple farmer demands by generating varieties with multiple traits—particularly if those traits are related to very specific consumption characteristics that may reflect subjective preferences—can be difficult and costly (David Beck, maize breeder, personal communication 2006).

maize variety to grow. The second consisted of an assessment of the extent to which each germplasm type supplies farmers with traits they value. Male and female farmers rated each variety in terms of their perception of the variety performance for each of the 19 traits as very good, good, poor, or very poor. There were instances in which varieties currently grown were not rated, because the farmer did not feel that he or she knew enough about their performance. Later we grouped ratings of varieties by maize types according to the definitions presented above.

Importance of Crop Characteristics for Farmers

Even though a large number of characteristics were rated, almost all farmers in both Oaxaca and Chiapas rated them as either very important or important. Table 7.7 presents the percentage of farmers who rated each characteristic as very important by gender for both study areas. Almost all characteristics were rated as very important by 50 percent or more of the farmers in both study areas. These high percentages suggest that the earlier focus groups were very accurate at identifying pertinent crop characteristics and that these farmers value multiple traits. To test whether any of these traits are particularly important to the poor, nonparametric correlations between the expenditure of the household and the ratings of importance were run for each trait. A significant negative correlation indicates that as expenditure decreases importance increases, that is, the trait becomes more important to the poor. Table 7.7 reports the statistically significant correlations as well.

The characteristics that were rated as very important by the greatest number of male farmers in Oaxaca are yield by weight, yield of dough to make tortillas, tolerance to drought, ease of shelling, and resistance to lodging. Yield by weight is a key trait for breeding. Yield of dough to make tortillas is a trait that is seldom taken into consideration by breeders. Drought and lodging are key sources of risk and vulnerability in maize production. As pointed out earlier, farmers in Oaxaca are still heavily oriented to subsistence farming, so yield of dough to make tortillas and ease of shelling are understandably key characteristics. The correlations showed that as poverty decreases, duration (growing cycle), good for sale, good for *elote* (corn on the cob), good for fodder, and yield by volume become more important. There were no traits that seem to be particularly important for poor male farmers. For women the traits most widely rated are resistance to lodging, yield of dough to make tortillas, *atole* (a traditional beverage made of maize) quality, tolerance for excess water, and *nixtamal* (the dough used to make tortillas) quality. Clearly, consumption characteristics seem more relevant for female than male farmers, as would be expected, because women are in charge of maize processing and preparation. Furthermore, these results indicate that an important concern for women is reduction of vulnerability. The correlations show that three traits are significantly more important for poor female farmers: tolerance to drought, resistance to rot, and

TABLE 7.7 Percentage of farmers who rated a characteristic as very important in Oaxaca and Chiapas, Mexico, by gender

	Coast of Oaxaca				Frailesca, Chiapas			
Characteristic	Men	Correlation coefficient[a]	Women	Correlation coefficient	Men	Correlation coefficient	Women	Correlation coefficient
Vulnerability								
Resistant to lodging	69.8		98.8		82.6		94.3	−.120*
Tolerant to drought	75.9		83.3	−.117*	75.2		72.2	
Tolerant to excess water	54.3		84.6		70.8		88.6	
Does not rot (good husk cover)	61.1		75.2	−.162**	68.9		80.4	
Duration (growing cycle)	49.4	.169**	80.9		62.1		82.3	
Resistant to pests	66		83.3	−.145**	69.6		80.4	
Resistant to insects in storage	58.6		75.9		61.5		80.5	
Produces something even in a bad season	58		75.9		64.6		76.7	
Good for sale	55.9	.181**	65.4		63.8		81.8	
Consumption-related								
Good for consumption	59.9		80.2		70.2		84.9	

Good for *atole*	59.3		91.4	68.9		90.6
Good *elote* for sale and consumption	50.6	.118**	69.8	60.2		74.2
Good for *antojitos*[b]	58.6		75.9	65.2		79.2
Easy to shell	70.4		76.5	42.9		73
Good for *nixtamal*	61.1		84.6	68.9		83.6
Good pasture	27.8	.155**	54.8	49.1	–.122*	64.8
Productivity						
Yield of dough to make tortillas	77.2		92	83.9		89.2
Yield by weight	84.6		67.9	89.4		67.1
Yield by volume	67.9	.120**	61.1	72.7		68.4
Number of households	162		162	161		158

NOTE: *, ** indicate correlations significant at the 10 and 5 percent levels, respectively.

[a] Nonparametric correlation between predicted expenditure and rating of importance. A negative sign indicates that the importance increases with poverty, and vice versa.

[b] *Antojitos* refers to an assortment of maize-based traditional snack foods.

resistance to pests. Clearly these traits are related to vulnerability, which seems to be more important to women than to men.

The characteristics that were rated as very important by the greatest number of male farmers in Chiapas are very similar to those for men in Oaxaca: yield by weight, yield of dough to make tortillas, resistance to lodging, tolerance to drought, and yield by volume. Only the importance of one trait is associated with the poor: good for fodder. For female farmers the traits most widely rated are also similar to those for women in Oaxaca. There is a consistent pattern of consumption characteristics being more relevant for women than for men. Thus even with the high level of commercialization—although marketability is considered more important than in Oaxaca—subsistence production is still relevant for women. Only the importance of resistance to lodging is associated with the poor, again a vulnerability factor.

Supply of Crop Characteristics by Maize Varieties

To examine systematically farmers' perceptions of the performance of the varieties available with respect to the characteristics they value, proportional odds regressions (Agresti 1996; Coe 2002) were run for all 19 traits identified in Table 7.7. The proportional odds model relates a dependent variable consisting of ordered response categories (for example, farmers' ratings of performance for a trait) to a set of independent variables (such as types of maize germplasm grown by farmers—defined in an earlier section—and other covariates explained below). The model was estimated independently for the 19 identified traits, separately for men and women, and individually for the two study areas. The results of this type of regression in this context are the ratio of the odds that farmers rated a maize category as superior compared to another maize category for a particular trait. In this regression, we included the predicted expenditure, used in the adoption section (see footnote 14) as a covariate to correct for differences in ratings associated with different levels of welfare. Furthermore, because women may not have participated directly in growing many varieties and hence may have very limited knowledge and experience of the variety, which could bias their ratings, a dummy variable specifying whether they actually had participated in growing the variety was also included in the regressions of female ratings. These results are presented in Tables 7.8 and 7.9 for Oaxaca and Chiapas, respectively. For simplicity these tables only display the characteristics for which there were statistically significant differences. The tables should be interpreted as follows: the category presented in the row was rated as superior to the category in the column for the characteristics described in the cell that results from their intersection. For example, in Table 7.8 male farmers rated creolized varieties superior to landraces for yield by weight and lodging, whereas landraces were rated superior to creolized varieties for ease of shelling and good for *nixtamal, elote,* and fodder. By comparing the characteristics described in cells that result from inverting the categories in the rows and the columns, one

TABLE 7.8 Comparisons of different types of germplasm with respect to traits with statistically significant different ratings, coast of Oaxaca, by gender

Categories in row rated as superior to categories in column	Improved varieties[a]	Creolized varieties	Landraces
Improved varieties		F Produces even in bad season*	Resistant to lodging** F Resistant to lodging**
Creolized varieties			Resistant to lodging*** Yield by weight*** F Resistant to lodging***
Landraces	Does not rot** Ease of shelling*** F Ease of shelling*	Ease of shelling**** Good for *nixtamal**** Good for fodder* Good for *elote**	

NOTES: Female ratings are preceded by an F. *, **, ***, **** indicate statistical significance at the 10, 5, 1, and .1 percent levels, respectively. The significance level was adjusted by the number of pairwise comparisons.

[a] Improved varieties include hybrids, recycled hybrids, and open-pollinated varieties, which were added together owing to the low number of observations by category.

can identify the trade-offs between two types of maize categories. Thus for male farmers, the trade-offs between landraces and improved varieties are resistance to ear rot and ease of shelling versus resistance to lodging.

RESULTS FROM OAXACA. Table 7.8 shows that for men in Oaxaca, there were statistically significant differences for seven of the 19 traits rated. There is no overall superior maize type; all types have advantages and disadvantages. Most advantages were associated with landraces; however, both improved and creolized varieties were superior with respect to resistance to lodging—a key vulnerability factor in the area. While landraces are considered to be superior for many traits, improved and creolized varieties provide a trait lacking in them—resistance to lodging. Furthermore, creolized varieties, although inferior to landraces for some consumption traits (good for *elotes, nixtamal,* and fodder, and ease of shelling) were superior for yield by weight. Clearly these maize types show some trade-offs between key traits. These results support the hypothesis that creolized varieties provide a combination of traits not provided by landraces or by improved varieties, and hence entail fewer trade-offs.

Furthermore, creolized seed is much cheaper. For example hybrid seed cost on average MX$17.44 per kilogram compared to MX$5.33 per kilogram for seed of creolized varieties, while seed of landraces costs MX$3.88 per kilogram. These findings help to explain the results of the qualitative study, in which farmers said that, although they considered seeds of improved varieties very

TABLE 7.9 Comparisons of different types of germplasm with respect to traits with statistically significant different ratings, Frailesca, Chiapas, by gender

Categories in row rated as superior to categories in column	Hybrids	Recycled hybrids	Open-pollinated varieties	Creolized varieties	Landraces
Hybrids		Good for *elotes*** Good for *antojitos*** Good for sale*	Good for sale*** Good for *antojitos*** Good for *atole** Good for *elote** Good for fodder*	Resistant to lodging* Good for *elote**	Resistant to lodging**** Good for sale***
Recycled hybrids	F Produces even in bad season*		F Produces even in bad season*		Resistant to lodging***
Open-pollinated varieties	Resistant to insects in storage***				Resistant to lodging*
Creolized varieties	Resistant to insects in storage**** F Resistant to insects in storage* F Resistant to pests*				
Landraces	Does not rot*** Resistant to insects in storage*** F Tolerant to excess water*** F Does not rot*			Does not rot**	

NOTES: Female ratings are preceded by an F. *, **, ***, **** indicate statistical significance at the 10, 5, 1, and .1 percent levels, respectively. The significance level was adjusted by the number of pairwise comparisons.

expensive, they would "make the sacrifice" and buy them if the improved varieties were truly superior (which they did not consider to be the case). The price differentials between seed of creolized varieties and landraces also illustrate that farmers perceive advantages in the former compared to the latter, as they are willing to pay a premium.

For women, there were statistically significant differences for only three of the rated traits. For two of these traits, resistance to lodging and ease of shelling, the results are similar to those of men. Only for yield reliability (that is, yields even in a bad year) did female farmers rate improved varieties higher than creolized varieties, unlike men, for whom no differences were observed. One would have expected creolized varieties to be rated higher in this respect, given that they have been grown longer in these areas and thus could be better adapted and more stable (low year-to-year variability). There is no clear explanation for these results, which merit further investigation.

As indicated in the methods section, a dummy was included in the female regressions to account for the actual experience of growing a variety. This variable was statistically significant for several traits[19] indicating—not surprisingly—that female farmers' actual experience with varieties influences their perceptions of varietal performance.

RESULTS FROM CHIAPAS. Table 7.9 shows that for male farmers in Chiapas, nine of the 19 traits rated showed statistically significant differences. There is no type that is superior for all traits, as in the case of Oaxaca; unlike the situation in that state, in Chiapas there is a wider range of maize types, and thus more comparisons were made. In general, men have a very positive opinion of hybrids. They rated hybrids higher than OPVs (in particular) and recycled hybrids for several traits (related to consumption and marketing characteristics). However, OPVs, creolized varieties, and landraces were rated higher than hybrids for resistance to insects in storage, a key trait for subsistence farmers. Landraces were also rated higher than hybrids with respect to resistance to ear rot. Resistances to insects in storage and to ear rot are closely linked to vulnerability, suggesting that landraces are valuable for addressing vulnerability issues. Overall, improved varieties were rated superior to landraces for resistance to lodging.

Farmers' ratings of creolized varieties do not indicate they perceive these varieties to have many or unique advantages, unlike in Oaxaca. However, the price of creolized seed is on average higher than that of landraces (MX$6.33 per kilogram versus MX$3.51 per kilogram) and much cheaper than the hybrid price (MX$20.25 per kilogram), suggesting farmers are willing to pay a premium for creolized varieties over landraces.

---

19. In Oaxaca, traits for which women had significantly different perceptions, depending on whether or not they had grown the variety, included resistance to lodging, resistance to pests, good for *nixtamal,* resistance to pests in storage, and yield of dough.

The data for women indicated statistically significant differences for only four of the traits rated. Women perceived hybrids to be inferior to landraces, creolized varieties, and even recycled hybrids. They perceived more and unique advantages in creolized varieties and landraces. Recycled hybrids were highly rated for yield reliability compared to other types of improved germplasm. As in the case of Oaxaca, however, there were many traits for which the variable indicating actual experience with a type of germplasm was significant; that is, women's actual experience with varieties influences their perceptions and how they rate traits.[20]

In summary, neither in Oaxaca nor in Chiapas is there an overall superior maize type; all types have advantages and disadvantages. In the subsistence-oriented farming systems of Oaxaca, landraces seem to be more advantageous, whereas in the commercially oriented systems of Chiapas, hybrids seem to have more advantages. Creolized varieties, although commonly planted by all poverty groups in both regions, are perceived as more advantageous in Oaxaca than in Chiapas. Nevertheless women in Chiapas do have a more positive perception of creolized varieties than of hybrids.

Case Study Perspectives on Impacts on Poverty and Well-Being

The case studies reveal several ways in which creolized maize contributes to the well-being of poor farmers in the study areas. Unlike the survey results, the case studies did not emerge with as many accounts of direct benefits from improved maize "from the bag," although commercial production using improved maize was observed among some farmers in Chiapas. The scarcity of positive feedback may be because even where improved maize was providing important economic benefits, problems still occurred, and people tend to express these when given a chance to talk about their experiences. Nevertheless, the benefits of creolized maize, where improved maize has changed over time, emerged strongly in both Oaxaca and Chiapas among farmers at all poverty levels.

Creolized maize seems to improve well-being mainly through a reduction of vulnerability. Poor farmers in both study areas depend on maize for their survival. Thus the introduction of germplasm that improves yields and reduces vulnerability to crop losses reduces vulnerability to food insecurity: "It is the food of our families, since we are all poor, we have no money to buy maize and if we don't plant it, what will we eat?" By reducing expenses needed for inputs, as well as reducing the cost of the seed itself, creolization also releases cash for other basic household expenses and reduces vulnerability to price and currency

20. In Chiapas, traits for which women had significantly different perceptions, depending on whether or not they had grown the variety, included resistance to lodging, resistance to pests, ease of shelling, resistance to insects in storage, yield reliability, yield of dough, and yield by volume.

fluctuations. Farmers expressed the idea that creolized seed combines the benefits of resistance and acclimation to local conditions provided by *criollos* or landraces with traits of improved seeds, such as yield, height, and wind resistance. Finally, the case studies support the survey findings that creolized varieties provide people with traits that they want and reduce trade-offs.

It is also worth noting that, associated with the perceptions of recycling and acclimation, there is a perception of security provided to farmers by "knowing" the seed, which was expressed repeatedly as being particularly important. Farmers need to see it perform before trying a new variety, even if it means using a second generation of seed. The trust placed in creolized varieties contributes to farmers' well-being in a subjective—but no less real—way by providing a sense of security, which is particularly important for poor and vulnerable farmers.

That the introduction of new germplasm has improved people's well-being is illustrated by the words of an informant from Nopala, Oaxaca: "It has given us results. Since we bought that seed many things began to improve for the people, because before we had to buy lots of maize around here . . . but now we buy less. And, last year I was even selling maize; this year we harvested less, but for September we will have new maize." Still, adopting different varieties does not seem to significantly change people's livelihood strategies. Rather, the risks involved with maize cultivation of any kind drive these strategies. It is not possible for poor people in either of the two study areas to meet their necessities (which depend increasingly on cash earnings) with the income obtained—if any—from growing maize. They also need to make investments beforehand to grow maize, which for most of the extreme and average poor case study informants has not been sustainable. Thus people say that it is not possible to live solely from maize cultivation and emphasize the difficulties related to cultivation: "No one can get rich here growing maize. . . . With the cost of fertilizers and liquids, our time in planting, processing, and transportation, if you do the numbers you see that you don't get anything back. Maize produces, but very little is left over. We are content just to be able to grow enough to eat." The more options people have for a better and safer income, the less maize they plant.

In spite of these problems and the limitations of maize production as a route to escape from poverty, our study reveals the enormous importance that maize continues to play in people's livelihoods, from ensuring food security to providing cash income for other basic needs. As one informant said, "We need it to live; without it we don't eat." For less-poor farmers engaged in commercial production, improved maize creates a better chance of prospering rather than just getting by. In both cases, there is no question that providing maize germplasm (through scientific improvement and creolization in the field) that increases yields and reduces risks will make a significant difference in people's well-being. Escape from poverty, however, requires a more comprehensive poverty-reduction strategy that is farther reaching than agricultural technology alone.

Conclusions

The coast of Oaxaca and the Frailesca, Chiapas, are highly contrasting regions. Poverty is pervasive, even in the more commercialized and developed Chiapas. Maize continues to play a key role in the livelihoods of the poor in both regions.

This chapter has shown that modern varieties, and particularly creolized varieties, are widely planted in the study areas of Oaxaca and Chiapas. We cannot establish a causal link between the adoption of improved germplasm and poverty alleviation in the sense of establishing whether those who adopted improved varieties are better off in terms of income, expenditure, or nutritional status compared to those who did not—we do not have a baseline study to compare the situation before and after adoption. However, we have shown the contribution of improved germplasm, and particularly of creolized varieties, to the well-being of poor farmers by examining how these types of germplasm supply traits they value. In Oaxaca, creolized varieties are perceived to provide traits that the landraces do not have and have fewer trade-offs than do improved varieties. Creolized seed is also cheaper than hybrid seed. Adoption patterns show that the poor plant the former. In Chiapas, hybrids and other improved varieties seem to be neutral; that is, the poor plant as much hybrid seed as do non-poor farmers, once one corrects for other factors. The impact of creolized varieties is less straightforward than in Oaxaca, but they are still widely planted. This observation suggests that in more commercial systems with a wider diversity of germplasm types, creolized varieties are not considered as advantageous (although this perception varies by gender), which contrasts with the situation in more isolated and subsistence-oriented systems. Linguistic and cultural factors, and agroecological factors to a much lesser extent, play a key role in decisions to adopt different types of maize in both study areas. The evidence supports our hypothesis about creolization and its role in farmers' maize agriculture in Oaxaca, but the case is much less clear in Chiapas. Creolized varieties seem to occupy a niche that shifts according to the availability of improved germplasm and the orientation of farmers' maize production.

Although farmers discuss varieties and their traits, farmers' distinctions between creolized seed and landraces are blurred: all seed that is not "from the bag" (improved varieties, in a sealed package) is widely referred to as *criollo*. Furthermore, improved varieties are said to be quickly converted into creolized ones. This process is perceived to occur through seed recycling, where seed is seen as "acclimating" to the land and therefore improving. Even when seed is acknowledged to degenerate through recycling, it is still a popular practice because of the high cost of new seed. Creolization is also seen to occur by planting different varieties near one another so that they cross. Some farmers deliberately plant varieties close together in the hope of achieving better characteristics in the newly created variety, while others cross varieties unintentionally. Farmers place great confidence in these creolized varieties, whether created by recycling or cross-pollination, because they have proven themselves over time and are seen as better adapted to local conditions.

In addition to selecting from one's own harvest, seeds are mainly obtained through informal social networks and, to a lesser extent, through government programs. Surprisingly, commercial seed outlets still play a very limited role. Social networks are key because they offer many options, are trusted, and most important, provide the opportunity for farmers to observe plants in the field before adopting. This need to see performance and reduce risk is true for all farmers, but particularly for the poorest. Maize is seen as a highly precarious undertaking, involving numerous risk factors. Thus varieties that are known—and those that reduce these risks—are important, especially to the poorest, most vulnerable farmers.

Government programs play a more important role in Chiapas than in Oaxaca, but they suffer from a lack of credibility in both study areas. Farmers' experiences with these programs have been problematic, including late delivery of seeds, restricted access to credit, absence of technical support, politicization of seed distribution, and quantity and quality requirements that the poorest farmers cannot meet. Experience with poor quality seed has left farmers suspicious of government seed and improved seed more generally. They also often do not trust advice about maize management practices or cannot afford to follow them. Improved experience with government programs could accelerate the benefits of improved maize in several ways.

Our research also illustrated the value of combining different research methodologies and disciplines. This combination strengthened the evidence presented and the ability to interpret results, as complementary insights were gained and similar conclusions were reached by applying different methods to study the same issues. The quantitative research using large sample sizes and statistical analysis enabled us to systematically assess adoption in relation to socioeconomic characteristics and farmers' practices, perceptions, and preferences. The qualitative case study research enabled us to discover many additional aspects of the rationales underlying farmers' practices with respect to selection and management of seed, including those related to vulnerability, poverty, perceptions of science, social networks, relationships with government, and politics. The results thus point to the critical importance of combining methods for impact assessment, as well as to the importance of dedicating adequate time to the integration of results from the different methods. Using findings from each method to strengthen analysis is also important, as is an iterative approach to data collection: results from one method are used to build instruments and are explored subsequently using another method.

The involvement of anthropologists from outside the CGIAR system and the willingness of CGIAR researchers to be critical enhance the credibility of the findings on impacts, positive and negative. The research could have been strengthened through the closer involvement of CIMMYT breeders in the research process, incorporating their insights throughout the process, focusing their attention on the issues in the research, and drawing out more actively the implications for breeding in the future. Breeders took part in informal conversations during the research, interacting with the social scientists on technical issues and during the seed "grow-out" phase of the research. Since then, the findings of the research have sparked a dialogue with breeders, and an exploration of methods to incorporate farmers' practices into the development of improved germplasm.

Several implications can be drawn from the results of this study. First, it is important to move away from the dichotomy of traditional versus modern varieties that is common in adoption and impact studies. As shown here, there are many different types of germplasm, with different advantages and disadvantages. All are influenced by different factors and have various impacts on farmers' well-being to the extent that they provide valued traits that respond to their needs and preferences. However, moving away from this simple dichotomy also entails methodological challenges that may require the use of multiple methodologies, including some that are not commonly used in adoption and impact studies—such as participatory and ethnographic methods and collection of maize samples from farmers.

A second implication, closely related to the above, is that we need to question the conventional adoption model for improved germplasm. This model assumes that the breeding process finishes once farmers have adopted a variety, that a variety once adopted should stay unchanged, and that if the variety does change, the changes are likely to be negative; therefore the seed should be replaced either with new seed of the original variety or of one that is even "better." Improved varieties do change in farmers' hands, and these changes are not necessarily negative; farmers may consider them positive. The changes are associated with farmers' selection and seed management practices. Rather than ignoring such changes, we should try to investigate ways to take advantage of them.

Third, rather than provide poor farmers with finished OPVs or hybrids, the research system could instead offer them improved crop populations containing desirable traits that are still quite diverse, with a view to further selection to reinforce what the farmers prefer. This practice should include some additional training of the farmers to make their selection more efficient (for example, not only to take ear characteristics into consideration for selection as is mainly currently done, but to include relevant plant characteristics as well). How to accomplish this goal is not yet clear, but it is an area that merits further research. For example, CIMMYT is exploring a method called "targeted allele introgression," which allows the incorporation of valuable traits (such as drought tolerance and storage pest resistance) from elite germplasm into local maize populations and builds on farmers' seed management practices (Bergvinson and García-Lara 2004).

Further work in this area of research should be particularly important to address the needs and conditions of the poor, because farmers' seed management practices in general and creolization in particular are more important among the poor, particularly in more subsistence-oriented systems. However, creolization may only function for open-pollinated crops[21] and may not be applicable for self-pollinated crops, such as wheat or rice, and certainly not for clonally propagated crops, such as potatoes. Thus the results of this research may not be applicable to other crops.

Fourth, there is a need for dialogue among scientists of different disciplines to analyze the implications of these results for breeding strategies and the diffusion of improved germplasm. Social scientists alone cannot introduce new practices. Breeders need to bring their technical expertise to judge the methods, feasibility, benefits, and costs for linking creolization into the breeding process.

Fifth, there is a need to go beyond a simplistic concept of yield as the yardstick of impact and look at the set of traits that farmers value, how those traits are being supplied by the germplasm available, and the trade-offs they entail. Decreasing these trade-offs has an important and positive impact on farmers' well-being, which is the particular value of creolized varieties in the systems that we studied. Even yield is a more complex concept than tons per hectare. As shown here, farmers have different concepts of yield, which are not necessarily correlated, for example, yield by weight, yield by volume, and yield of dough to make tortillas.

Sixth, extension strategies should better understand local innovation and adaptation of improved varieties. Extension agents should not assume that an "improved" variety is automatically superior, especially for all characteristics that matter to farmers. An improved variety may be indeed superior for some traits but not for others, hence the value of local adaptation and creolization. There may be a role for extension in terms of strengthening the capacity of farmers to innovate and adapt improved varieties to their needs and circumstances, not just to promote adoption. For example, farmers can be trained to understand maize reproduction better, to support their capacity to creolize improved varieties. This is another area that merits further research.

Seventh, researchers and extension agents should be aware of farmers' actual practices with regard to management and recycling of improved and creolized seeds. Actual practice is determined by farmers' resource base, beliefs, and access to and trust in different sources of information. Such awareness provides insight into the usefulness of different varieties under various conditions and the likely outcomes of introduction, adoption, and creolization.

---

21. Open pollination occurs when pollen from one plant fertilizing another is the dominant process; self-pollination occurs when pollen from the same plant is responsible for fertilization.

Eighth, the implications of being poor for farmers, the importance given to different traits, and the constraints they face are not the same in subsistence- and commercially oriented systems. For example, improved germplasm, particularly hybrids, are better able to benefit the poor in a commercially oriented system, but have a much more limited value in a more subsistence-oriented and isolated system. An a priori classification of areas by the dominant orientation of maize production should be very useful in targeting agricultural research to address the needs of the poor.

The implications of creolization for agricultural research organizations may be different for an institution with a global mandate, such as CIMMYT, than for a national research organization, such as INIFAP. For CIMMYT it may be important to establish the extent to which adaptation of improved varieties by farmers also occurs and is appreciated in other maize regions of the developing world. For example, it is estimated that about 60 percent of the maize area in Latin America and 64 percent of that in Sub-Saharan Africa is planted in farmer-saved seed (Morris 2002). Such numbers suggest that creolization could be important in these regions, but clearly this is an empirical question. Creolization is also a process that may be worth studying to understand the potential impacts of transgenic maize varieties among the poor in areas where farmers recycle seed (see Bellon and Berthaud 2006). Furthermore, it should be within the purview of CIMMYT to explore and develop innovative methods to improve the process of creolization for the benefit of poor farmers whenever this process is used (for example, targeted allele introgression). Clearly CIMMYT cannot and should not work on such methods alone, but in close partnership with national programs that focus on improving the livelihoods of poor maize farmers that use this process. One role of INIFAP could be to assess the extent that creolization is an important and widespread process among small-scale Mexican farmers and, if this process is as important for the poor as our study suggests, to develop and implement new methodologies for improving the efficiency of the process in delivering germplasm that is relevant for the poor.

Finally, the results suggest that tools used by poverty alleviation programs are useful for a broad targeting of agricultural research. By focusing our research efforts on areas of high and very high marginality, we can target the research to address the needs and issues relevant to the poor as a first step. Once the targeted regions have been identified, the study of people's asset bases, perceived risks, beliefs and experiences, social networks, and the local political economy—and the relationships among them—is necessary to discover likely patterns of adoption and impact. Such study can best be accomplished through a combination of conventional survey and participatory and ethnographic methods. Our research has shown that results can be achieved with reasonable expenditures of time and resources—and is worth the effort if helping poor farmers is a central objective.

Acknowledgments

The authors thank the following people for their valuable contributions to this project: Miriam Lopez Lara and Javier Rodríguez for carrying out the case study fieldwork; José Alfonso Aguirre Gómez, Dagoberto Flores, and Irma Manuel Rosas for conducting the technical focus groups; Jeff White and Eduardo Martínez for the geographic information system; David Beck for information on maize breeding; Satwant Kaur and John Engels for editorial assistance; Juan Carlos de Loera for inputting the survey data; and Ginette Mignot and Liliana Santamaria for administrative assistance.

References

Adato, M., and R. Meinzen-Dick. 2003. Assessing the impact of agricultural research on poverty and livelihoods. *Quarterly Journal of International Agriculture* 42 (2): 149–167.

Adesina, A., and M. Zinnah. 1993. Technology characteristics, farmers' perceptions and adoption decisions: A tobit model application in Sierra Leone. *Agricultural Economics* 9 (4): 297–311.

Agresti, A. 1996. *An introduction to categorical data analysis.* New York: Wiley.

Aguirre-Gómez, J. A. 1999. Análisis regional de la diversidad del maíz en el Sureste de Guanajuato. Ph.D. thesis, Universidad Nacional Autónoma de México, México, D.F.

Barkley, A., and L. Porter. 1996. The determinants of wheat variety selection in Kansas, 1974 to 1993. *American Journal of Agricultural Economics* 78 (1): 202–211.

Bellon, M. R. 1996. The dynamics of crop infraspecific diversity: A conceptual framework at the farmer level. *Economic Botany* 50 (1): 26–39.

Bellon, M. R., and J. Berthaud. 2001. In-situ conservation of maize diversity, gene flow, and transgenes in Mexico. Paper presented at the OECD Conference on LMOs and the Environment, Raleigh-Durham, N.C., November 27–30.

———. 2006. Potential impacts of transgenic varieties on maize diversity in traditional Mexican agricultural systems. *Agriculture and Human Values* 23 (1): 3–4.

Bellon, M. R., and J. Risopoulos. 2001. Small-scale farmers expand the benefits of improved maize germ plasm: A case study from Chiapas, Mexico. *World Development* 29 (5): 799–811.

Bellon, M. R., and J. E. Taylor. 1993. "Folk" soil taxonomy and the partial adoption of new seed varieties. *Economic Development and Cultural Change* 41 (4): 763–786.

Bellon, M. R., M. Adato, J. Becerril, and D. Mindek. 2005. *Impact of improved germplasm on poverty alleviation: The case of Tuxpeño-derived materials in Mexico.* Mexico, D.F.: Centro Internacional de Mejoramiento de Maíz y Trigo (CIMMYT). <www.cimmyt.org/english/docs/impacts/impmaize_05.pdf> (accessed October 2006).

Bergvinson, D., and S. García-Lara. 2004. Genetic approaches to reducing losses of stored grain to insects and diseases. *Current Opinion in Plant Biology* 7 (4): 480–485.

Coe, R. 2002. Analyzing ranking and rating data from participatory on-farm trials. In *Quantitative analysis of data from participatory methods in plant breeding,* R. Bellon and J. Reeves, eds. Mexico, D.F.: Centro Internacional de Mejoramiento de Maíz y Trigo (CIMMYT).

CONAPO-PROGRESA (Consejo Nacional de Población y Programa de Educación, Salud y Alimentación). 2000. *Indices de marginación 1995.* México, D.F.: Consejo Nacional de Población y Programa de Educación, Salud y Alimentación.

Edmeades, S., M. Smale, M. Renkow, and D. Phaneuf. 2004. Variety demand within the framework of an agricultural household model with attributes: The case of bananas in Uganda. Environment and Production Technology Division Discussion Paper 125. Washington, D.C.: International Food Policy Research Institute.

Foster, J., J. Greer, and E. Thorbecke. 1984. A class of decomposable poverty measures. *Econometrica* 52 (3): 761–765.

Guevara-Sanginés, A., C. Muñoz Piña, G. Estrada Díaz, and N. Acosta Romero. 2000. Manual para la Evaluación de Impactos sobre el abatimiento de la Pobreza a Partir de la Inversión en Proyectos Ambientales en Pequeñas Poblaciones Rurales. Working Paper 500-03. Mexico, D.F.: Departamento de Economía, Universidad Iberoamericana.

Hintze, L. H., M. Renkow, and G. Sain. 2003. Variety characteristics and maize adoption in Honduras. *Agricultural Economics* 29 (3): 307–317.

Louette, D., A. Charrier, and J. Berthaud. 1997. In situ conservation of maize in Mexico: Genetic diversity and maize seed management in a traditional community. *Economic Botany* 51: 20–38.

Morris, M. L. 2002. *Impacts of international maize breeding research in developing countries, 1966–98.* Mexico, D.F.: Centro Internacional de Mejoramiento de Maíz y Trigo (CIMMYT).

Morris, M. L., and M. A. Lopez-Pereira. 1999. Impacts of maize breeding research in Latin America 1966–1997. Mexico, D.F.: Centro Internacional de Mejoramiento de Maíz y Trigo (CIMMYT).

Morris, M. L., J. Risopoulos, and D. Beck. 1999. Genetic change in farmer-recycled maize seed: A review of the evidence. CIMMYT Economics Working Paper 99-07. Mexico, D.F.: Centro Internacional de Mejoramiento de Maíz y Trigo (CIMMYT).

Pressoir, G., and J. Berthaud. 2004. Population structure and strong divergent selection shape phenotypic diversification in maize landraces. *Heredity* 92 (2): 95–101.

Skoufias, E., B. Davis, and J. R. Behrman. 1999. An evaluation of the selection of beneficiary households in the education, health, and nutrition program (PROGRESA) of Mexico. Final report. Washington, D.C.: International Food Policy Research Institute.

Smale, M., M. R. Bellon, and A. Aguirre. 2001. Maize diversity, variety attributes and farmers' choices in southeastern Guanajuato, Mexico. *Economic Development and Cultural Change* 50 (1): 201–225.

Wood, D., and J. Lenné. 1997. The conservation of agrobiodiversity on-farm: Questioning the emerging paradigm. *Biodiversity and Conservation* 6 (1): 109–129.

8 National and International Agricultural Research and Rural Poverty: The Case of Rice Research in India and China

SHENGGEN FAN, CONNIE CHAN-KANG, KEMING QIAN, AND K. KRISHNAIAH

Agricultural research has played an important role in agricultural production and productivity growth in many developing countries. High-yield varieties (HYVs) released by national and international agricultural research centers have substantially increased crop production in many Asian countries. The rural poor have benefited directly from income increases as a result of production growth. In addition, rapid agricultural growth stimulated broader economic development that led to the regional economic boom of the 1980s and 1990s (Rosegrant and Hazell 2001). Thus rural poverty also declined through these indirect effects in the region, and the predicted food shortage never occurred. Although there have been many studies on the effects of the green revolution on production and productivity growth in the 1970s and 1980s (for example, Hayami and Ruttan 1985; Hazell and Ramasamy 1991), the question today is whether these national and international efforts will continue to have high payoffs in terms of further growth in agricultural production. In addition, what role the Consultative Group for International Agricultural Research (CGIAR) centers have played as a partner in this process has not been well documented. Moreover, there have been few attempts to link agricultural research investments to rural poverty reduction.[1] This study is designed to help fill these gaps using the case of rice in India and China. The study measures the impact at national levels, taking account of the important ways, direct and indirect, in which the poor can be affected. Information on the poverty effects of agricultural research investments can help national and international policymakers mobilize resources and set priorities for agricultural research in the future.

India and China are the two most populous countries in the world, together accounting for more than 38 percent of the world's total population and almost 50 percent of its rural residents. In spite of recent rapid economic growth in both countries, many people still live below the official national poverty line. India has an estimated 200 million and China 30 million rural people below the poverty

1. Evenson and Gollin (2002) estimated economic returns to varietal improvement of CGIAR research.

line. However, if the poverty line of US$1 per day measured in purchasing power parity (PPP) is used, China would have substantially more poor than the official figure. Using this line, China had more than 100 million rural poor in 1998 (World Bank 2000a).[2]

Rice is a major staple food crop for many developing countries, not only as a main source of calories but also as an important source of income and employment for many farmers, particularly those in poor households. For developing countries as a whole, rice accounted for 34 percent of land area planted in cereal and 47 percent of cereal production in 2000. Rice is, in fact, the dominant cereal in China and India, occupying 35 and 45 percent, respectively, of the total area planted in cereal in 2000. For that same year, rice accounted for 45 and 57 percent of total cereal production in China and India, respectively.

China and India are the two leading rice-producing countries and have been so since 1961, the first year that data became available from FAO (2002). In 2001, they jointly produced 53 percent of the world's rice on 48 percent of its area devoted to rice. In China and India, rice is the most important food crop, accounting for about 30 percent of food energy intake (FAO 2002).

The International Rice Research Institute (IRRI) has been collaborating with China and India for the past several decades. The major modes of collaboration have been joint research and exchanges of human resources, scientific information, and germplasm. For this study, we have selected rice in these two countries to evaluate the total benefits from varietal improvement research, and we attempt to partition the benefits attributable to IRRI and others and estimate the contribution of rice-breeding research to poverty reduction.

In contrast to the traditional econometric approach proposed by Griliches (1957), our study uses extensive data on the adoption and performance of the rice varieties used by Chinese and Indian farmers to evaluate the total benefits from rice varietal improvement research. We then rely on genetic or pedigree information to analyze how international agricultural research has contributed to productivity gains in Chinese and Indian rice production. Finally, we use the calculated benefits, together with poverty impact parameters reported in recent International Food Policy Research Institute (IFPRI) studies, to assess the contribution of domestic and international rice research to poverty reduction.

Rural Poverty in China and India

Headcount ratio, the percentage of the population falling below the poverty line, is the most widely used measure of poverty incidence. The poverty line used in India is defined as 49 rupees per month at 1973–74 prices (Datt and Ravallion

2. India's national poverty line is very close to the international poverty line of US$1 per day. For example, Datt and Ravallion (1997) reported that one-third of India's population lived in poverty in the mid-1990s if the US$1 per day 1993 PPP is used. The official poverty rate was 35 percent from 1992 to 1997 (World Bank 2000b).

1997). This poverty line is equivalent to US$0.965 per person per day measured in 1993 PPP, only slightly below the US$1 per day widely used for cross-country comparison by the World Bank and other organizations. China's official poverty line is equivalent to US$0.66 per day measured in 1985 PPP (World Bank 2000a). Using these official poverty lines, the incidence of poverty has declined dramatically over the past several decades in both countries. In India, rural poverty fluctuated between 50 and 65 percent in the 1950s and early 1960s before beginning a steady decline from about two-thirds of the rural population in the mid-1960s to one-third of it in the late 1980s. Rural poverty increased to about 40 percent in the early 1990s when economic policy reforms were initiated. Recent official data show that the poverty rate declined to 27 percent in 1999.

The long downward trend in poverty in rural India from 1967 to 1999 coincided with several important developments. The rapid adoption of HYVs, together with improved irrigation and the use of fertilizer, sharply increased agricultural production and productivity. This change in technology was a direct result of increased government investment in agricultural research and extension, infrastructure, irrigation, and education during the 1960s, 1970s, and 1980s. The increase in government investment also improved nonagricultural employment opportunities and increased wages, which contributed to further reductions in rural poverty (Fan, Hazell, and Thorat 2000; Fan, Zhang, and Zhang 2002).

For the past two decades, China has achieved remarkable progress in reducing rural poverty. Following rural reforms, per capita income increased from 220 yuan in 1978 to 522 yuan in 1984 (1990 prices), an average growth rate of 15 percent per year. The income gains were shared widely enough to cut the poverty rate by more than half. By 1984, only 11 percent of the rural population was below the official poverty line, compared to 33 percent in 1978. Because of equitable land distribution, income inequality as measured by the Gini coefficient increased only slightly despite the sharp income increase observed between 1978 and 1984 (Fan, Zhang, and Zhang 2002).

From 1985 to 1989, rural income continued to increase, but at a much slower pace, averaging 3 percent per year, mainly because of the stagnation of agricultural production after the reforms. By the end of 1984, the effects of fast agricultural growth on rural poverty were largely exhausted. Rural income distribution became less egalitarian, and the Gini coefficient rose from 0.264 in 1985 to 0.301 in 1989 (State Statistical Bureau 1990). As a result, the number of poor increased from 89 million in 1984 to 103 million in 1989 when measured by the official poverty line. Only in 1990 did rural poverty begin to decline again. The number of rural poor dropped from 103 million in 1989 to 34 million in 2000, equivalent to an average reduction of 9 percent per year.[3] The

3. If poverty is measured using the international poverty line of US$1 per day (constant 1985 PPP dollars), then China still had more than 100 million rural poor and 20 million urban poor in 1998 (World Bank 2000a).

above discussion suggests that agricultural growth, including that spurred by agricultural research, plays a key role in reducing rural poverty.

Rice Research and Rice Production

For thousands of years farmers in Asia have improved their rice yields by selecting and saving seed from the higher-yielding plants in local fields. Modern national and international rice breeding programs have developed more formal and structured methods in crossing and selecting improved rice varieties. The international exchange of genetic resources in various forms (landraces and advanced lines) has become an important feature of modern rice breeding.[4]

IRRI's rice breeding program began in October 1961, and in the following year 38 crosses were made. The variety IR8, released in 1966, changed the face of Asian agriculture, with yields ranging from six to eight tons per hectare in experimental fields compared to three to four tons per hectare with older varieties. IRRI crosses grew in number and complexity over time, and by 1975, 29 IR varieties had been released. Breeding research gave greater emphasis to insect and disease resistance and adaptability to unfavorable environments, resulting in greater geographic spread, higher yields, and improved yield stability. In addition, newer varieties grew faster, meaning that they used less water, were exposed to field hazards for a shorter period of time, and facilitated multiple cropping.

Rice research in India has a long history and has been one of the top priorities of the government-supported research program. Core activities of varietal development and related activities are performed by several research institutions: (1) the Directorate of Rice Research (DRR) and its funded centers (about 54 of them) located across the country in all states; (2) the Central Rice Research Institute (CRRI) in Cuttack, Orissa, and its substations; and (3) a half-dozen institutes affiliated with the Indian Council of Agricultural Research (ICAR). The state universities, such as those in Tamil Nadu, Andhra Pradesh, West Bengal, and Punjab, are also conducting rice research.

The introduction of semidwarf varieties from IRRI to India occurred in 1964, when C. Subramaniam, minister of Food and Agriculture, visited IRRI and was given seeds of new rice varieties that included TN-1. By 1966, IR8 and other IRRI lines were being tested in various experimental fields in India. Shortly after their introduction, these IRRI varieties were crossed with local varieties. By 1998 about three-quarters of the area devoted to rice in India was sown in HYVs (Indiastat 2002).

---

4. The history of international rice research draws heavily from various IRRI publications and Dalrymple (1986); the evolution of Chinese and Indian rice research programs is taken from those countries' respective government documents.

Conventional rice breeding began in China in 1906. However, systematic and well-targeted breeding using rigorous methods did not start until 1919, when the Nanjing Higher Agricultural School and Guangzhou Agricultural Specialized School set up breeding programs. Following the establishment of the People's Republic in 1949, the government paid greater attention to rice breeding. The development of the rice-breeding program is characterized by three stages. During the first stage, from 1950 to the beginning of the 1960s, great efforts were made in the selection, evaluation, and use of local rice varieties. The second stage of rice breeding, from the beginning of the 1960s to the beginning of the 1970s, focused on the breeding of dwarf varieties.[5] The third stage is characterized by the development of hybrid rice, in which China was a pioneer. Research on hybrid rice in China began in the mid-1960s, and in 1976 China became the first country to commercially use hybrid rice varieties.[6] Since then, the area planted in hybrid rice has increased steadily. In 1981, hybrid rice accounted for 23 percent of total rice production, but two decades later it accounted for 61 percent of total production.[7]

The more formal IRRI involvement in China's rice breeding program began in the 1970s, although IR8 was introduced and tested in Guangdong in 1967. In the early 1970s, a delegation of Chinese officials visited the Philippines and was given a bag of rice seed developed at IRRI. This event marked the first formal cooperation between IRRI and China.

As a result of these national and international efforts, rice crop production in both China and India has increased substantially for the past several decades. From 1961 to 2001, rice production grew at an average of 2.7 percent per year in India and by 2.6 percent per year in China, much higher than their respective population growth rates of 2.1 and 1.6 percent. Much of the increase in rice production was a result of gains in yield. In India, yield increase accounted for 77 percent of the total increase in rice production, while in China almost all the production increase came from yield increase. In India, yield doubled from 1.5 tons per hectare in 1961 to 3.0 tons per hectare in 2001, while in China yield tripled from 2.1 to 6.3 tons per hectare over the same period (Table 8.1). The development of improved or modern rice varieties in conjunction with irriga-

5. After a farmer found a dwarf plant (only 70 centimeters tall) in 1956, Chinese scientists began the breeding program that led to the development of the first dwarf HYV of rice, Guang Chang Ai, in 1957, a few years before the foundation of IRRI (Shen 1980; Dalrymple 1986). Guang Chang Ai is an Indica variety, and its offspring were quickly adopted in southern China. The first semidwarf Japonica variety introduced to China in 1957 was Nongken 58, a selection from a Japanese variety, which was crossed with various local varieties.

6. In 1974, Professor Yuan Long Ping, from the Hybrid Rice Research Center in Hunan, and his team successfully developed the first hybrid rice variety.

7. China has never officially published rice output by type. The shares reported here are calculated by the authors using area-by-variety data from the Ministry of Agriculture.

TABLE 8.1 Trends in area planted in rice, production, and yield

Item	1961	1970	1980	1990	2000	2001	Growth rate (%)
Area harvested (million ha)							
India	34.7	37.6	40.2	42.7	44.8	44.5	0.6
China	27.0	33.1	34.5	33.5	30.3	28.6	0.02
World	115.5	133.1	144.6	146.9	154.1	151.5	0.58
Production (million tons)							
India	53.5	63.3	80.3	111.5	129.4	131.9	2.71
China	56.2	113.1	142.9	191.6	189.8	181.5	2.62
World	215.7	316.4	396.8	518.2	600.6	592.8	2.54
Yield (kg/ha)							
India	1,542	1,685	2,000	2,613	2,890	2,964	2.09
China	2,079	3,416	4,144	5,717	6,264	6,350	2.60
World	1,867	2,377	2,745	3,529	3,897	3,912	1.95

SOURCE: FAO (2002).

tion and the greater use of modern inputs (such as fertilizer and pesticides) has been instrumental in achieving these yield increases.

Research Benefits and the Contribution of International Research

In this section, we quantify the economic impact arising from the development of improved rice varieties in India and China. We begin by estimating the total benefits from rice varietal improvement research, irrespective of the sources of the gains. Next we use genetic or pedigree information on each variety planted in the two countries to assess the contribution of IRRI to these benefits.

Estimation of Benefits

The economic benefits from rice varietal improvement research result mostly from the productivity gains that farmers experience after adopting improved varieties. Typically, measuring these benefits is based on comparing a "with research" scenario to a counterfactual scenario (Pardey et al. 1996, 2002; Heisey and Morris 2002). The first step toward measuring these benefits is to determine the gain in yield resulting from the development and adoption of HYVs. To isolate the genetic contribution of improved varieties to yield increase from other factors, we collected experimental yield data of adopted rice varieties in India and China.[8] Experimental yields have the advantage of holding many of the variables influencing yields constant, and hence may provide a good approximation of the genetic contribution to yield gains. Empirical evidence shows that absolute yields achieved in experimental trials are higher than those in farmers' fields. However, it is uncertain whether relative yield gains in trials are also greater (Pardey et al. 1996; Heisey and Morris 2002). Here we assume that the proportional gains achieved in experimental trials are representative of the proportional gains realized by farmers. Using the experimental yield data, we selected numeraire varieties specific to each country. The numeraire should be a variety that was widely adopted in either China or India before the establishment of their respective rice research programs, and that has been grown as a control variety at research stations ever since. We then compute the yield premium of newer adopted varieties against the numeraire variety.[9] Suppose that before variety B was released, it was tested against the numeraire variety A. The yield premium of variety B is given by

$$P_B = (Y_B / Y_A) - 1,$$

where P_B is the yield premium of variety B, Y_B the yield of variety B, and Y_A is the yield of the numeraire variety A. As the check variety used in experimental trials changes over time, we use the chain rule to link back to the numeraire variety A. Thus before variety C was released, variety B was used as a check variety. The yield premium of variety C over the numeraire variety A is given by

$$P_C = ([Y_C / Y_B']) \times [Y_B / Y_A]) - 1.$$

Note that Y_B and Y_B' are not equal, because they are the yields of the same variety tested at different times. Although the yield premium gives the relative gain in yield, the absolute yield gain of variety C against the numeraire variety A is estimated as

$$\Delta Y_C = Y_C - Y_A = P_C \times Y_A.$$

The benefits for each variety are calculated by multiplying the yield gain by price, and again by area sown to the variety. For a given region and year, the total benefits B are simply the sum of those for all varieties and can be written as

$$B = \Sigma_i \Delta Y_{ir} A_{irt} AP_{rt},$$

8. Experimental yield data for China were obtained from the Chinese National Rice Research Institute in Hangzhou, China. Data for India are from the coordinated trials of the All India Coordinated Rice Improvement Program (AICRIP) of the Directorate of Rice Research. Different levels of input are often used in conducting the yield experiments between two varieties and therefore, the average yield advantage is calculated from different levels of input use.

9. Pardey et al. (1996) used a similar procedure.

where A_{irt} represents the area of variety i in region r at time t, and AP_{rt} is the average of the counterfactual and actual producer price of rice at region r at time t. The counterfactual price captures the price-reducing effect of improved rice varieties, that is, what the price of rice would have been in the absence of the development and adoption of improved rice varieties. Under unitary demand and supply elasticities, the proportional shift in supply translates to the same proportional shift in prices. Assuming that the price of rice in 2000 was under a perfect market, we estimated the counterfactual and average price series as

$$CP_{rt} = P_{r2000}(1 + k_{rt})$$
$$AP_{rt} = (CP_{rt} + P_{r2000})/2,$$

where CP_{rt} is the counterfactual price at region r and time t, P_{r2000} is the price of rice in region r in 2000, and k_{rt} is the supply shift in region r and time t. Under neutral technical change with fixed-factor proportions, the percentage increase in experimental yield P_{irt} translates into an equal, proportional, rightward shift in supply (Alston, Norton, and Pardey 1995, 339).

Three major types of rice are planted in China, namely, Indica, Japonica, and hybrid. Therefore it is necessary to choose a numeraire variety specific to each type of rice. The numeraire variety we chose for conventional Indica rice is Bao Tai Ai, a variety released in 1959 by the Yulin Regional Agricultural Experiment Station in Guangxi. Due to data limitations, we choose Nongken 58, a variety introduced from Japan in the 1950s, as our numeraire for Japonica varieties. As all early hybrid varieties had an IRRI parent, the numeraire we chose for hybrid rice is Zhen Zhu Ai, a conventional Indica variety that does not have any IRRI ancestry.[10] These numeraire varieties were all widely adopted and used as breeding materials for subsequent varieties.

For India, we choose a numeraire variety specific to each state. These numeraire varieties were local varieties widely adopted by farmers in the early 1960s before the introduction of IR8 to India. The numeraire varieties used for each state are Andhra Pradesh, HR67; Assam, Latisail; Bihar, N136; Gujurat, Mashuri; Haryana, Himachal, and Punjab, Jhona349; Karnataka, SR26 B; Kerala, Ptb 10; Madya Pradesh, Safri17; Maharashtra, Ratnagiri1; Orissa, T141; Tamil Nadul, CO25; Uttar Pradesh, Sarjoo49; and West Bengal, NC1263.[11]

Figure 8.1 compares rice farm yield and experimental yield achieved in India and China. Figure 8.1A shows that farm yield doubled from 1.5 to 3 tons per hectare in India from 1961 to 2001. In China, the observed increase in yield was even more significant, tripling from 2.1 tons per hectare in 1961 to 6.3 tons per hectare in 2001. Compared with farm yield, experimental yield increased substantially less over time in both India and China (see Figure 8.1B). This is

10. This choice was recommended to us by Professor Yuan Long Ping at the China National Hybrid Rice Research Center, Changsha, Hunan, China.

11. Our source of experimental yield data in India was AICRIP.

FIGURE 8.1 Average farm field yield and experimental yield in India and China

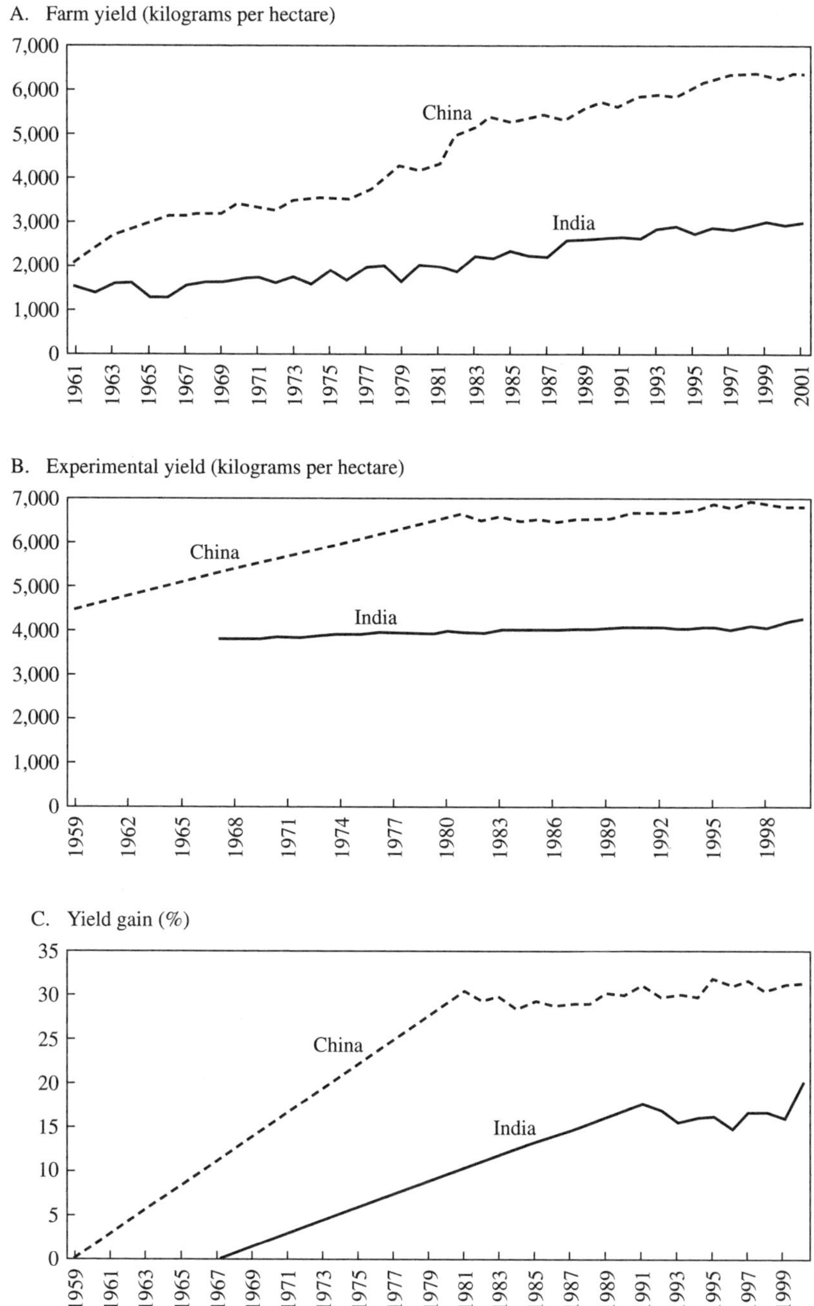

SOURCES: Industry yield compiled by authors from FAO (2002); authors collected experimental yield data and compiled experimental yield and yield gain.

because the increased use of inputs such as fertilizer also contributed to farm yield, while the increased use of inputs has been controlled for in the experimental tests.

On average, experimental yield increased from 3.8 to 4.3 tons per hectare from 1967 to 2000 in India. For comparison purposes, total factor productivity (TFP) for the Indian agricultural sector as a whole barely budged in the 1970s but grew quickly in the 1980s and 1990s (Fan, Hazell, and Thorat 2000). In contrast, Evenson, Pray, and Rosegrant (1998) found that growth in TFP for the Indian crop sector slowed down during the 1980s. In China, after a rapid increase from 4.5 tons per hectare in 1959 to 6.6 tons per hectare in 1981, experimental yield increased little in the 1980s and 1990s, ranging from 6.5 to 6.8 tons per hectare. Similar to these trends, Rozelle et al. (2003) found that the TFP for rice increased little from the mid-1980s to the mid-1990s.

Figure 8.1C shows the average yield gain over the numeraire variety. In China, the gain in yield resulting from new varieties accelerated from 1959 to the early 1980s and plateaued afterward. In India, the average gain in yield increased sharply from 1967 to the early 1990s, remained constant in the following years, and increased again in the late 1990s. Overall the yield gain realized in China was higher than in India. In 2000 the average gain in yield with respect to the numeraire was 31 and 20 percent in China and India, respectively.

Table 8.2 presents the estimated benefits from rice research reported in constant 2000 prices. In India the benefits from rice research varied from US$3.9 billion in 1991 to US$3.6 billion in 2000. In China the benefits from rice research amounted to US$5.2 billion in 2000. The source of these benefits changed significantly over time. In 1981 Indica rice accounted for 72 percent of the total rice research benefits, whereas Japonica and hybrid rice accounted for 4 and 24 percent, respectively. In 2000 72 percent of the rice research benefits were attributed to hybrid rice, whereas the share of Indica rice declined to only 16 percent and Japonica rice accounted for 12 percent.

India's research benefits as a share of total rice production value ranged between 20 and 24 percent between 1991 and 2000 (Table 8.3). In China, rice research benefits accounted for a similar share of rice production value, averaging 20.1 percent in 1981 and 17.1 percent in 2000.

Benefits attribution

The use of IRRI varieties by the national agricultural research system falls within the following categories: (1) direct use of IR varieties under either IR names or local names, (2) direct use of IR breeding lines or crosses under either IR numbers or local names, and (3) use of IR varieties or lines as parents in local breeding programs. To gain some insight into IRRI's impact in China and India, we first examined the share of area devoted to rice sown in varieties that have IRRI ancestry (Table 8.4). In China, the share increased from 23 percent in 1981 to a peak of 65 percent in 1991, then declined to nearly 20 percent

TABLE 8.2 Annual benefits from rice research

	China					India	
	Indica	Japonica	Hybrid	All rice	Agricultural research expenditures	All rice	Agricultural research expenditures
1981	3,833	187	1,304	5,324	237	na	na
1982	4,674	187	928	5,789	246	na	na
1983	3,810	203	1,329	5,342	306	na	na
1984	3,225	204	1,917	5,347	349	na	na
1985	3,501	262	1,547	5,311	342	na	na
1986	3,293	305	1,520	5,118	347	na	na
1987	2,584	296	1,818	4,698	328	na	na
1988	2,566	362	2,540	5,468	384	na	na
1989	2,583	461	2,487	5,531	399	na	na
1990	2,474	433	3,378	6,284	361	na	na
1991	1,342	506	2,963	4,812	387	3,930	300
1992	1,944	718	3,352	6,014	454	3,916	299
1993	1,494	747	3,099	5,340	473	3,907	294
1994	1,805	682	3,194	5,681	506	3,842	310
1995	1,108	593	3,676	5,377	503	4,012	325
1996	1,581	632	4,163	6,376	522	3,587	333
1997	1,277	1,262	4,574	7,113	483	4,233	352
1998	1,284	907	4,658	6,849	573	4,217	361
1999	1,153	651	4,317	6,121	660	4,020	455
2000	849	650	3,729	5,228	na	3,615	na

NOTES: All data are in millions of 2000 U.S. dollars; na, data not available.

in 2000 in 1997. Table 8.4 also reveals that the impact of IRRI in China occurred mostly through the use of IRRI varieties as breeding material rather than through direct adoption.

Moreover, IRRI contributed mostly to hybrid rice, whereas practically none of the Japonica varieties were bred with IRRI materials. In 1997 50 percent of hybrid, 31 percent of Indica, and only 0.5 percent of Japonica varieties had an IRRI ancestor in their pedigree. In India IRRI's impact is found in both the direct adoption of IRRI varieties and the use of breeding materials from IRRI. In 2000 the area planted in varieties with IRRI ancestry (including direct adoption) accounted for nearly 60 percent of total rice area in India, and about 14 percent of the varieties adopted were IRRI-released.

To attribute the shares of the rice benefits to IRRI, we followed the method described in Pardey et al. (1996), who developed various rules to attribute benefits to a specific research or breeding program, in this case to IRRI research.

TABLE 8.3 Rice research benefits as a share of production value

	China				India
	Indica	Japonica	Hybrid	All rice	All rice
1981	24.5	3.8	21.8	20.1	na
1982	24.1	3.2	20.1	19.4	na
1983	21.2	3.5	17.9	17.2	na
1984	21.4	3.2	17.7	16.5	na
1985	23.6	4.2	16.3	17.4	na
1986	23.3	4.9	14.4	16.5	na
1987	21.0	4.6	14.3	15.0	na
1988	25.0	6.3	17.3	17.8	na
1989	24.7	6.6	16.3	17.0	na
1990	26.1	6.4	18.9	18.5	na
1991	17.7	7.5	15.9	14.6	23.9
1992	24.3	10.5	17.9	17.9	22.0
1993	23.1	10.9	16.6	16.7	21.5
1994	23.4	10.3	18.3	17.9	22.4
1995	18.3	10.8	16.9	16.2	21.9
1996	21.7	8.4	20.1	17.9	19.5
1997	23.0	13.6	20.7	19.2	21.7
1998	21.6	9.9	21.6	18.7	21.1
1999	19.7	7.0	20.1	16.7	21.1
2000	19.5	7.7	21.1	17.1	22.7

NOTES: All data are percentages; na, data not available.

These rules take into consideration various factors involved in varietal development, such as the recent versus the earlier research and breeding efforts versus heritability of traits. The binary-parents rule gives full credit to IRRI if the two parents of a variety or any of its ancestors were IRRI-released. If only one parent was IRRI-released or had IRRI ancestry, then the variety was considered 50 percent IRRI. The all-antecedents rule assigns equal weights to the variety and each of its ancestors. Thus, if we trace the pedigree back to the grandparent level, the variety and each of its ancestors is given a weight of one-seventh if released by IRRI. The geometric rule assigns higher weight for the recent generations and lower weight for the early generations. The all-credit-to-last-cross rule takes only the last cross into account. Thus, if the variety was released by IRRI, it gets all credit; otherwise, it gets none. Finally, the any-ancestor rule gives credit to IRRI if a variety or any of its ancestors was released by IRRI. The all-credit-to-last-cross rule and the any-ancestor rule represent polar cases: the former is the most conservative rule and the latter is the least conservative.

TABLE 8.4 Percentage of area devoted to rice that is planted with IRRI ancestors

	China			India		
Year	Direct adoption	With IRRI ancestry	Total IRRI	Direct adoption	With IRRI ancestry	Total IRRI
1982	0.2	23.9	24.1	na	na	na
1983	0.1	29.3	29.4	na	na	na
1984	0.1	36.0	36.1	na	na	na
1985	0.0	38.7	38.7	na	na	na
1986	0.2	45.3	45.5	na	na	na
1987	0	49.6	49.6	na	na	na
1988	0	58.8	58.8	na	na	na
1989	0	56.2	56.2	na	na	na
1990	0	62.6	62.6	na	na	na
1991	0	64.9	64.9	23.2	38.8	62.0
1992	0	58.9	58.9	34.7	34.0	68.7
1993	0	54.7	54.7	21.0	41.6	62.6
1994	0	53.0	53.0	25.0	30.3	55.3
1995	0	53.6	53.6	20.8	37.3	58.1
1996	0	41.1	41.1	24.4	35.3	59.8
1997	0	36.8	36.8	21.9	41.7	63.5
1998	0	30.5	30.5	18.7	44.5	63.3
1999	0	27.2	27.2	15.3	44.8	60.1
2000	0	18.7	18.7	14.4	43.9	58.3

NOTES: IRRI, International Rice Research Institute; na, data not available.

Using these various attribution rules, we present in Table 8.5 the contribution of IRRI to the total benefits from rice varietal improvement research in India and China. IRRI accounted for a sizable share of rice research benefits in India. With the any-ancestor rule, IRRI accounted for 81 percent of the rice research benefits in 1991 and for 63 percent in 2000. With the most conservative scenario (all-credit-to-last-cross rule), IRRI's contribution was still important, accounting for 63 percent of the research benefits in 1991 and for 12 percent in 2000. According to the binary-parents, all-antecedents, and geometric rules, IRRI's contribution to research benefits ranged from 18 to 77 percent from 1991 to 2000.

In contrast, the share of the rice benefits attributable to IRRI was smaller in China. IRRI's varieties were mostly used as breeding materials in China and were not directly adopted by farmers. As a result, the all-credit-to-last-cross rule gives overall 0 percent of the research benefits to IRRI. With the any-ancestor rule, IRRI's share of research benefits was 23 percent in 1981, increasing to 69

TABLE 8.5 Rice research benefits attributed to IRRI under alternative attribution rules (%)

	China					India				
Year	Binary parents	All antecedents	Geometric	All credit to last cross	Any ancestry	Binary parents	All antecedents	Geometric	All credit to last cross	Any ancestry
1981	14.1	7.6	5.1	0.0	23.2	na	na	na	na	na
1982	13.8	7.4	5.0	0.2	22.8	na	na	na	na	na
1983	17.4	10.5	6.7	0.2	28.9	na	na	na	na	na
1984	20.9	13.3	8.2	0.1	36.8	na	na	na	na	na
1985	21.5	13.0	7.5	0.1	36.5	na	na	na	na	na
1986	25.9	12.9	7.0	0.2	41.7	na	na	na	na	na
1987	31.9	15.0	7.4	0.0	50.0	na	na	na	na	na
1988	34.6	14.6	7.0	0.0	57.9	na	na	na	na	na
1989	31.4	12.3	5.7	0.0	54.2	na	na	na	na	na
1990	36.1	14.4	6.6	0.0	62.1	na	na	na	na	na
1991	39.2	15.0	6.8	0.0	68.9	75.3	40.5	55.2	63.4	81.0
1992	33.0	12.5	5.8	0.0	59.5	77.2	41.0	56.4	65.7	81.8
1993	31.2	11.2	5.2	0.0	57.4	56.5	27.9	36.8	40.3	67.4
1994	28.7	10.2	4.8	0.0	53.3	57.0	28.4	36.8	40.9	64.9
1995	27.6	9.1	4.1	0.0	50.2	44.7	19.8	22.1	20.5	58.6
1996	21.9	7.3	3.3	0.0	40.5	42.6	20.2	22.5	19.8	55.7
1997	18.8	6.1	2.8	0.0	35.5	42.0	20.4	21.1	17.0	57.1
1998	16.8	5.3	2.4	0.0	31.7	48.3	24.0	24.3	19.4	63.8
1999	16.2	5.0	2.2	0.0	29.9	44.9	19.3	18.1	13.2	63.0
2000	11.9	3.8	1.7	0.0	21.6	46.8	20.9	18.6	11.8	63.5

NOTES: All data are percentages; IRRI, International Rice Research Institute; na, data not available.

percent in 1991, but declining gradually to 22 percent in 2000. With the geometric rule, IRRI's contribution to total benefits ranged from 1.7 to 8.2 percent over the 1981–2000 period compared with 12–39 percent with the binary-parents rule and 4–15 percent with the all-antecedents rule.

Table 8.6 compares the benefits and costs of IRRI's research. The benefits attributed to IRRI using the geometric rule are presented next to IRRI's total budget and China's and India's contribution to IRRI. The geometric attribution is one of the most conservative rules, taking into account not only the recent crosses but also past breeding efforts. More weights assigned to the recent crosses than the earlier ones attribute more benefits to the national agricultural research

TABLE 8.6 Annual international rice research benefits and costs

	Research benefits contributed by IRRI		Expenditures by IRRI		
Year	China	India	Total	China's contribution	India's contribution
1981	270,402	na	38,942	na	na
1982	290,109	na	40,761	na	187
1983	356,711	na	38,350	na	195
1984	440,074	na	40,429	150	188
1985	396,607	na	45,592	146	218
1986	356,467	na	42,435	171	178
1987	346,393	na	45,243	69	173
1988	383,977	na	41,395	67	166
1989	317,536	na	47,010	64	129
1990	415,284	na	51,668	62	124
1991	328,615	2,167,777	46,224	60	119
1992	348,260	2,206,824	48,616	93	117
1993	277,479	1,436,881	50,993	103	114
1994	270,443	1,415,077	44,631	100	112
1995	221,254	887,621	44,008	98	219
1996	211,383	807,302	42,877	96	187
1997	196,548	892,439	36,736	95	158
1998	165,085	1,022,552	36,310	na	na
1999	136,553	729,510	35,875	na	na
2000	88,924	671,972	32,600	130	158

SOURCES: Research benefits compiled by the authors. IRRI expenditures from 1981 to 1997 are from the CGIAR secretariat; 1998–2000 expenditures are from the CGIAR 1999 financial report and the 2000 annual report, respectively. China's and India's contributions to IRRI from 1982 to 1997 are from IRRI (2002a, 2002b); China's and India's contributions to IRRI in 2000 are from IRRI (2000).

NOTES: All data are in thousands of 2000 U.S. dollars. Only the very conservative attribution rule, geometric, was used here. IRRI, International Rice Research Institute; na, data not available.

system than to IRRI. Even using this conservative rule, the benefits from IRRI's research in India and China well exceed both countries' contributions. In 2000, benefits attributed to IRRI are 684 times China's funding contribution to IRRI, and they are more than 4,000 times that of India. The benefits from IRRI research in China were nearly threefold greater than IRRI's budget, whereas in India the benefits were twentyfold greater than IRRI's total budget. Total benefits attributed to IRRI from China and India are $761 million in 2000. This amount is twice as large as CGIAR's annual budget.

Impact on Poverty

New technology resulting from agricultural research can help alleviate poverty in several ways. First, following the releases of new and improved cultivars, farmers can produce more output at the same cost (or the same level of output at a lower cost), which directly improves farmers' incomes (Kerr and Kolavalli 1999). Second, the diffusion of modern varieties result in lower food prices, as demonstrated in several studies, such as Ruttan (1977), Lipton and Longhurst (1989), and more recently Datt and Ravallion (1997). This reduction in food prices is critical, given that the poorest people spent a large share of their income on food. Third, the productivity consequences of improved varieties results in greater demand for labor and increased wages. Hossain (1988), for example, studied the effects of technological progress in rice cultivation in Bangladesh and found that the poor benefited from the new technology as a result of greater employment opportunities as well as the upward pressure on wage rates in the labor market. This finding concurs with many past studies, such as Jayasuriya and Shand (1986), Quizon and Binswanger (1986), Basant (1987), Acharya (1989), and David and Otsuka (1994).

The benefits arising from rice varietal improvement research are distributed between producers and consumers. Producers gain from expanded production due to reduced production cost. However, they may lose income because of depressed sales prices. The net gain by producers can be either positive or negative. For consumers, their gain will always be positive because of lower prices. In our study, we focus on the impact on rural poor. (The benefits to urban poor can be equally large, as indicated in Chapter 9 of this volume.) Therefore, our estimates on the impact on rural poverty reduction are conservative. We use the following steps to estimate the impact of national and international rice varietal improvement research on rural poverty reduction. First, we calculate the marginal impact on rural poverty reduction of an increase in agricultural production value. This measure gives the reduction in the number of rural poor people per additional unit of agricultural production value. The parameters needed are reported by two recent IFPRI publications (Fan, Hazell, and Thorat 2000; Fan, Zhang, and Zhang 2002). Second, we calculate the total reduction in the number of rural poor people that occurs because of rice varietal improvement research by considering the estimated research benefits as the additional increase in agricultural production value. Finally, we use IRRI's share of total rice research benefits estimated from the geometric attribution rule to estimate the poverty reduction impact attributed to IRRI. These are lower-bound estimates, as the geometric rule is one of the most conservative rules.

Fan, Hazell, and Thorat (2000) estimated a system of econometric equations to calculate the impact of different types of government spending on agricultural growth and rural poverty reduction in India using state-level data for 1970–93. The model is structured to enable the identification of the various channels through which different types of government expenditures affect the poor. The study distinguishes between direct and indirect effects of agricultural growth due to agricultural research. The direct effects arise in the form of benefits the poor receive from higher income through growth in agricultural production. The indirect effects come from increased rural wages and employment and changed food prices. This approach has two advantages. First, both direct and indirect effects of agricultural growth can be estimated. In addition to the direct effects on farmers' incomes from increased output, price effects as well as effects on rural wages and nonfarm employment have also been included. Second, other types of investment, such as infrastructure, education, and health, are also included to avoid at least some of the potentially upward-biased estimates of research investment impact.

The estimated poverty equation in the cited system shows that with every 1-percent increase in agricultural production or productivity growth, the total number of rural poor in India is reduced by 0.241 percent as a result of all direct and indirect effects. Using this total elasticity, we can calculate the marginal impact of an additional unit in agricultural production value on rural poverty reduction. Multiplying this marginal poverty impact by the estimated productivity benefits from rice research gives the total reduction in the number of rural poor people due to rice variety improvement research. Table 8.7 shows the estimated results for India. The reduction in the number of rural poor people as a result of rice varietal improvement research varied from 4.95 million in 1991 to 4.81 million in 1997 and then declined to 3.06 million in 1999. This annual reduction expressed as a percentage of total rural poor ranges from 2.12 percent in 1991 to 1.81 percent in 1999.

Turning to the impact of IRRI varietal improvement research on rural poverty reduction, Table 8.7 shows that in 1991, a reduction of 2.73 million in the number of rural poor is attributable to IRRI's research. In 1999, the estimated reduction in the number of rural poor attributable to IRRI varietal improvement research was 0.56 million. We also calculated the reduction in the poor per US$1 million of IRRI spending (Table 8.7).[12] In India, for every US$1

12. A more complete analysis would have allowed for the lagged relationships between agricultural research expenditures and their productivity increases by calculating research stocks from

TABLE 8.7 Impact on rural poverty of rice research in India

Year	Rural poor (million)	Reduction in number of poor from rice research (million)	Reduction as a percentage of total poor (%)	Reduction in number of poor from IRRI research (million)	Reduction in number of poor per US$1 million of IRRI spending
1991	233	4.95	2.12	2.73	59,040
1992	237	5.12	2.16	2.89	59,379
1993	242	4.90	2.03	1.80	35,372
1994	274	5.29	1.93	1.95	43,629
1995	252	4.81	1.91	1.07	24,203
1996	251	4.39	1.75	0.99	23,033
1997	249	4.81	1.93	1.01	27,590
1998	212	4.23	1.99	1.02	28,221
1999	169	3.06	1.81	0.56	15,490

NOTE: IRRI, International Rice Research Institute.

million invested by IRRI, 59,040 individuals moved above the poverty line in 1991, and 15,490 in 1999.

Similar to the India study, Fan, Zhang, and Zhang (2002) developed and estimated a simultaneous-equation model to estimate the effects of different types of government expenditure in China using provincial-level data for 1970–97. From their estimated poverty equation, the total elasticity of poverty reduction with respect to agricultural output growth is 1.924 percent. As for India, we use this elasticity to calculate the reduction in the number of poor people per unit increase in agricultural production value and the reduction in the number of poor people from IRRI rice varietal improvement research.

The total reduction in rural poor through rice research in China has been much larger than that in India (Table 8.8). In 1981, 23 million individuals escaped poverty as a result of rice varietal improvement research. However in 1999, only 1.53 million rural poor came out of poverty because of rice research. In relative terms, escape from poverty through rice research as a proportion of the total number of rural poor was 12 percent in 1981 and 5 percent in 1999. Table 8.8 also shows that the reduction in the number of poor people brought about by IRRI's varietal improvement research declined from 1,016,000 in 1981 to 30,000 in 1999. Finally, the reduction in the number of poor people per

past investment data and using estimated lagged structures (Fan, Hazell, and Thorat 2000; Fan, Zhang, and Zhang 2002). However, we do not have enough years of rice expenditure data to undertake these calculations here.

TABLE 8.8 Impact on rural poverty of rice research in China

Year	Rural poor (million)	Reduction in number of poor from rice research (million)	Reduction as a percentage of total poor (%)	Reduction in number of poor from IRRI research (million)	Reduction in number of poor per US$1 million of IRRI spending
1981	194	23.07	11.89	1.02	26,083
1982	140	16.23	11.60	0.70	17,259
1983	123	12.06	9.80	0.70	18,224
1984	89	7.54	8.48	0.54	13,443
1985	96	7.85	8.17	0.51	11,211
1986	97	7.24	7.46	0.44	10,416
1987	91	5.71	6.27	0.37	8,197
1988	86	5.92	6.88	0.37	8,883
1989	103	7.63	7.41	0.39	8,229
1990	97	7.15	7.37	0.42	8,104
1991	95	5.20	5.47	0.32	6,828
1992	90	5.89	6.54	0.30	6,224
1993	80	4.40	5.50	0.20	3,978
1994	70	3.57	5.10	0.15	3,362
1995	65	2.85	4.39	0.10	2,345
1996	58	2.98	5.13	0.09	2,022
1997	50	2.77	5.53	0.07	1,828
1998	42	2.15	5.12	0.05	1,254
1999	34	1.53	4.51	0.03	839

NOTE: IRRI, International Rice Research Institute.

US$1 million of IRRI spending was 26,083 in 1981. Because of rapidly diminishing rural poverty, the reduction in the number of poor people for every US$1 million spent by IRRI declined to 839 in 1999.[13]

Conclusions

Using varietal adoption and performance data, in this chapter we calculated the total benefits from rice varietal improvement research in China and India for the past two decades. We then used genetic or pedigree information to partition

13. A recent study by Thirtle, Lin, and Piesse (2003) has estimated that there is need for US$179 of investment in agricultural research and development to help one person out of poverty. Our estimates show that India would have required US$17 in 1991 and US$65 in 1999, and China would have required US$38 in 1981 and US$1,191 in 1999 to help one person escape poverty.

the total benefits between these two countries and IRRI. Finally, we used reported elasticity of poverty reduction with respect to agricultural output growth to assess the effects of national and international research on poverty reduction in rural India and China.

The results indicate that rice varietal improvement research has contributed tremendously to increased rice production in both countries. In China research benefits as a share of rice production value range from 14 to 20 percent.[14] In India, they range from 20 to 24 percent. In both countries the benefits produced just from rice research are, on average, 10 times greater than their respective total agricultural research investments.

Without research investments in rice, the number of poor would be much higher today.[15] For every US$1 million invested at IRRI in 1999, more than 800 and 15,000 rural poor moved above the poverty line in China and India, respectively. A similar or even larger poverty impact is observed in Indonesia, Vietnam, and Bangladesh, although formal analyses have not yet been done in these countries.

However, most of these benefits are the results of research conducted in the 1960s, 1970s, and 1980s. For both China and India, the increase in experimental yield slowed in the 1990s. One of the reasons is the lack of agricultural research investment at both the national and international levels. As a percentage of agricultural gross domestic product, agricultural research investment in both countries was relatively low, 0.3 percent for China and 0.4 percent for India. For other low-income Asian countries, the percentages are in the range of 0.5–1. For developed countries, the range is as high as 2–4 percent.

IRRI's budget has been severely cut in recent years. Its budget of $32.6 million in 2000 was the lowest in 20 years, and was only 63 percent of its peak of $51.6 million (measured in 2000 prices) in 1990. The declining effects of IRRI's research on poverty reduction in India and China have several explanations and implications. First, rapid poverty reduction in both China and India will naturally lead to a smaller marginal impact on poverty reduction from rice research. Second, IRRI's research now has lower production effects than it did before. Table 8.6 clearly shows that economic benefits from IRRI's rice research have declined after the mid-1990s. Therefore, even if India and China had the same number of poor as before, poverty reduction effects would have to decline. This observation implies that there are probably better investments than rice research to reduce poverty. As China's poor have increasingly been concentrated in west-

14. This result is consistent with the findings of Fan and Pardey (1997), who concluded that about 20 percent of the total production value from 1965 to 1993 is from the increased agricultural research investment.

15. In separate studies, Fan (2003) and Fan, Fang, and Zhang (2003) concluded that the effects of agricultural research on urban poverty are as large as those on rural poverty, and agricultural research may play an even larger role in helping the urban poor in the future, as more poor will be concentrated in the urban centers.

ern China, where there is little rice production, the poverty reduction impact of rice research is naturally low. In this case, research in other commodities or even investment in education and infrastructure may have higher returns.

For IRRI's research, if its objective is to maximize poverty reduction, it may have to shift to other countries where the poverty rates are still high and to other sectors where the poor can directly benefit. Thus new country and sector priorities have to be set.

Acknowledgments

The authors are grateful for research support provided by the International Center for Agricultural and Rural Development of the Chinese Academy of Agricultural Sciences. Funding from the Standing Panel on Impact Assessment of the CGIAR, the Natural Science Foundation of China (No. 70525003), and the Australian Centre for International Agricultural Research is acknowledged.

References

Acharya, S. 1989. Agricultural wages in India: A disaggregated analysis. *Indian Journal of Agricultural Economics* 44 (1): 121–139.

Alston, J. M., G. W. Norton, and P. G. Pardey. 1995. *Science under scarcity: Principles and practice for agricultural research evaluation and priority setting.* Ithaca, N.Y.: Cornell University Press.

Basant, R. 1987. Agricultural technology and employment in India: A survey of recent research. *Economic and Political Weekly* 22 (1,2): 1348–1364.

Dalrymple, D. G. 1986. *Development and spread of high-yielding rice varieties in developing countries.* Washington, D.C.: USAID.

Datt, G., and M. Ravallion. 1997. *Why have some Indian states performed better than others at reducing rural poverty?* FCND Discussion Paper 26. Washington, D.C.: International Food Policy Research Institute.

David, C., and K. Otsuka. 1994. *Modern rice technology and income distribution in Asia.* Boulder, Colo.: Lynne Rienner.

Evenson, R. E., and D. Gollin. 2002. The green revolution: An end-of-century perspective. Paper presented at the conference, Why Has Impact Assessment Research Not Made More of a Difference? February 4–7, 2002, San José, Costa Rica.

Evenson, R. E., C. E. Pray, and M. W. Rosegrant. 1998. *Agricultural research and productivity growth in India.* IFPRI Research Report 109. Washington, D.C.: International Food Policy Research Institute.

Fan, S. 2003. Agricultural research and urban poverty in India. *Quarterly Journal of International Agriculture* 42: 63–78.

Fan, S., C. Fang, and X. Zhang. 2003. Agricultural research and urban poverty: The case of China. *World Development* 31(4): 733–741.

Fan, S., P. Hazell, and S. Thorat. 2000. Government spending, agricultural growth and poverty in rural India. *American Journal of Agricultural Economics* 82 (4): 1038–1051.

Fan, S., and P. Pardey. 1997. Research, productivity, and output growth in Chinese agriculture. *Journal of Development Economics* 53: 115–137.

Fan, S., L. Zhang, and X. Zhang. 2002. *Growth and poverty in rural China: The role of public investment.* IFPRI Research Report 125. Washington, D.C.: International Food Policy Research Institute.

FAO (Food and Agriculture Organization of the United Nations). 2002. FAOSTAT. <http://apps.fao.org> (accessed July 2002).

Griliches, Z. 1957. Hybrid corn: An exploration in the economics of technological change. *Econometrica* 25 (4): 501–522.

Hayami, Y., and V. Ruttan. 1985. *Agricultural development: An international perspective.* Baltimore: Johns Hopkins University Press.

Hazell, P., and C. Ramasamy. 1991. *The green revolution reconsidered.* Baltimore: Johns Hopkins University Press for the International Food Policy Research Institute.

Heisey, P. W., and M. L. Morris. 2002. *Practical challenges to estimating the benefits of agricultural R&D: The case of plant breeding research.* Paper presented at the annual meeting of the American Agricultural Economics Association (AAEA), July 28–31, 2002, Long Beach, Calif.

Hossain, M. 1988. *Nature and impact of green revolution in Bangladesh.* IFPRI Research Report 67. Washington, D.C.: International Food Policy Research Institute.

Indiastat. 2002. Datanet India Private Limited. <www.indiastat.com> (accessed July 2002).

IRRI (International Rice Research Institute). 2000. *Annual report 2000.* Manila: International Rice Research Institute.

———. 2002a. *Facts about cooperation: People's Republic of China and IRRI.* Manila: International Rice Research Institute. <http://www.irri.org/media/facts/pdfs/china.pdf> (accessed July 2002).

———. 2002b. *Facts about cooperation: India and IRRI.* Manila: International Rice Research Institute. <www.irri.org/VIS/facts/india.pdf> (accessed July 2002).

Jayasuriya, S., and R. Shand. 1986. Technical change and labor absorption in Asian agriculture: Some emerging trends. *World Development* 14: 415–428.

Kerr, J., and S. Kolavalli. 1999. *Impact of agricultural research on poverty alleviation: Conceptual framework with illustration from the literature.* EPTD Discussion Paper 56. Washington, D.C.: International Food Policy Research Institute.

Lipton, M., and W. R. Longhurst. 1989. *New seeds and poor people.* Baltimore: Johns Hopkins University Press.

Pardey, G. P., J. M. Alston, J. E. Christian, and S. Fan. 1996. *Summary of a productive partnership: The benefits from U.S. participation in the CGIAR.* EPTD Discussion Paper 18. Washington, D.C.: International Food Policy Research Institute.

Pardey, G. P., J. M. Alston, C. Chan-Kang, E. C. Magalhaes, and S. A. Vosti. 2002. *Assessing and attributing the benefits from varietal improvement research: Evidence from Embrapa, Brazil.* EPTD Discussion Paper 95. Washington, D.C.: International Food Policy Research Institute.

Quizon, J., and H. P. Binswanger. 1986. Modeling the impact of agricultural growth and government policy on income distribution in India. *World Bank Economic Review* 1: 103–148.

Rosegrant, R. W., and P. Hazell. 2001. *Transforming the rural Asian economy: The unfinished revolution.* Hong Kong: Oxford University Press.

Rozelle, S., S. Jin, J. Huang, and R. Hu. 2003. The impact of investments in agricultural research on total factor productivity in China. In *Crop variety improvement and its effect on productivity,* R. E. Evenson and D. Gollin, eds. Cambridge: CAB International.

Ruttan, V. W. 1977. The green revolution: Seven generalizations. *International Development Review* 19 (4): 13–16.

Shen, J.-H. 1980. Rice breeding in China. In *Rice improvement in China and other Asian countries.* Manila: International Rice Research Institute and Chinese Academy of Agricultural Sciences.

State Statistical Bureau. 1990. *China development report.* Beijing: China Statistical Press.

Thirtle, C., L. Lin, and J. Piesse. 2003. The impact of research-led agricultural productivity growth on poverty reduction in Africa, Asia, and Latin America. *World Development* 31 (12): 1959–1975.

World Bank. 2000a. *China: Overcoming rural poverty.* Washington, D.C.: World Bank.

———. 2000b. *India: Policies to reduce poverty and accelerate sustainable development.* Washington, D.C.: World Bank.

9 Agricultural Research and Urban Poverty in China and India

SHENGGEN FAN

Many studies have shown that investments in agricultural research can yield favorable economic returns (Alston et al. 2000) and contribute to significant reductions in rural poverty (Kerr and Kolavalli 1999; Fan, Hazell, and Thorat 2000; Hazell and Haddad 2001; Fan, Zhang, and Zhang 2002). The links between agricultural research and food price benefits for consumers have also been quantified, using the consumer surplus as a welfare measure (Akino and Hayami 1975; Mellor 1975; Scobie and Posada 1978; and Pinstrup-Andersen 1979).[1] But little work has been done on quantifying the specific impact of agricultural research on urban poverty reduction, even though rapid urbanization is increasing the incidence of urban poverty in developing countries (Haddad, Ruel, and Garret 1999; Ravallion 2000). This chapter helps fill that gap and reports on an econometric study of the links between past expenditures on agricultural research and urban poverty reduction in China and India.

Both countries have been very successful in increasing agricultural production and reducing rural poverty in recent decades, and both now face worsening urban poverty. The two countries have also intervened heavily in their food markets over the years in ways that affect market prices, and hence influence the main channel by which agricultural research affects the urban poor.

The chapter is organized as follows. I first review the historical trends in agricultural research investment in China and India, followed by a brief discussion of changes in agricultural production, food prices, and urban poverty. I then present a conceptual framework and model for the analysis on how agricultural research affects the urban poor in the two countries, and then discuss the estimation procedures and results. I conclude with some policy implications.

This chapter draws heavily on Fan (2003) and Fan, Fang, and Zhang (2003).

1. The effects of rural growth on both rural and urban poverty reduction have also been quantified by Ravallion and Datt (1996) for India; Fan, Zhang, and Zhang (2002) for China; and Deininger and Squire (1996) and Irz et al. (2001) for cross-country analyses.

Investment Trends in Agricultural Research

Government spending on agricultural research in China and India has increased significantly over the past four decades, but not without substantial year-to-year variations (Tables 9.1 and 9.2). In China, investment in agricultural research was quite modest during the first five-year (1953–57) plan, averaging 72 million yuan per year (all values in 1990 prices). During the Great Leap Forward (1958–60), expenditures on agricultural research increased dramatically to 497 million yuan per year, but then fell to 425 million yuan per year during the following three years. Research expenditure increased modestly to 643 million yuan during the Cultural Revolution (1966–76), and then increased steadily thereafter until 1994. Since then, agricultural research expenditures have shown little increase.

As a percentage of agricultural gross domestic product (AgGDP), agricultural research investment was a relatively low, 0.12 percent during the first five-year plan, but it increased dramatically to 0.54 percent during the Great Leap Forward. The percentage has gradually declined to below 0.4 percent in recent years. Thus government investment in agricultural research has increased substantially in absolute terms for the past several decades but has declined relative to the size of the agricultural sector.

In India, investment in agricultural research was quite modest during the 1960s, ranging from 1.6 to 1.9 billion rupees (Rs; all values in 1995 prices). During the 1970s, expenditures on agricultural research increased dramatically to 4.0 billion Rs around 1980, more than doubling in the decade. During this period many agricultural universities and national research institutions were set up (Evenson, Pray, and Rosegrant 1998). These were the driving force behind the green revolution that more than doubled the yields of rice and wheat within

TABLE 9.1 Public investment in agricultural research in China

Year	Agricultural research expenditure (constant 1990 million yuan)	Percentage of agricultural gross domestic product
1953–57	72	0.12
1958–60	497	0.54
1961–65	425	0.57
1966–76	643	0.43
1977–85	1,348	0.44
1986–90	1,725	0.39
1991–94	2,099	0.39
1995–97	2,203	0.32

SOURCES: Fan and Pardey (1992), Fan and Pardey (1997), and State Statistical Bureau and State Science and Technology Commission (1992–2004).

TABLE 9.2 Public investment in agricultural research in India

Year	Research expenditures (million 1990 Rs)	Research expenditures (million 1990 international dollars or PPPs)	Percentage of agricultural gross domestic product
1964	1,629	378	na
1965	1,581	367	0.21
1966	1,869	434	0.25
1967	1,590	369	0.18
1968	1,684	391	0.19
1969	1,879	436	0.20
1970	1,902	441	0.20
1971	1,886	438	0.21
1972	1,973	458	0.22
1973	1,741	404	0.17
1974	2,504	581	0.26
1975	3,178	737	0.33
1976	3,471	805	0.38
1977	3,965	920	0.38
1978	4,407	1,022	0.43
1979	4,148	962	0.45
1980	3,982	924	0.38
1981	4,128	958	0.39
1982	4,292	995	0.41
1983	4,695	1,089	0.40
1984	4,978	1,155	0.43
1985	4,572	1,061	0.39
1986	5,115	1,186	0.44
1987	6,011	1,394	0.50
1988	6,517	1,512	0.48
1989	6,507	1,509	0.46
1990	7,085	1,643	0.48
1991	6,873	1,594	0.46
1992	6,754	1,567	0.44
1993	7,280	1,689	0.44
1994	7,246	1,681	0.42
1995	7,293	1,692	0.43

SOURCE: Agricultural research expenditures were obtained from the State Planning Commission, Government of India.

NOTES: The gross domestic product deflator was used to deflate expenditures to 1995 prices. We then used the 1995 exchange rate based on purchasing power parity (PPP) to convert expenditures into 1995 international dollars. na, data not available.

a decade. During the 1980s, research expenditures continued to increase to 7.0 billion Rs in 1990. But in the 1990s, research expenditure increased only modestly to 7.3 billion Rs by 1995.

As a percentage of AgGDP, agricultural research investment in India was relatively low at 0.20 percent during the 1960s, but it increased dramatically to more than 0.40 percent in the 1970s. In the 1980s, the percentage continued to rise, to a peak of 0.50 percent in 1987. But the percentage has gradually declined to below 0.43 percent in recent years. Similar to China, government investment in agricultural research has increased in absolute terms over the past decade but has declined relative to the size of the agricultural sector.

Trends in Agricultural Production and Food Prices

Between 1970 and 2002, China's AgGDP increased nearly sixfold, while it doubled in India. Total factor productivity (TFP) also improved over the same period, by 85 percent in China and 52 percent in India (Table 9.3). These figures include all agricultural output and hence are influenced by agricultural diversification. Production and consumption of livestock and horticultural products have grown faster than food staples in both countries, and hence part of the AgGDP increase reflects a shift to higher-value products.

Urban food prices (in real terms) have remained stagnant in India since 1970, a remarkable achievement, given the huge increase in the population and the diversification of the national diet into higher-value foods. Urban food prices also changed little in China during 1970 to 1984 despite rapid population growth, but have increased by about 50 percent since then (Table 9.3). Again, this trend reflects a rapid shift toward a more diversified diet, but it also shows the effect of policy reforms in China that allowed agricultural prices to increase from the low levels at which they were formerly held. Food prices would have been much higher in both countries had the investments in agricultural research not been made. But how much higher and with what effects on poverty are questions that require the kind of quantitative analysis developed later in this chapter.

Urban Poverty

In China, the incidence of urban poverty is low compared to that of rural poverty (Table 9.4). Using a poverty line of US$1.0 income per capita per day measured in 1985 purchasing power parity (1985 PPP dollars), the incidence of rural poverty was 4.6 percent in 1998, and the number of rural poor was 42.1 million. In contrast, the incidence of urban poverty was only 2.06 percent and the number of urban poor was 6.32 million (Table 9.4), or about 5 percent of the nation's total poor when the US$1.00 per day poverty line is used.

TABLE 9.3 Growth in production and productivity and urban food price indices in India and China

Year	China			India		
	AgGDP	Total factor productivity	Urban food price index	AgGDP	Total factor productivity	Urban food price index
1950	na	na	100	100	na	na
1951	na	na	97	101	na	na
1952	100	100	97	105	na	na
1953	105	97	100	113	na	na
1954	109	97	101	116	na	na
1955	118	104	100	115	na	na
1956	126	105	101	121	na	na
1957	124	103	104	116	na	na
1958	127	116	104	128	na	na
1959	108	107	104	126	na	na
1960	95	87	104	135	na	na
1961	106	74	109	135	na	na
1962	110	77	109	132	na	na
1963	124	80	114	135	na	na
1964	139	85	116	148	na	na
1965	161	89	117	131	na	na
1966	176	93	118	130	na	na
1967	178	95	120	149	na	na
1968	179	97	120	149	na	na
1969	188	98	119	158	na	na
1970	209	95	119	169	100	100
1971	216	94	117	166	99	98
1972	216	92	118	158	91	99
1973	237	96	116	169	99	102
1974	246	100	116	167	96	103
1975	256	101	115	188	109	102
1976	255	99	114	177	104	96
1977	246	98	114	195	113	98
1978	262	103	114	200	115	96
1979	313	106	114	174	99	95
1980	326	108	114	196	112	96
1981	362	116	114	207	118	97
1982	414	125	114	205	116	96
1983	455	131	116	225	129	97
1984	508	143	118	228	125	96
1985	511	151	123	230	128	94
1986	531	152	123	229	124	95
1987	586	156	126	226	126	95

TABLE 9.3 *Continued*

	China			India		
Year	AgGDP	Total factor productivity	Urban food price index	AgGDP	Total factor productivity	Urban food price index
1988	625	151	130	260	148	100
1989	634	153	128	264	140	98
1990	712	162	125	275	138	99
1991	703	164	126	271	138	100
1992	715	169	128	287	144	103
1993	740	174	128	298	146	101
1994	848	176	169	313	152	102
1995	951	183	179	311	na	104
1996	1,036	192	177	341	na	104
1997	1,055	190	172	332	na	103
1998	1,107	na	167	353	na	104
1999	1,126	na	na	354	na	101
2000	1,127	na	na	353	na	na
2001	1,174	na	na	377	na	na
2002	1,231	na	na	357	na	na

SOURCES: Fan (2003) and Fan, Fang, and Zhang (2003).

NOTE: AgGDP, Agricultural gross domestic product; na, data not available.

TABLE 9.4 Poverty in China

		Incidence of urban poverty for a given poverty level (%)			Number of urban poor for a given poverty level (million)			Rural poverty (US$1.00/day)	
Year	Urban per capita income (yuan)	US $1.00/ day	US $1.50/ day	US $2.00/ day	US $1.00/ day	US $1.50/ day	US $2.00/ day	Number (million)	Incidence (%)
1992	2,191	2.09	13.74	35.66	5.22	34.32	89.06	80.1	8.8
1994	2,686	2.73	13.18	29.49	7.46	36.00	80.55	70	7.6
1995	2,828	1.65	10.28	25.73	4.70	29.26	73.22	65	7.1
1996	2,879	1.69	8.41	23.31	4.92	24.50	67.92	58	6.3
1997	3,001	2.00	9.21	21.36	5.98	27.53	63.85	49.6	5.4
1998	3,078	2.06	8.86	19.58	6.32	27.17	60.04	42.1	4.6

SOURCE: Rural poverty data from Rural Survey Organization of State Statistical Bureau (2004).

NOTES: Per capita income is measured in 1992 prices. Total consumption expenditures are used for poverty measures.

However, there are good reasons to use a higher poverty line when measuring urban poverty. One prominent reason is the much higher cost of living for urban than rural residents. Consequently, in this study, I also use poverty lines of US$1.50 and US$2.00 per capita per day. These increments lead to significant increases in the estimated number of urban poor in 1998, from 6.32 million when using the US$1.00 poverty line to 27.17 million and 60.04 million, respectively, when using the US$1.50 and US$2.00 poverty lines.

The incidence of urban poverty changed little between 1992 and 1998 when the US$1.00 poverty line is used, though the number of poor increased from 5.22 million to 6.32 million. When the higher poverty lines are used, there is a dramatic decline in urban poverty between 1992 and 1998. This shift suggests that many people who lived just above the US$1.00 poverty line in 1992 have moved to higher income levels, whereas the poorest of the poor living below the US$1.00 line have not benefited much at all.

One important characteristic of the urban poor in China is the high share of total consumption expenditure they spend on food. If the US$2.00 per capita per day poverty line is used, then in 1998 the urban poor spent about 58 percent of their total expenditures on food, compared to 50 percent for the average urban population. Clearly the urban poor would suffer more than most from higher food prices.

In India, both rural and urban poverty rates were high in 1970, with 57 percent of the rural population and 47 percent of the urban population living below the official poverty line (Table 9.5).[2] Due in large part to rapid agricultural growth, the rural poverty rate declined to 45 percent by the mid-1980s (Ahluwalia 1985). The urban poverty rate also declined to 36 percent. In addition to growth in urban income, the decline in real food prices relative to nonfood prices may have played a large role in this reduction. From the mid-1980s to 1987, rural poverty continued to decline to 39 percent, but urban poverty changed very little. The reduction in rural poverty during this period is mainly because of the development of rural nonfarm employment and increases in rural wages. The so-called "trickle-down" benefits of agricultural growth for the rural poor were almost nonexistent, as both agricultural production and productivity growth were largely stagnant (Fan, Hazell, and Thorat 2000). The impact of agricultural growth on urban poverty through lower food prices was also absent.

Rural and urban poverty declined relatively rapidly from the end of the 1980s to the early 1990s, largely because of the rapid increases in agricultural production and productivity (Fan, Hazell, and Thorat 1999). The growth in agricultural production and productivity may have also contributed to urban poverty

2. India's national poverty line is very close to the international poverty line of US$1 per day. For example, Datt and Ravallion (1998) reported that one-third of India's population was below the poverty line in the mid-1990s if the US$1 per day 1993 PPP is used. The official poverty rate is 35 percent from 1992 to 1997 (World Bank 2000).

TABLE 9.5 Poverty in India, 1970–95

Year	Rural poverty rate (%)	Urban poverty rate (%)	Urban poor (million)	Rural poor (million)	Share of urban poor (%)
1970	57.61	47.16	51.69	256.53	16.77
1971	54.84	44.98	51.12	248.99	
1972	na	na	na	na	na
1973	55.36	45.67	55.97	260.99	17.66
1974	55.72	47.96	61.07	267.46	18.59
1975	na	na	na	na	na
1976	na	na	na	na	na
1977	na	na	na	na	na
1978	50.60	40.50	59.95	259.54	18.77
1979	na	na	na	na	na
1980	na	na	na	na	na
1981	na	na	na	na	na
1982	na	na	na	na	na
1983	45.31	35.65	62.36	253.06	19.77
1984	na	na	na	na	na
1985	na	na	na	na	na
1986	na	na	na	na	na
1987	38.81	34.29	67.73	232.36	22.57
1988	39.60	35.65	72.55	241.10	23.13
1989	39.06	36.60	76.71	241.77	24.09
1990	34.30	33.40	72.06	215.79	25.03
1991	36.43	32.76	72.72	232.89	23.79
1992	40.00	33.50	74.36	259.76	22.26
1993	36.66	30.51	71.63	241.73	22.86
1994	41.00	33.50	80.88	274.36	22.77
1995	37.15	28.40	70.54	252.15	21.86

SOURCES: Rural and urban poverty rates from Datt (1998); the number of rural and urban poor was calculated by the author using rural and urban population data from FAO (2003).

NOTE: na, data not available.

reduction by keeping food prices low. In fact, the relative food price index dropped by 2 percent during this period.

In summary, whenever there is higher growth in agricultural production and productivity, rural poverty declines. But urban poverty also falls when agricultural growth is high.

Econometric Model

To analyze the links between agricultural research and urban poverty, an econometric model was developed in which an agricultural production function, price determination function, and urban poverty equation are simultaneously determined. Many poverty determinants—such as income and its distribution, production or productivity growth, and prices—are generated from the same economic process as poverty and hence must be specified as endogenous to avoid estimation biases. Also, because agricultural research investments affect poverty through changes in food prices, it is difficult to capture this link using a single-equation approach.

Although there are inevitably some differences in variables that could be enumerated and used in the two country studies, the basic model is the same and can be represented by

$$Y = h(FARMINP, RES, IRRI, INFRA, EDUC, RAIN, X), \quad (1)$$

$$FP = g(Y, GDP, POP, WPI, S), \quad (2)$$

$$UP = f(FP, M, GINI, Z). \quad (3)$$

Equation (1) is an agricultural productivity function. The dependent variable Y is either agricultural labor productivity measured in constant prices (China) or TFP (India). Given that the dependent variable is agricultural output for China, then such farm inputs (*FARMINP*) as land, fertilizers, and machines per unit of labor are included on the right-hand side of the function, but these are not relevant when the dependent variable is TFP, as for India. Other variables include *RES,* which is a research stock variable (a function of current and lagged government spending on agricultural research); *IRRI,* the percentage of the total cropped area that is irrigated; *INFRA,* measures of the stock of infrastructure, such as roads, electricity, and telecommunications; *EDUC,* measures of the stock of education; and *RAIN,* annual rainfall.

Institutional changes and policy reforms have made important contributions to growth in agricultural and nonagricultural production and poverty reduction in both countries. There is no need to estimate these contributions for the purposes of this study, but to reduce possible estimation biases that may arise from neglecting them, we added year and regional dummies X to capture year-specific institutional and policy changes as well as the effects of any remaining agroclimatic factors on growth in agricultural production. This specification is more flexible than Fan (1991) and Fan and Pardey (1997), who used time-period dummies for longer periods to capture the effects of institutional change on production growth.

Equation (2) models the determination of food prices *FP.* Food prices are measured as a ratio of food prices to nonfood consumer prices. Growth in agricultural production or TFP (Y) increases the supply of agricultural products and hence is expected to contribute to lower food prices. Per capita gross domestic product *GDP* and population size *POP* are used to capture demand-side factors in the food markets. Food prices in both countries may also be affected by international markets *WPI,* although both countries intervened in their food markets during the period of study. Variable *S,* which consists of a set of regional level dummies, is intended to capture the effect of all other factors on changes in food prices.

Equation (3) models the determinants of urban poverty *UP.*[3] Urban poverty is expected to be positively related to food price changes relative to nonfood prices *FP* and to inequality in urban incomes *GINI,* and negatively related to the per capita income of urban residents *M.* Variable *Z* (which constitutes year and province dummies) is included to capture the effects of all other omitted variables.[4]

Data and Model Estimation

China

The urban poverty and income variables were constructed from China's urban household survey. The urban household survey is conducted annually by the State Statistical Bureau to monitor changes in urban household expenditures and consumption. Forty to fifty thousand households were surveyed annually between 1992 and 1998. We were able to obtain access to 10 percent of the total sample, taken from one representative city in each province.

To obtain appropriate poverty measures, we first had to convert our chosen poverty lines (US$1.00, 1.50, and 2.00 per capita per day, measured in 1985 PPP dollars) into local currency at nominal prices. We first converted the poverty line from 1985 PPP dollars into Chinese currency based on the 1985 PPP exchange rate. Then we used the Chinese consumer price index to calculate the national poverty lines at current prices. Finally, province-level poverty lines were calculated by adjusting for differences in the cost of living by province.[5]

To measure urban poverty *UP,* we used the percentage of the urban population that had less than the chosen poverty line when measured in 1985 PPP. Baseline results were obtained using a poverty line of US$1.50 per capita per day, but we also ran the model using other poverty lines to check the sensitivity of the results. We chose a baseline poverty line of US$1.50 because it is broadly comparable to the widely used US$1.00 poverty line for rural areas.

---

3. To simplify the presentation, I have omitted subscripts to indicate observations in year t and at the province level, the subscripts t–1, . . ., t–j, which indicate observations in years $t-1, \ldots, t-j$.

4. It is true that nonfood price also affects the welfare of the poor. But the urban poor spend 60–80 percent of their income on food, and therefore food price change would have substantial impact on their welfare. In addition, we have controlled for nonfood price by deflating their income *M* to real terms and focusing on the effects of food price changes.

5. China did not start radical price reforms until 1984. Before that, prices were strictly controlled by state governments and were allowed to vary by only a few percentage points across provinces. We therefore assumed that price levels were the same for all provinces in 1984.

The average per capita income of the urban population *M* was calculated from the urban household expenditure survey, using the urban consumer price index as a deflator. The food price variable *FP* was measured as the food procurement price index relative to the urban consumer price index. The *GDP* variable is gross domestic product measured in constant prices. The variable *POP* is the combined population of urban and rural areas.

The agricultural production used for China was specialized to

$$Y = h(LAND, AK, RES, IRRI, SCHY, ELEC, ROADS, RTR, RAIN, X). \quad (4)$$

Agricultural production inputs are measured as follows: *LAND,* arable land only; *AK,* agricultural capital; and *IRRI,* the percentage of in total arable land that is irrigated. The variables *Y, LAND,* and *AK* are divided by labor input, so the equation becomes a labor productivity determination equation. For education *SCHY,* we used data on the percentage of the population with different education levels to calculate the average years of schooling, assuming 0 years for illiterate and semiliterate persons, 5 years for those with primary school education, 8 years for junior high school education, 12 years for a high school education, 13 years for those with professional school education, and 16 years for persons with college education. The road variable *ROADS* is defined as road density, measured as length of roads in kilometers per thousand square kilometers of geographic area. The rural telecommunication *RTR* variable is the number of rural telephones per thousand rural residents, and rural electrification variable *ELEC* is measured as rural electricity consumption per rural resident.

Public investment in agricultural research and development (R&D) is reported in the total national science and technology budget. The sources of agricultural R&D investment are from different government agencies. Science and technology commissions at different levels of government allocate funds to national, provincial, and prefecture institutes primarily as core support. These funds are primarily used by institutes to cover researchers' salaries and benefits and administrative expenses. Project funds come mainly from other sources, including departments of agriculture, research foundations, and international donors. Recently revenues generated from commercial activities (development income) have become a particularly important source of revenue for the research institutes. The research expenditures reported in this study include only those expenses used to directly support agricultural research. The data reported here were taken from Fan and Pardey (1992) and various publications from the State Statistical Bureau and the State Science and Technology Commission. Research expenditures and personnel numbers include those from research institutions at national, provincial, and prefecture levels and from agricultural universities. Input and output data are taken from various statistical yearbooks of the State Statistical Bureau and Ministry of Agriculture. Road density and education levels are taken from various issues of *China Transportation Yearbook, China Population Yearbook,* and *China Education Yearbook.*

India

State-level data from 1970 to 1995 were used in the model estimation. Most of the data are taken from the official sources of the Indian Government (Fan, Hazell, and Thorat 2000). The head-count ratio data used in this analysis were constructed by Datt and are published in World Bank (2000). Datt used the poverty line originally defined by the Planning Commission, and more recently endorsed by the same agency, which is based on a nutritional norm of 2,400 calories per person per day. It is defined as the level of average per capita total expenditure at which this norm is typically attained and is equal to a per capita monthly expenditure of Rs 57 at October 1973–June 1974 all-India urban prices.[6] The mean income and Gini coefficients are also taken from Datt (1998).

The agricultural productivity used for India was specialized to

$$TFP = h(RES, IR, ROADS, PVELE, LITE, GCSHEL, GERDEV, GCSSL, RAIN). \quad (5)$$

Our TFP growth index is the ratio of an aggregated output index to an aggregated input index. The following variables are included in the equation: *RES,* research stock, which is a function of current and lagged government spending in agricultural research and extension; *IR,* percentage of irrigated cropped area in total cropped area; *ROADS,* road density; *PVELE,* percentage of villages electrified; *LITE,* literacy rate of the rural population; *GCSHEL,* capital stocks of government investments in health; *GERDEV,* rural development; *GCSSL,* soil and water conservation; and *RAIN,* annual rainfall. Because of concerns that the measure of TFP used may be sensitive to the cost data used in aggregating inputs, a primal approach was also tried. By first estimating a production function for Indian agriculture using district level data, production elasticities for such key inputs as land, labor, fertilizer, machinery, and animals were obtained and then used to construct an estimate of TFP growth at the state level. The results were similar to those obtained by using the cost shares (a dual approach). But the dual approach is preferred here because the elasticities used in the primal approach do not vary by states.

The road density variable is defined as the length of road per unit of geographic area. Education is measured as the literacy rate, defined as the percentage of literate people in the total rural population above 7 years old. The irrigation variable is defined as the percentage of the total cropped area under

6. It is rather difficult to calculate back how much of this poverty line is equivalent to international dollars, measured based on purchasing power parity (PPP). The official exchange rate in 1973–74 is 7.74 rupees per dollar, which implies that the poverty line is 57/7.74/30 = 0.245 dollars per day. There was no PPP exchange rate available back then. But we do know that the ratio between the official and PPP exchange rates in 1995 is 0.21. Using this ratio, the poverty line is $1.16, measured in 1995 international dollars. But assuming the same ratio in 1973 and 1995 is not realistic and therefore this conversion should be used with caution.

irrigation. The electrification variable measures the percentage of all villages that have access to electricity. These variables were aggregated from district level data, which were obtained from the Planning Commission through the National Center for Agricultural Policy and Economics Research, New Delhi.

The food price variable is measured as the change in food prices relative to nonfood prices in urban areas. Gross domestic product and population data are from World Bank (2004) database. The world food price index is a weighted average price index for rice, wheat, and maize in the international market, and the international prices of these commodities are taken from FAO (2003).

Functional Form and Estimation Technique

We used double-logarithmic functional forms for all equations in the model. More flexible functional forms, such as the translog or quadratic, impose fewer restrictions on estimated parameters, but many coefficients are not statistically significant because of multicollinearity problems among the many interaction variables. For our system-level estimation, we used the full information maximum likelihood technique.

In China, because the urban poverty data by province are only available for six years (1992 and 1994–98), a two-step procedure was used in estimating the full equations system. The first step involved estimating the production and price functions for 1970 to 1998 and calculating predicted values of *FP* at the provincial level using the estimated parameters. The second step then involved estimation of the poverty equation using the predicted values of the *FP* variable and available poverty data for 1992 and 1994–98. The advantage of this procedure is that it uses all the information available for estimating the production function and food price equations and therefore increases the reliability of the estimates. It also avoids endogeneity problems with many of the variables in the poverty equation.

Lags and Distributions of R&D Investments

Government investments in R&D can have long lead times in affecting agricultural production, as well as long-term effects once they kick in. One of the thornier problems to resolve when including agricultural research investments in a production function concerns the choice of appropriate lag structure. Most past studies use stock variables, which are usually weighted averages of current and past government expenditures on R&D. But what weights and how many years' lag should be used in the aggregation are currently under hot debate.[7] Be-

7. Alston, Craig, and Pardey (1998) argue that research lag may be much longer than previously thought, perhaps even infinite. But this argument may be less relevant for most developing countries, as their national agricultural research systems are much younger and their research tends to be more applied and hence has shorter useful life than is true for developed nations.

cause the shape and length of these investment lags are largely unknown, we use a free-form lag structure in our analysis; that is, we included current and past government expenditures on R&D in the production function. Then we used statistical tools to test and determine the appropriate length of lag for R&D expenditure.

Various procedures have been suggested for determining the appropriate lag length. The adjusted R^2 and Akaike's Information Criteria (AIC) are often used by many economists (Greene 1993). In this study, we simply used the adjusted R^2. Because the R^2 value estimated from a simultaneous equations system does not provide the correct information on the goodness of fit of the estimated model, we used the adjusted R^2 from a single-equation approach to the production function equation. The optimal lag length is determined by the length of lag that maximizes the adjusted R^2. The AIC is similar in spirit to the adjusted R^2 in that it rewards goodness of fit, but it penalizes for the loss of degrees of freedom. The lag determined by the adjusted R^2 approach is 17 years for China and 13 years for India.

Another problem related to the estimation of the lag structure is that the independent variables ($RDE, RDE_{-1}, RDE_{-2}, \ldots, RDE_{-i}$) are often highly correlated, making the estimated coefficients statistically insignificant. Several ways of tackling this problem have been proposed. The most popular approach is to use what are called "polynomial distributed lags" (PDLs). In a PDL, the coefficients are all required to lie on a polynomial of some degree *d*. In this study, we used PDLs of degree 2. In this case, we only needed to estimate three instead of $i + 1$ parameters for the lag distribution. For more detailed information on this subject, refer to Davidson and MacKinnon (1993). Once the lengths of lags are determined, we estimate the simultaneous equations system with the PDLs and appropriate lag length for research investment.

Estimation Results for China

The estimated model for China is presented in Table 9.6. Two sets of results are reported for the poverty determination equation, corresponding to poverty lines of US$1.50 and US$2.00 per capita per day. Most of the estimated coefficients are statistically significant at the 5 percent confidence level (one-tail test) or better. Since we used double-logarithmic functional forms, the estimated coefficients are in elasticity form.

The estimated agricultural productivity equation (1) confirms that agricultural research, improved roads, irrigation, telecommunications, access to electricity, and education all contributed significantly to agricultural productivity over the sample period. The coefficient reported for agricultural R&D is the sum of the past 17 years' coefficients from the PDL distribution. The significance test is the joint *t*-test of the three parameters of the PDL.

TABLE 9.6 Estimates of the simultaneous equation system for China

				Equation											R^2
(1)	*Y*	=		0.516 *LAND* (16.25)*	+	0.104 *AK* (2.98)*	+	0.085 *RES* (3.97)*	+	0.097 *ROAD* (3.53)*	+	0.412 *IRRI* (16.39)*	+	0.458 *SCHY* (3.58)*	0.903
			+	0.071 *RTR* (5.26)*	+	0.039 *ELEC* (0.79)*	+	0.123 *RAIN* (4.92)*							
(2)	*FP*	=	−	0.430 *Y* (3.01)*	−	0.023 *GDP* (−0.16)	+	0.386 *POP* (1.50)	+	1.620 *WPI* (1.11)					0.798
(3a)	*UP*	=	−	3.54 *M* (−16.35)*	+ +	1.04 *GINI* (2.69)*	+	1.69 *FP* (4.37)*							0.908
(3b)	UP_2	=	−	2.13 *M* (−9.43)*	+	0.529 *GINI* (1.83)*	+	1.41 *FP* (5.11)*							0.916

NOTES: * indicates statistical significance at the 5 percent level; *t*-values in parentheses. The variables *Y, LAND,* and *AK* are divided by labor, so the equation is a labor productivity determination equation. The coefficient for *RES* is the sum of the coefficients for the past 17 years, and the *t*-value of the coefficient is the joint *t*-value of the coefficients for the past 17 years. The dependent variable *UP* in equation (3a) is the incidence of urban poor using the US$1.50/day poverty line, whereas the independent variable UP_2 in equation (3b) is the incidence of urban poverty using the US$2.00/day poverty line. The year and regional dummies are included in the regressions but are not reported here owing to space limitations. See the text for definitions of the variables.

The estimated food price equation (2) indicates that increases in agricultural output do exert a strong downward pressure on food prices with an elasticity of 0.43. However, per capita GDP and total population size have statistically insignificant impacts on agricultural prices. World food prices also have an insignificant impact on domestic food prices, indicating that past price policies have acted to buffer domestic prices from world price movements.

The estimated poverty equations (3a and 3b) show that food prices have a very significant impact on urban poverty, and this result holds for both the poverty lines used. For every 1 percent decline (increase) in food prices, urban poverty is reduced (increased) by 1.69 percent when the poverty line is US$1.50, and by 1.41 percent when the poverty line is US$2.00. Growth in per capita income has also contributed significantly to rapid reductions in urban poverty, while a worsening income distribution in urban areas has worked to increase urban poverty.

Estimation Results for India

The estimated model is presented in Table 9.7. Because we used double-logarithmic functional forms, the estimated coefficients are in elasticity form. The estimated agricultural productivity equation (1) confirms that agricultural research, improved roads, irrigation, access to electricity, education, and rainfall all contributed significantly to agricultural production over the sample period. The coefficient reported for agricultural R&D is the sum of the past 13 years' coefficients from the PDL distribution. The significance test is the joint *t*-test of the three parameters of the PDL.

The estimated food price equation (2) indicates that increases in agricultural output do exert a strong downward pressure on food prices with an elasticity of 0.231. Per capita GDP and total population size have positive, but statistically insignificant, impacts on agricultural prices. World food prices have a significant impact on domestic food prices, indicating that domestic urban food prices are linked with the international market.

The estimated poverty equation (3) shows that food prices have a very significant impact on urban poverty. For every 1 percent decline (increase) in food prices, urban poverty is reduced (increased) by 0.35 percent. Growth in per capita income has also contributed significantly to rapid reductions in urban poverty, while a worsening income distribution in urban areas has worked to increase urban poverty.

Contribution of Agricultural Research to Urban Poverty

By totally differentiating equations (1)–(3), the impact of government investment in agricultural R&D in year $t - i$ on poverty at year t can be derived as

$$dUP/dRDE_{-i} = (\partial UP/\partial FP)(\partial FP/\partial Y)(\partial Y/\partial RDE_{-i}). \qquad (6)$$

TABLE 9.7 Estimates of the simultaneous equation system for India

					Equation											R^2
(1)	*TFP*	=	–0.026	+	0.255 *RES*	+	0.215 *IR*	+	0.242 *ROADS*	+	0.062 *PVELE*	+	0.708 *LITE*	+	0.012 *GCSHEL*	
			(–0.78)		(1.82)*		(1.83)*		(2.43)*		(0.60)		(1.95)*		(0.39)	
				+	0.022 *GERDEV*	+	0.0015 *GCSSL*	+	0.272 *RAIN*						0.301	
					(0.63)		(0.37)		(5.47)*							
(2)	*FP*	=	0.025	–	0.231 *TFP*	+	0.112 *GDP*	+	0.034 *POP*	+	0.271 *WPI*					0.363
			(2.22)*		(–3.03)*		(1.56)		(1.67)		(8.03)*					
(3)	*UP*	=	7.07	–	1.637 *M*	+	1.003 *GINI*	+	0.350 *FP*							0.911
			(21.15)*		(–23.89)*		(15.15)*		(1.78)*							

NOTES: The estimated first equation is from Fan, Hazell, and Thorat (2000). * indicates statistical significance at the 5 percent level; t-values in parentheses. The coefficient for *RES* is the sum of the coefficients for the past 13 years, and the t-value of the coefficient is the joint t-value of the coefficients for the past 13 years. The year and regional dummies are included in the regressions but are not reported owing to space limitations. See the text for definitions of the variables.

Aggregating the total effects of all past government expenditures on R&D over the lag period gives the sum of marginal effects for any particular year. This value is equivalent to the marginal impact of a change in the "stock" of R&D investment at time *t,* where the stock *RES* is measured as

$$\text{RES}_t = a_t RE_t + a_{t-1} RE_{t-1} + \ldots + a_{t-17} RE_{t-17}, \tag{7}$$

where RE is government spending on agricultural R&D and the a_{t-i} coefficients are the estimated parameters in the production or productivity function equation.

China

When the poverty line of US$1.50 per capita per day is used, the estimated elasticity of urban poverty to agricultural research is –0.064.[8] That is, for every 1 percent increase in agricultural research investment, urban poverty declines by 0.064 percent. But with a poverty line of US$2.00, the elasticity declines to –0.053. Lowered food prices due to agricultural research accounted for 18 percent of poverty reduction over 1992–98 with a poverty line of US$1.50 but 30 percent with a poverty line of US$2.00.

Using these elasticities and the values of the relevant variables for specific periods of time, we calculated the number of poor urban people raised above the poverty line for an additional 10,000-yuan increase in the stock of agricultural research investment. Similarly, we were able to calculate the total number of urban poor who were raised above the poverty line each year as a result of actual investments in agricultural research. The results are shown in Table 9.8.

Using the results obtained with the US$1.50 poverty line, each additional 10,000-yuan increase in the 1992 stock of agricultural research raised 6.08 urban people above the poverty line. This figure had declined to 3.96 for increases in the 1998 stock of agricultural research. Given actual levels of investment in agricultural research, then, 6.27 million urban people were raised above the poverty line in 1992 and 2.96 million in 1998. This decline in poverty impact since 1992 suggests that agricultural research investments may have been even more effective in helping the urban poor prior to 1992. Unfortunately, we do not have urban poverty data from earlier years to test this proposition.

The incremental poverty reduction effects are much larger when the US$2.00 poverty line is used instead. In this case, every 10,000-yuan increase in the 1992 stock of agricultural research investment lifted 12.7 urban people out of poverty, and a similar increase in the 1998 stock of agricultural research investment lifted 7.9 urban people out of poverty. The total number of urban

8. Thirtle, Lin, and Piesse (2003) calculated a poverty reduction elasticity with respect to agricultural R&D of 0.119, which is much larger than what we obtained for both India (0.021) and China (0.064). Our estimates are for urban poverty, whereas those of Thirtle, Lin, and Piesse are for general poverty in which rural poor account for the majority. It is expected that poverty reduction elasticities of agricultural R&D are greater for rural poverty than for urban poverty.

TABLE 9.8 Impact of agricultural research on urban poverty, China

	Number of poor reduced per 10,000 yuan for a given poverty level		Total number of poor reduced (million)		Poor reduced as a share of total urban poor (%)	
Year	US $1.50/day	US $2.00/day	US $1.50/day	US $2.00/day	US $1.50/day	US $2.00/day
1992	6.08	12.78	6.27	13.19	18.27	14.81
1994	5.27	9.88	4.51	8.45	12.53	10.49
1995	4.27	9.1	3.32	7.09	11.35	9.68
1996	3.31	7.73	2.59	6.03	10.57	8.88
1997	5.05	9.86	4.01	7.83	14.57	12.26
1998	3.96	7.91	2.96	5.91	10.89	9.84

people raised above the poverty line by actual research expenditures is also much higher; 13.2 million in 1992 and 5.9 million in 1998.

The results obtained here for the urban poor are quite comparable with similar calculations of the impact of agricultural research investments on the rural poor. For example, Fan, Zhang, and Zhang (2002) have estimated that for every 10,000-yuan increase in the stock of agricultural research investment, 7.8 rural people were raised out of poverty in 1997. The large impact on rural poverty comes not only from increased agricultural productivity, but from greater nonfarm employment as a result of agricultural and nonfarm sector linkages.

India

The estimated elasticity of urban poverty to agricultural research is –0.021. That is, for every 1 percent increase in agricultural research investment, urban poverty declines by 0.021 percent. Using this elasticity and the values of the relevant variables for specific periods of time, we were able to calculate the number of poor urban people raised above the poverty line for an additional one-million-Rs increase in the stock of agricultural research investment. Similarly, we can calculate the total number of urban poor who were raised above the poverty line each year as a result of actual investments in agricultural research. The results are shown in Table 9.9.

Each additional million-Rs increase in the 1970 stock of agricultural research investment lifted 196 urban people out of poverty. This figure had declined to 72 people by 1995. Given actual levels of investment in agricultural research, then, 1.21 million urban people were raised above the poverty line in 1970 and 1.70 million in 1995. This result suggests that, although the marginal impact of agricultural research on urban poverty reduction is declining, the total number of rural poor who became non-poor as a result of agricultural research actually increased over time.

TABLE 9.9 Impact of agricultural research on urban poverty, India

Year	Number of poor reduced per million Rs (1995 price)	Total number of poor reduced (million)	Poverty reduction as a share of total urban poor (%)
1970	196.26	1.21	2.33
1971	215.87	1.32	2.59
1973	229.58	1.30	2.32
1974	166.07	1.35	2.21
1978	102.47	1.46	2.43
1983	103.03	1.57	2.53
1987	85.10	1.66	2.44
1988	73.73	1.56	2.14
1989	74.52	1.57	2.04
1990	69.99	1.61	2.24
1991	69.49	1.55	2.12
1992	79.86	1.75	2.36
1993	64.61	1.52	2.11
1994	68.66	1.61	1.99
1995	72.11	1.70	2.39

The results obtained here for the urban poor are quite comparable with similar calculations by Fan, Hazell, and Thorat (2000) of the impact of agricultural research investments on the rural poor (Fan 2003). For example, for every increase of one million Rs in the stock of agricultural research investment, 84.5 rural people were raised out of poverty in 1995. As in China, the large impact on rural poverty arises not only from the direct impact of increased agricultural productivity on the poor, but from indirect nonfarm employment effects.

Conclusions

This study has estimated the impact of agricultural research investments on urban poverty in China and India using time series data and an econometric modeling approach. The model explicitly tracks the causal links between agricultural research investments and subsequent production or productivity increases in agriculture and how these increments impact food prices and the incidence of urban poverty.

The results show that agricultural research has played an important role in reducing urban poverty in both countries. Without investments in agricultural research, urban poverty in China and India would be much higher today. In China, each 10,000-yuan increase in the stock of agricultural research investment raises about as many urban as rural people above the poverty line. The strength of this impact has declined over time, as per capita incomes have risen and food has become a less dominant item in most households' budgets. But with rapid urbanization, agricultural research will still need to play a key role in supplying adequate food at affordable prices to ensure that urban and rural poverty remain low.

In India, each million-Rs increase in the stock of agricultural research investment also raises about as many urban as rural people above the poverty line. With rapid urbanization, agricultural research will still need to play a key role in supplying adequate food at affordable prices to ensure that urban and rural poverty remain low.

But agricultural research investment in China and India has stagnated in recent years, both in terms of real expenditure and as a share of AgGDP. By the late 1990s, government investment in agricultural research as a percentage of AgGDP was only about 0.3 percent in China and 0.4 percent in India. These are extremely low compared with the 2–4 percent typical of many developed countries, and are even lower than 0.5–1 percent found in many other developing countries. One possible result of this stagnation in investment was that both rural and urban poverty declined at a slower rate in the 1990s than in the 1970s and 1980s.

Today the urban poor account for a quarter of the total poor in both China and India. It is projected that more than half the Indian population will reside in urban cities by 2030, and the poor will be urbanized faster than the general population (Ravallion 2000). China is expected to be even more urbanized than India, with 40 percent of its population currently residing in urban centers and a projected 60 percent urbanized in 2030 (Cao 2003). China and India have achieved great successes in feeding their large and growing populations and in reducing both rural and urban poverty during recent decades through government investments in agricultural research, rural infrastructure, and education. But neither country can afford to become complacent. Continued government support for these investments is still needed, otherwise food insecurity, malnutrition, poverty, and social conflict will shadow many people in both countries for a long time to come.

Acknowledgments

The author is grateful for research support provided by the International Center for Agricultural and Rural Development of the Chinese Academy of Agricultural Sciences. The author also thanks Cheng Fang and Xiaobo Zhang for allowing me to use some parts of our collaborative work in preparing this chapter. Peter Hazell encouraged the merging of the China and India case studies and provided editorial assistance. Funding from the Standing Panel on Impact Assessment of the CGIAR, the Natural Science Foundation of China (No. 70525003), and the Australian Centre for International Agricultural Research is acknowledged.

References

Ahluwalia, M. S. 1985. Rural poverty, agricultural production, and prices: A reexamination. In *Agricultural change and rural poverty,* J. Mellor and G. Desai, eds. Baltimore: Johns Hopkins University Press.

Akino, M., and Y. Hayami. 1975. Efficiency and equity in public research: Rice breeding in Japan's economic development. *American Journal of Agricultural Economics* 57 (1): 1–10.

Alston, J., B. Craig, and P. Pardey. 1998. *Dynamics in the creation and depreciation of knowledge, and the returns to research.* EPTD Discussion Paper 35. Washington, D.C.: International Food Policy Research Institute.

Alston, J., C. Chan-Kang, M. Marra, P. Pardey, and T. Wyatt. 2000. *A meta-analysis of rates of return to agricultural R&D, Ex Pede Herculem?* IFPRI Research Report 113, Washington, D.C.: International Food Policy Research Institute.

Cao, G.-Y. 2003. *The future population of China: New projections.* Laxenburg, Austria: International Institute for Applied Systems Analysis (IIASA).

Datt, G. 1998. *Poverty in India and Indian states: An update.* FCND Discussion Paper 47. Washington, D.C.: International Food Policy Research Institute.

Datt, G., and M. Ravallion. 1998. Why have some Indian states done better than others at reducing rural poverty? *Economica* 65: 17–38.

Davidson, R., and J. MacKinnon. 1993. *Estimation and inference in econometrics.* New York and London: Oxford University Press.

Deininger, K., and L. Squire. 1996. A new data set measuring income inequality. *World Bank Economic Review* 10: 565–591.

Evenson, R., C. Pray, and M. Rosegrant. 1998. *Agricultural research and productivity growth in India.* IFPRI Research Report 109. Washington, D.C.: International Food Policy Research Institute.

Fan, S. 1991. Effects of technological change and institutional reform on production growth in Chinese agriculture. *American Journal of Agricultural Economics* 73: 266–275.

———. 2003. Agricultural research and urban poverty in India. *Quarterly Journal of International Agriculture* 42: 63–78.

Fan, S., and P. G. Pardey. 1992. *Agricultural research in China: Its institutional development and impact.* The Hague: International Service for National Agricultural Research.

———. 1997. Research, productivity, and output growth in Chinese agriculture. *Journal of Development Economics* 53: 115–137.

Fan, S., C. Fang, and X. Zhang. 2003. Agricultural research and urban poverty: The case of China. *World Development* 31 (4): 733–741.

Fan, S., P. Hazell, and S. Thorat. 2000. Government spending, agricultural growth and poverty in rural India. *American Journal of Agricultural Economics* 82 (4): 1038–1051.

Fan, S., L. Zhang, and X. Zhang. 2002. *Growth, inequality, and poverty in rural China: The role of public investment.* IFPRI Research Report 125. Washington, D.C.: International Food Policy Research Institute.

FAO (Food and Agriculture Organization of the United Nations). 2003. FAOSTAT. <http://faostat.fao.org/default.htm> (accessed July 2003).

Greene, W. H. 1993. *Econometric analysis.* New York: Prentice-Hall.
Haddad, L., M. T. Ruel, and J. L. Garret. 1999. *Are urban poverty and undernutrition growing? Some newly assembled evidence.* FCND Discussion Paper 63. Washington, D.C.: International Food Policy Research Institute.
Hazell, P., and L. Haddad. 2001. *Agricultural research and poverty reduction.* IFPRI 2020 Vision Discussion Paper 36. Washington, D.C.: International Food Policy Research Institute.
Irz, X., L. Lin, C. Thirtle, and S. Wiggins. 2001. Agricultural productivity growth and poverty alleviation. *Development Policy Review* 19 (4): 449–466.
Kerr, J., and S. Kolavalli. 1999. *Impact of agricultural research on poverty alleviation: Conceptual framework with illustrations from literature.* EPTD Discussion Paper 56. Washington, D.C.: International Food Policy Research Institute.
Mellor, J. W. 1975. *The impact of new agricultural technology on employment and income distribution—Concepts and policy.* Occasional Paper 2. Washington, D.C.: U.S. Agency for International Development.
Pinstrup-Andersen, P. 1979. Modern agricultural technology and income distribution: The market price effect. *European Review of Agricultural Eco*nomics 6 (1): 17–46.
Ravallion, M. 2000. *On the urbanization of poverty.* World Bank Policy Research Paper 2586. Washington, D.C.: World Bank.
Ravallion, M., and G. Datt. 1996. How important to India's poor is the sectoral composition of economic growth? *World Bank Economic Review* 10 (1): 1–25.
Scobie, G., and R. Posada. 1978. The impact of technological change on income distribution: The case of rice in Colombia. *American Journal of Agricultural Economics* 60 (1): 85–92.
State Statistical Bureau. 2004. *Poverty Monitoring Report of Rural China.* Beijing: China Statistical Publishing House.
State Statistical Bureau and State Science and Technology Commission. 1992–2004. *China science and technology statistical yearbook.* Beijing: China Statistical Publishing House.
Thirtle, C., L. Lin, and J. Piesse. 2003. The impact of research-led agricultural productivity growth on poverty reduction in Africa, Asia, and Latin America. *World Development* 31 (12): 1959–1975.
World Bank. 2000. *India: Policies to reduce poverty and accelerate sustainable development.* Washington, D.C.: World Bank.
———. 2004. *World development indicators 2004.* Washington, D.C.: World Bank. CD-ROM.

10 Impacts of Agricultural Research on Poverty: Synthesis of Findings and Implications for Future Directions

RUTH MEINZEN-DICK AND MICHELLE ADATO

This chapter reviews the results of all seven case studies presented in the volume in terms of the evidence pertaining to the impacts of technology on poverty. We begin by examining the rate, pattern, and determinants of adoption and the ways in which dissemination approaches affect adoption, and then turn to the evidence on quantitative and qualitative impacts on productivity, income, and welfare of poor households and individuals. The impact of agricultural technology on poverty is affected by the rate and pattern of the adoption of that technology. If an agricultural technology is not adopted, it is unlikely to have an effect. But the pattern of adoption (who adopts it, when, and for what) is also likely to affect the distribution of benefits and costs. Therefore the case studies explicitly investigated the rate and pattern of adoption of the new technologies, with particular attention to the interaction with vulnerability, assets, and mediating institutions. Early on, it became clear from the qualitative work that the rate and pattern of adoption is affected by who is doing the dissemination, the methods used, and how people respond to them. The impacts of the technology are direct (for those who adopt) and indirect (for those who adopt and for some of those who do not), and are mediated by vulnerability, the institutional environment, and social status. Some benefits and costs are quantifiable and others are not. Expected benefits and costs, in turn, affect the likelihood of adoption, thus completing the circle (though complicating the identification of causality).

Finally, we turn to implications of this study for methods of impact assessment and implications for how agricultural research can have a greater effect on poverty. These implications include the priorities and conduct of the agricultural research, the partnerships formed for innovation and dissemination, and developing a learning culture within research systems.

Rate, Pattern, and Determinants of Technology Adoption

Farmers experience directly the effects of agricultural research when they adopt the resulting technologies.[1] It is thus important to examine facilitating factors

1. As noted in Chapter 1, the term "technologies" describes the output of agricultural research,

and barriers to adoption to see how the benefits are distributed between better-off and worse-off households, and between different household members. This examination helps identify factors that might include or exclude various groups from direct benefits. Three main sets of factors are likely to affect adoption: (1) whether the technologies are anticipated by potential adopters to increase or decrease their production, profits, and vulnerability; (2) whether the farmers have the requisite assets to make technology adoption worthwhile; and (3) the nature of mediating institutions, including the extent to which they represent the interests of poor people, and people's attitudes toward the institutions. Permeating all three areas are relationships of gender, class, and power that can help explain the status of individuals with respect to vulnerability, assets, and institutions, and whether they benefit from the technologies. Table 10.1 summarizes factors affecting adoption in the micro-level case studies.

Vulnerability

One of the most striking characteristics of poor people's lives is not just their low income, but their vulnerability to many hazards, including loss of income, health, and even basic safety (Hoddinott and Quisumbing 2003a,b; Skoufias 2003). Many of the case studies found that concerns about vulnerability—and whether agricultural research will increase or decrease it—were significant determinants of adoption. In assessing constraints and outcomes, the case studies examined both self-subjective and external, "objective" assessments of vulnerability as potential factors affecting adoption. Subjective assessments of risk affect directly behavior and adoption decisions because they incorporate what potential adopters believe are risks, and how farmers trade off risks against potential benefits. This is not to say that poor people will not adopt technologies they perceive as risky if the perceived benefits are high enough, but rather that average or potential outputs are not always the main factor considered or prioritized.

Some of the new technologies in the case studies increased vulnerability in some respect. This trend was particularly evident for polyculture fishponds in Bangladesh, in which owners reported losing their season's investment—as well as fear of losing their investment—because of many different factors, including diseases of the fish and even poisoning of the pond by others. In Mexico new maize varieties are a source of uncertainty in terms of how they will perform under the farmer's particular conditions. Many farmers do not adopt until they have seen a variety growing on fields nearby. In both Mexico and Zimbabwe, dependence on the market to get improved seed each year is another source of vulnerability for people who may not always have the cash necessary to buy seed or cannot count on traders having good-quality seed. Such

but it is broadly defined to include not only physical technologies but also germplasm and management practices.

TABLE 10.1 Factors affecting technology adoption in the five micro-level case studies

Case studies	Adoption
Bangladesh (rice)	Assets: main asset required to adopt MVs was water control. Institutions: recognizing the need to liberalize imports of small water pumps overcame this potential asset constraint to adoption.
Bangladesh (fishponds and vegetables)	Vulnerability: concerns inhibiting adoption included disease of fish, deliberate poisoning of pond. Vulnerability: vegetable production reduced physical vulnerability of women as they do not have to go outside the homestead to undertake agricultural activities. Assets: group ponds tried to overcome lack of private ownership of natural capital as a constraint to adoption.
Mexico (maize)	Vulnerability: vulnerabilities perceived in trying new varieties of improved maize without observing performance, and in certain traits of hybrids and landraces. Vulnerability: creolization reduced vulnerability. Institutions: low level of trust in government seed and assistance; high trust in social networks.
Zimbabwe (maize)	Vulnerability: dependence on market for improved seed increased vulnerability because of unreliable market access and cash flow problems. Vulnerability: concern over accusations of witchcraft from observing neighbors' fields or sharing information on yields and income. Assets: men's access to financial assets and formal marketing institutions made them more likely than women to adopt HYVs. Women preferred OPVs, where seeds and markets are accessed through their informal networks, giving them a degree of independence from male control. Institutions: seed companies were resented for promoting the new varieties by withdrawing older ones.
Kenya (SFR)	Vulnerability: adoption of labor-intensive SFR increased susceptibility to labor shortages in the context of AIDS. Vulnerability: SFR reduced concerns about "spoiling the soil." Assets: biomass transfer did not require much landownership. Assets: education was not necessary for adoption-specific knowledge to be transferred. Institutions: mixed experience with groups—some increases in capacities, power, and social cohesion, but also existing power relationships that work against the poor were reproduced in the context of the new technology adoption function of groups.

NOTES: MV, modern variety; HYV, high-yield variety; OPV, open-pollinated variety; SFR, soil fertility replenishment.

conditions favor open-pollinated varieties (OPVs), which can be replanted without a significant reduction in yield, over hybrids that must be purchased every season. Poor seed quality from the distributing nongovernmental organization (NGO) was also a constraint to vegetable production in Bangladesh. Zimbabwean farmers' perceptions of their vulnerability to witchcraft—or their fear of being accused of using witchcraft as a result of showing too much interest in their neighbors' field practices—provided a constraint to diffusion of new technology through farmers learning from their neighbors. The institutional arrangements of group fishponds in Bangladesh introduced another source of vulnerability: production became susceptible to intragroup conflict. Worry about increased vulnerability in the context of diseases that reduce the labor available for farm activities was also cited as a factor restricting the adoption of labor-intensive technologies (such as the case of soil fertility replenishment [SFR] in Kenya), especially in areas with a high prevalence of AIDS.

Other technologies reduced vulnerability. Creolized maize varieties were seen as more resistant to local stresses. Agroforestry alternatives to chemical fertilizer reduced cash input requirements and farmers' concerns about "spoiling the soil." Modern rice varieties changed the seasonal pattern of rice production in Bangladesh, thereby reducing the length of the "hungry season" before the first major harvest of the year. Women in Bangladesh valued the vegetable program because it increased productive employment around the homestead, thereby reducing their vulnerability to harassment from going outside the homestead for employment.

The implication of these findings is that agricultural research must look beyond increasing average productivity if the goal is for the poor to adopt and fully benefit from the technologies. For example, stable yields may be more important than higher but more volatile yields. Agricultural research now pays considerable attention to adaptation of technology to biophysical sources of vulnerability (for example, drought and pest resistance, rice varieties for deep water conditions in Bangladesh), but the institutional, social, and economic factors that increase vulnerability are not always considered. Dealing with these issues might require technologies that reduce dependence on purchased inputs or a focus on strengthening supporting institutions, especially those that facilitate access to effective risk-coping instruments. These include formal sector credit, crop insurance, and safety nets as well as microfinance institutions, and even customary mutual assistance societies. Where the sources of vulnerability are strongly related to farmers' assessments of vulnerability that may be changed through access to information, information campaigns may be helpful in reducing concerns and enhancing people's sense of security in adoption.

Assets

Poor people generally have fewer assets than the non-poor. Thus agricultural technologies that require a high level of assets to adopt are more likely to exclude the poor from direct benefits. All agricultural production requires some combination of assets, but those technologies that build upon the assets that the poor are likely to have are more likely to be adopted by the poor than those that require large lump-sum investments.

Rice in Bangladesh presents a classic case of a technology that does not require a large number of assets to adopt. Although land is needed to grow rice, modern varieties (MVs) can be adopted on any size holding. However, MVs require more labor than traditional varieties (TVs). As a result, smaller farmers had a higher adoption rate than did larger farmers, as they were likely to have relatively more access to more motivated (that is, family) labor. Because MVs did not require long-term investment, even tenants could adopt. However, water control is usually required, which favored those farms at higher elevations, provided they had irrigation pumps (physical assets). Here the government's policy to liberalize imports of small pumps reduced the lumpiness of irrigation investments, and the expansion of water markets increased smallholder access to water control, enabling them to adopt MVs.

Comparing the outcomes of the vegetable and fishpond experiences in Bangladesh shows how the asset threshold requirements of a technology affect whether it is adopted primarily by the rich or by the poor. Improved vegetables were disseminated to poor women, who could grow them on their homestead land. As even households with no agricultural land have some homestead land, very poor families could participate. In contrast, one of the fishpond programs focused on those with private fishponds, who tended to be non-poor. Moreover, homestead land is more under women's control. Farmland (including fishponds) is more likely to be under men's control. Hence the vegetable program reached women, whereas control of output of the private fishpond program went mostly to men. Given the gendered nature of poverty in Bangladesh, these differences in control over assets and technology are important.

The agroforestry program in Kenya and the group fishpond program in Bangladesh provided alternatives to large private landholdings for technology adoption. In Kenya, small farms might not have enough land to devote to trees for SFR, but biomass transfer allowed even those with little land to cut leaves from shrubs growing alongside roads and other public land to use on their fields. This practice, however, required considerable labor, which the poor could not always supply. Group fishponds substitute social capital (a proxy for which is the strength and functioning of group membership) for ownership of natural capital (land), thereby allowing landless women to adopt the technology, provided the groups could be sustained. However, difficulties with the technology itself or the organizations disseminating the technology could cause groups to fall apart.

Although natural capital assets like land and water are the most obvious factors affecting the decision to adopt technologies, other assets also play a major role. Financial capital is needed for any purchased inputs, which favors those with savings, credit, or remittances. Physical capital includes not only pump sets, but also access to roads and other infrastructure, which affects access to markets and even to information. Social capital may also play a role, as in the group fishponds (for example, the propensity for collective action), groups for collective nurseries and technology dissemination in the Kenyan agroforestry case (for example, the transfer of information), or social networks for seed exchange (for example, to reduce transaction costs and increase trust) in Mexico. Again, differences in control over assets within the household may be significant. In Zimbabwe, men were more likely to adopt hybrids because they had more access to cash and markets, whereas women's social networks gave them an advantage in obtaining seed for OPVs.

Human capital includes both labor and knowledge. The poor are often assumed to be "labor surplus," but the case studies indicate that this is not always true. Lack of able-bodied adults was often cited as a reason that households were poor, and this situation will increase with the spread of AIDS and other diseases. Furthermore, poor households are often involved in multiple livelihood activities, thereby reducing their labor availability for intensive farming operations. Although this shortage of labor may exclude extremely poor households from adopting new technologies, the quantitative analyses of the five micro-oriented case studies did not find labor scarcity to be a major reason for the poor not to adopt technologies. However, the widespread diversification of livelihood strategies found across our case studies, and across wealth categories within each case, signals an important change in the role of agriculture, to which research institutions must adapt. Instead of full-time farm families, we find households with multiple activities (and in the Kenya case, lack of interest in farming among young people). These conditions can create constraints to some kinds of agricultural intensification, for example, through time constraints and lack of continuity of presence on the farm, but they also create opportunities, such as access to cash for input purchases, risk diversification, and information.

Quantitative analyses often use schooling attainment as a proxy for the knowledge dimension of human capital. By combining qualitative and quantitative analyses, the micro-oriented case studies showed where the level of formal schooling was an inadequate measure of the knowledge needed to adopt and how formal education might indeed play a role. In Kenya, schooling attainment did not have a significant effect, because disseminating institutions made efforts to explain the technology in the simplest possible terms. The Zimbabwe study revealed generational differences in the way youth and their parents obtained information about new technologies: the older generation relied on direct observation and practical experience, whereas the youth relied on advertisements, contact with extension agents, and more theoretical learning.

Attention to the assets needed to adopt particular technologies can help agricultural research provide direct benefits to the poor. Even within the household, considerable differences in control over assets, between men and women, and younger and older generations, can affect who adopts and benefits from the technologies. The micro-oriented case studies indicate that aspects of the technology itself or the accompanying policies and institutions that lower the amount of land, education, or cash required to adopt the technology or allow substitution of one asset for another (for example, collective action or labor for land) can help the poor to adopt the technologies. For effective poverty reduction, it is essential for agricultural research and technology dissemination to engage with other sectoral programs that help to build the assets of the poor, including education, health, infrastructure, and microfinance.

Mediating Institutions

Policies, institutions, and social and political processes shape how people gain access to various assets and use them to create livelihood strategies. The relevant institutions here involve a combination of governmental policies, governmental and nongovernmental systems for agricultural extension, cultural norms, power relations, gender roles, land tenure, markets for inputs and outputs, labor relations, public goods, financial services, local governance, and local organizations. Although this synthesis cannot capture the diversity of effects discussed in the individual case studies, we highlight some key factors here.

It is not only agricultural policies that influence the adoption and impact of agricultural research. In Bangladesh, liberalization of imports led to increased availability of small pump sets, which was key to the widespread adoption of MVs. A range of political processes in Zimbabwe has mediated the impacts of agricultural technology in several ways, and in turn, technology was politicized. Postindependence resettlement projects provided people with land that, together with technology packages, facilitated adoption. However, the eventual decline in governmental investment in agriculture, first in maize breeding and then in the national Agritex extension service, and the increasing role of the private sector, directed the priorities for agricultural research and extension toward the needs of larger-scale, non-poor producers. A shift in Agritex priorities toward cash-croppers and the volatile political climate of recent years have lead to a mistrustful atmosphere, where some farmers in one of the study areas viewed the phasing out of the older, "more reliable" varieties and replacing them with the newer varieties as a conspiracy between Agritex and the private sector to discredit the government.

Power relations relevant to adoption play out between farmers and outsiders, within communities, and even within households. Traders who supply seed, private sector breeders, and government or other extension agents can have considerable power over smallholders, pushing the adoption of, or restricting access to, particular technologies (for example, Seed Co withdrawing favored hybrid varieties in Zimbabwe so that farmers felt forced to adopt new varieties). Intracommunity power relations in Kenyan villages were reproduced among farmer groups organized for agroforestry promotion. In both the Zimbabwean maize and Bangladesh fish and vegetable cases, men were dominant in the households, but women could strengthen their standing if they had control over some aspects of production via informal social networks or NGO efforts.

Culture also mediates the experience of agriculture, making new technologies more attractive or constraining people's ability to take advantage of them. Staple foods are often laden with cultural meaning and values (for example, maize in Mexico and rice in Bangladesh). Notions of an ideal good farmer in Kenya reflect and drive people's aspirations and perceptions of what they should strive for, even if myriad constraints confound their achievements. The belief in witchcraft in Zimbabwe—and more specifically, the fear of being accused of witchcraft—affected the sharing of information on maize performance among farmers. Restrictions on women's mobility in Bangladesh affect their ability to move freely outside the home, where most technology use and dissemination take place. In Mexico, participation in religious festivals is important for social status and drives poor farmers to harvest their maize early and sell the grain before the price reaches its maximum. This social requirement confers advantages to a diversity of maize varieties that can be harvested at different times.

Although specific cultural norms are not generalizable, the importance of considering how agricultural research and technologies interact with culture at different levels to affect adoption or outcomes certainly is. Who adopts the new technologies (whether women or men, elites or poorer classes, members of one ethnic group or another) is affected by culture. Fortunately, there is a wealth of ethnographic studies available that can provide a starting point for understanding the norms and values pertaining to agriculture in a given region. Many agricultural scientists with backgrounds from rural areas are also aware of some of these factors and can be encouraged to recognize their importance, rather than thinking of them as "primitive" notions.

Technology Dissemination Pathways

Dissemination pathways—how people learn about or obtain a technology—play a fundamental role in affecting who learns about new technologies and who adopts. The various case studies examined very different dissemination methods. Methods have diversified away from sole reliance on extension that uses government agents to visit individual farmers. Though these methods still exist (and were still popular with farmers in Kenya), dissemination now involves mass media (for example, radio in Zimbabwe) and a wide array of methods in which farmers are trained collectively and farmers train one another. In Zimbabwe and Kenya these methods include farmer field days, demonstration units, seminars, meetings, chief's *barazas,* and training for youth in schools. Because of their emphasis on innovation in dissemination methods, the government, NGOs, and the World Agroforestry Centre (ICRAF) in western Kenya include farmer exchanges and the formation or use of farmers', women's, and church organizations for dissemination. In Bangladesh dissemination of fishponds and vegetables mainly occurred through training and credit offered by government and NGOs. "Model farmers" and/or "adaptive research farmers" (often better-off farmers), who serve as examples to others and adapt new technologies to local conditions, were important in the Bangladesh rice, Zimbabwe, and Kenya cases. In Mexico formal dissemination was mainly limited to governmental distribution of seed and provision of advice, with some participation of the private sector. In all five cases, informal methods of exchange and learning among farmers played a large role in dissemination (with the exception of one region of Zimbabwe).

Collectively the findings reinforce the notion that there is no one best method for dissemination. Rather, a diversity of methods is preferred by farmers, and indeed is needed to reach different types of farmers. Although international and national agricultural research centers may not have the mandate of disseminating the technologies, if they want the results of their research to reach the poor, thought should be given to outreach channels and the kinds of partnerships to be developed with government agencies, private sector, NGOs, and farmers' associations. Thus it is important to conduct sufficient research ex ante on potential dissemination options—and on the local culture and power relations in which they will be embedded—before determining the most appropriate means of dissemination. In addition the findings

1. highlight the importance of trust in facilitating or hindering effective dissemination and affecting subjective assessments of vulnerability from adoption;
2. illustrate the extensive use of formal local organizations—NGOs, user groups, and community-based organizations in general—in sharing and screening information;
3. demonstrate the widespread use of informal social networks for sharing of experiences; and
4. confirm the potential of farmer participation in the technology development process as a way of enhancing dissemination.

Trust

In both Bangladesh cases and in Mexico, the case studies found a low level of confidence in public agencies and public officials in general, including those responsible for dissemination of agricultural technologies. In the two Bangladesh case studies, governmental extension agents are seen as uninterested and not reaching the poor, especially poor women: "The government officers are just there for their own interests. They sit in their offices but they don't come to us" (Chapter 4, p. 124).

In Zimbabwe, trust in the government was high during dissemination of the first generation of maize in the early 1980s, because the government was dedicated to providing an enabling environment for small-scale commercial production and devoted resources accordingly. However, in the later period, several factors combined to lower farmers' assessments of the government. First, government is seen to have narrowed its concerns to better-off farmers. Second, in one region Agritex officers are perceived to have only impractical "book" knowledge of maize cultivation (though younger farmers tended to trust the knowledge of these officers more than did older farmers, who trusted their own experience more). Third, recent political instability has created more distrust in general. Women were not given resettlement land in their own right, and they have been excluded from Zimbabwean government dissemination channels, with men operating in the public sphere, attending dissemination activities, and otherwise taking responsibility for commercial maize production. Women expressed preference for OPVs, for which they obtain seed and sell maize through their informal networks and which do not require obtaining loans for fertilizer, because women do not have access to these credit markets. Government did not provide extension for OPVs, and in fact it was unlawful to plant these varieties for many years.[2]

In Mexico government extension services are widely criticized for arriving late or not at all. "There is no faith in the government now, because they don't come through with what they promise. . . . The support comes so late that nothing can be done" (Chapter 7, p. 253). In addition, government seed is seen to be of bad quality and not worth paying for, and the distribution of seed and agricultural support has been politicized; that is, they are rewards for political support. These findings underscore the point that negative perceptions endure over time, and understanding local history is important to understanding adoption and impact. People are influenced by what came before—for example, in Mexico bad experience with government seeds, and in Zimbabwe, loan defaults where the purchase of fertilizer was followed by drought.

Even where new technologies or systems have resolved earlier problems, people are often not willing to take another chance. Because history and experience are not readily brought out through the livelihoods framework, it is important to use this framework in conjunction with other modes of analysis, or introduce additional useful concepts as needed. If government remains a major source of dissemination of agricultural technologies, then the generalized lack of trust in government found across the case studies is problematic. Only in Kenya was distrust of government not a significant issue, where the

2. The particularly volatile political climate of recent years has undoubtedly further complicated relationships between the government and farmers, but determining the relation between these developments and improved maize was beyond the scope of this study. It also would have been difficult for fieldworkers to explore this issue directly. Even without this line of questioning, they were forced to leave the field a month early because of rumors related to their perceived political objectives that threatened their safety.

Ministry of Agriculture and Rural Development (MoARD)—heavily involved in the dissemination of SFR technologies—was rated highly by many farmers. This exception may be because MoARD has developed innovations in dissemination, involving networks of catchment committees. In Kenya successful efforts were made by government (and NGOs and ICRAF) to disseminate to women. Such specific dissemination efforts are important because where there are intrahousehold differences in control over resources, who has access to technologies matters for individual welfare outcomes. Combining qualitative and quantitative data helps to understand this dynamic: the survey found that women were active adopters of improved fallows in pilot villages, but men were the main adopters in nonpilot villages, where male-headed and wealthier households were significantly more likely to adopt. The qualitative study showed that this difference was due to the greater efforts to reach women and poor households in the pilot sites. In the nonpilot villages, men were more likely to learn about the technologies because they had more social connections, ability to travel, and exposure to other development work through exchange visits facilitated by different organizations within and outside the village.

NGOs had a better reputation than government among farmers who had experience with them, particularly in Bangladesh, where NGOs play a large role in dissemination. Poor groups in both Bangladesh cases reported being reached by NGOs. However, in the vegetable and fishpond case, some farmers said the very poor were excluded due to insufficient resources, and that lack of social connections and education discouraged the participation of very poor people in such organizations. It is clear that access to assets and power enable non-poor farmers to join and influence organizations. In both cases NGOs were not viewed entirely favorably; rather, their performance was highly variable in terms of competence, integrity, and operating style. Some were seen to be particularly unfair with regard to giving credit, by disbursing more easily to favored people. They were also said by some very poor women to treat people unequally: "they only give seeds and loans to people with whom they have a good relationship" (Chapter 4, p. 125). In the vegetable and fishpond study, participation in NGOs was limited by having small children at home and by small household size, because participating in organizations takes time. Nevertheless, the women who did participate were the major beneficiaries of these programs. Membership in NGOs was found to increase women's confidence because of the solidarity of the group, the new status and freedom of movement, and heightened political consciousness, as reflected in both the focus groups and the survey results. However, when a fishpond did fail because of inadequate NGO supervision, it was felt to be very disempowering for participating women. Thus for an NGO to help the poor requires attention not only to the technologies, but also to organizational issues, including the operation of the NGO itself and the farmer groups with which it works.

The other study in which NGOs featured prominently was in Kenya, where groups organized by NGOs and other institutions were said to have provided new social solidarity and confidence among some participants. The Kenya Woodfuel Agroforestry Programme, run by an NGO, got the highest review of the SFR disseminating organizations. However, NGOs in Kenya were also criticized for providing insufficient support and leaving too early. The timing of a disseminating institution's decision to exit should be carefully assessed, as this was a widespread local concern: "what limits full implementation is that [farmers] are usually left before [they are] standing on their feet" (Chapter 5, p. 178).

The private sector is also involved in dissemination, mainly in Zimbabwe and Mexico, where they are involved in maize seed distribution. In all cases featuring the private sector, they were said to be concerned with the needs of larger, commercial or "successful" farmers, and less concerned with the needs of poor farmers. In Zimbabwe, the private sector played a large role in dissemination of the second generation of maize in the 1990s, focusing on maize traits of most concern to commercial farmers rather than the preferences of poor farmers. In Mexico, the private sector does not feature strongly, except that it is seen to provide better-quality seed than the government, though less affordable to the poor. The small farmers interviewed in the Mexico case study suspected the motives of the companies, which affected whether farmers accept advice from this sector: "They tell us that the hybrids will not produce from one year to the next. But I think that this is a lie, because the seed companies are making money" (Chapter 7, p. 254). Banking is another private sector that was viewed unfavorably by poor farmers in Mexico, requiring collateral that they do not have and blocking their access to other credit by holding farmers' farming certificates due to outstanding loans or defaults.

The Consultative Group on International Agricultural Research (CGIAR) centers and national agricultural research systems (NARS) were rarely identified by farmers as disseminators. As indicated above, farmers identify government extension services, NGOs, and the private sector as the disseminating organizations with which they have experience. The one exception was in Kenya, where ICRAF is widely recognized and highly regarded in the wide pilot dissemination area. The only criticism from farmers was for the system of adaptive research farmers. ICRAF was seen as giving too much attention to these farmers, and in one village it was criticized for leaving too soon.

Finally, choices of varieties are often made from among what is available, regardless of whether the sources are trusted or the varieties desirable. In Mexico and Zimbabwe, farmers explained that they often take what they can get. According to a Zimbabwean farmer, "We adopted the new seed varieties because our trusted variety R201 is no longer available. If it comes back from wherever it is, we will go back and grow it" (Chapter 6, p. 209).

Local Organizations

One innovation in dissemination methods involves the use of local organizations or groups. These are intended to increase the efficiency of dissemination through reaching multiple farmers, building capacity through training groups to train others, and empowering farmers through engaging them in collective endeavors (particularly for women who might not otherwise have this access). The three technologies where group-based methods were widely used were the vegetables and fishponds in Bangladesh and SFR in Kenya. In both Bangladesh and Kenya, groups met some of the above objectives with respect to efficiency, capacity, and empowerment in various ways. For example, in the Bangladesh fishpond study, one NGO disseminated fish technology to households with sufficient resources to own private fishponds, while another was able to reach the poor by facilitating the formation of groups of landless or land-poor women that could collectively rent a pond.

However, working with groups also proved complicated and problematic, as local power relationships and other social dynamics tend to be reproduced in organizations. For example, in one area of Bangladesh, only a quarter of the group members received training, other groups misappropriated funds (in part because of insufficient supervision by the NGO), and some never functioned well as a unit. Other problems raised in Bangladesh were the perceptions that groups unfairly favored some people, that many people cannot join a group because groups or ponds are not available, or people are reluctant to join. Women in particular may be reluctant to join or to leave the home for group activities. However, when they did join, women felt empowered in several ways.

The Kenya dissemination methods went the furthest in terms of innovation, in the concentration of different institutions on different methods, and in the use of local organizations. For this reason, more attention is given in this section to the Kenya case as an illustration of the benefits and drawbacks of these dissemination innovations. All the villages studied in the qualitative research on dissemination in Kenya used different forms of local groups. These groups were intended not only to disseminate technology but to strengthen human and social capital such that farmers can sustain the dissemination process inside the village and ultimately expand it to others. In practice, Kenya groups received mixed assessments, with many problems similar to those encountered in Bangladesh.

In Kenya groups were seen as a relatively important source of information. In one case, poor women said that "committee members participated very much in organizing and mobilizing farmers" (Chapter 5, p. 180). However, women also experienced problems, such as a low level of participation in groups, either because of self-exclusion, exclusion by group members, or the failure of groups to conduct dissemination with other farmers as envisioned. In general women were more positive in their evaluation of groups than were men, particularly about women's groups. Poor and non-poor women alike said that domination by men in the mixed groups reduces women's participation and learning, suggesting the importance of having separate groups for men and women. In one village, some women's, church, and welfare groups were also agriculture groups that contributed food to funerals. This joining of group functions can support social capital and address people's priorities, especially in the context of widespread HIV and AIDS. Existing groups that incorporate dissemination tended to be more active and sustainable than new groups formed solely for this purpose.

Groups in the Kenya case had mixed impacts on social capital. Some groups said that the extension activities had brought their community closer together, for example, in discussing and exchanging information about the various technologies. However, local groups also introduced or exacerbated social tensions and politics. One or more of the following issues were reported in all except one village: failure to reach farmers outside the group, uneven distribution of resources, domination of groups by farmers of greater wealth or social status, conflicts over resources, rivalry among leadership, mismanagement of funds, the ability of some to amass wealth through the process, and domination by elites. However, poor farmers did acquire some power through the process. In one village, for example, poor men said that farmers made demands on the committee and the committee in turn made demands for extension services on the government. The lesson to be derived from these mixed experiences is that group-based methods, like other development efforts that involve community participation, may have payoffs that make them worth pursuing, but they require careful attention to group dynamics and power relations to achieve effective and equitable outcomes. There is a growing literature (see Ostrom 1990; Baland and Platteau 1996; Place et al. 2002) on conditions for collective action, which can be applied to identify where group-based approaches are likely to be effective (for example, where agriculture is important to livelihoods, group sizes are manageable, a history of cooperation exists, and social divisions are not too great). Similarly, research on gender and participation highlights the importance of both formal and informal rules and incentives for ensuring women's effective participation (see Guijt and Shah 1998; Agarwal 2001; Meinzen-Dick and Zwarteveen 2001). Such rules may be just a matter of allowing both male and female household heads to be members or choosing a time and place to meet that is convenient for women; or they may involve more complex processes, such as increasing women's confidence to participate or structuring organizations in such a way that women feel free to speak out.

Informal Social Networks

Despite the importance of institutions of government, NGOs, the private sector, agricultural research institutes, and local organizations, the most consistently important dissemination institutions across the case studies were informal social networks. In Zimbabwe and Mexico, for example, farmers extensively use informal networks to get the maize seeds they preferred. Women in particular used networks to obtain OPVs that many prefer but cannot acquire officially. In Mexico and in the Bangladesh rice study most seeds come from farmers' own harvests or exchanges with neighbors. In Mexico farmers trust these seeds far more than those from the government, because they trust these informal relationships and they are able to see a new maize variety perform in the field before taking the risk of planting it. As one farmer explained (Chapter 7, p. 252):

> Sometimes the maize is unknown and you don't trust to buy it. Rather, you go with your people because you see that the crop grows well and the ears are pretty. So you ask if they have some stored and you buy a bit for planting. With the seed from the stores, there is no confidence. . . . You have to see it growing in the fields of your neighbors who have grown that variety. If not, you don't buy it.

As with trust in formal institutions, trust among farmers is important for informal methods of dissemination to function well. In Kenya conversation and observation of others' fields are key sources of dissemination. In fact, the emergence of new testers in areas of Kenya where external assistance had largely been withdrawn suggests that large concentrations of early users leads to large concentrations of new testers. However, these informal farmer-to-farmer methods cannot always be assumed to work. In one region of Zimbabwe, people deny learning by observing the fields of neighbors, because showing too much interest in your neighbors' fields can provoke accusations of witchcraft. Similar fears limit people's willingness to discuss yields and crop income with others. In this region, information from neighbors is viewed with suspicion, and information is currently disseminated in a fragmentary fashion in a climate of distrust. In this environment, informal farmer-to-farmer dissemination should not be heavily relied upon. These are important dynamics to understand prior to developing a dissemination strategy, which requires qualitative research, which in turn requires building sufficient trust to obtain information on sensitive topics such as witchcraft. This issue is not insignificant: 71 percent of the qualitative sample in this region believed that magic enhances agricultural skills. Even those who said they do not believe in magic still sometimes took measures to protect their fields from bad magic (Bourdillon, Hebinck, and Hoddinott 2002).

Farmer Participation in Breeding and Adaptation

Participatory processes in breeding have become popular in many CGIAR centers, although there were not many cases of this among our studies. The Kenya and Bangladesh studies reported that new technologies and management systems were being tested in farmers' fields and that there were forums for obtaining feedback from farmers. In Kenya this process took place through ICRAF's and CARE's use of adaptive research farmers (ARFs), individuals who are selected for testing technology under local conditions (they are supposed to be selected with widespread community participation, but this did not always occur). In terms of feedback for researchers, the study found that this system is important for adapting technology to local conditions. At the same time, the system seems to have some adverse impacts on social capital. ARFs were sometimes unpopular: they were blamed for not teaching others and were said to be using their position to gain power and prestige over other farmers, and they were resented for the amount of attention they received from outsiders: ARFs "are frequently visited and make others feel left out and different from the preferred farmers" (Chapter 5, p. 182). As in the case of local organizations, technology dissemination articulates with local social systems and can have adverse effects on social capital and on how local people respond to outside organizations in the future. This cascade of effects again underscores the importance of ex ante research to understand and better plan for such social dynamics.

In the villages studied in Mexico, there was no evidence of farmer participation in the formal work of plant breeders. However, the Mexico study is the best example of farmer breeding and experimentation, in the sense that—intentionally or by accident—farmers were continuously crossing maize and developing creolized varieties with traits that they valued. As explained by one farmer (Chapter 7, p. 255):

> A year ago, I planted the one we call "*tablita*" in one plot and in another together with another variety. But if I cross it now with 526 it produces half yellowish grains and the ear is a little bit narrower . . . but it became stronger. That is what we want—to cross a *criollo* with a variety to make it more resistant, so that it doesn't rot much.

Whereas CIMMYT (Centro Internacional de Mejoramiento de Maíz y Trigo/International Maize and Wheat Improvement Center) and INIFAP (Instituto Nacional de Investigaciones Forestales, Agricolas y Pecuarias) had bred mainly for height and yield, the case study of farmers' informal crossing of varieties revealed the many other traits that farmers value with respect to their vulnerability context, asset base, and cultural preferences. The challenge for plant breeders in Mexico is to understand these adaptation processes to better learn about what farmers want. It may be that for open-pollinated crops, instead of trying to develop finished cultivars, it would be better to develop a strategy of releasing improved crop populations containing desirable traits and disseminating more information about how farmers themselves can cross these materials with their own genetic stock to select for combinations of traits.

In other cases, farmer participation in selection of materials or even setting of priorities for new varieties could lead to more useful (or more trusted) materials, but it is critical to identify which farmers are able to give input. The Zimbabwe example shows that breeding for large farmers may not meet the needs of smallholders. If only men are included in the participatory research groups, then women's preference for varieties that have certain traits for cooking and taste, or that reduce specific vulnerabilities, may be missed.

This approach is consistent with the agricultural knowledge and information systems approach promoted by Röling (1988) that focuses on a system of actors and how they interacted to generate, transform, transmit, and utilize knowledge and information. Thus, rather than a pipeline or linear transfer of technology paradigm for agricultural research, in which international centers develop materials that are given to national research programs for adaptation and then passed on to extension systems to deliver to farmers, there is a move toward an innovations systems approach of developing partnerships among farmers, their organizations, universities, agriculture-based industries, public and private sector researchers, extension and training institutions, agricultural press, and information services (Hall et al. 2001).

Impacts on Poverty, Vulnerability, and Well-Being

Agricultural research can address poverty through direct effects on farm households that adopt the resulting technologies and through the indirect effects on the wider population. Some direct impacts, such as changes in agricultural productivity and farm income, are generally easier to measure; these have been the focus of much impact research. Other direct effects, such as social capital formation and changes in power relationships, are less easy to assess and are rarely evaluated. Direct effects are experienced by the adopters, but adoption may have a much broader set of indirect impacts on adopters and nonadopters. Indirect effects include lower food prices, more off-farm employment opportunities, and higher wage rates in those nonfarm activities; they may also include negative indirect effects, such as displacement of women's activities, the inability of nonadopting poorer farmers to compete, social tensions exacerbated by dissemination activities, or environmental degradation. As households diversify away from agriculture as the mainstay of their livelihoods, indirect effects are likely to become more important. The seven case studies provide evidence on a range of both direct and indirect impacts on poverty outcomes (for a summary of these effects, see Table 10.2). However, it would be useful to have further research to evaluate the relative importance of direct and indirect impacts of different technologies and dissemination methods.

Direct Impacts

AGRICULTURAL PRODUCTIVITY AND INCOMES. Of the five micro-oriented case studies, rice in Bangladesh shows the largest productivity impacts of agricultural research. Average yields increased 2.4 percent per year, from 1.52 tons

TABLE 10.2 Indirect and direct effects of agricultural technology adoption in the seven case studies

Case study	Direct effects	Indirect effects
India	Large productivity increases in rice	Large impacts on poverty—both rural and urban
China	Large productivity increases in rice	Large impacts on poverty—both rural and urban
Bangladesh (rice)	Large productivity increases due to improved varieties; income increases constrained by farm size and low price Declining soil fertility	Low price of rice important for net food purchasers Increases in employment opportunities in agriculture Improvement in working conditions in agriculture Decrease in availability of wild green leafy vegetables
Bangladesh (fishponds and vegetables)	Improved productivity of fishpond and vegetable production Small impacts on income of poor because technologies formed small part of household livelihoods and some of the private fishpond owners not poor to begin with Increased empowerment of women, when technology directed to them	Increased availability of vegetables in study sites Social capital strengthened by some groups disseminating the technology, but weakened when groups fell apart Diffusion of vegetable technologies
Mexico (maize)	Yields increased due to improved varieties, but perceived as more variable—hence creolization as an intermediate solution (reduced variability) Not perceived as route out of poverty but as providing a solid base from which to diversify Maize perceived as essential to being able to feed one's family, especially for the poorest farmers	Widespread diffusion and adaptation of improved maize via creolization led to reduction of trade-offs between traits of improved maize and landraces; increased ability to predict performance
Zimbabwe (maize)	Production and productivity gains of 20 percent Better-off farmers able to convert productivity-driven income gains to asset accumulation, increasing resilience to shocks	Built networks for information and technology demonstration for men

TABLE 10.2 *Continued*

Case study	Direct effects	Indirect effects
Kenya (SFR)	Doubling of maize productivity compared to no SFR; significantly better economic returns than with no SFR Spillovers in knowledge: improved understanding of soil fertility issues on whole farm Increased access to technology for women	Social capital strengthened by some groups disseminating the technology, particularly women's groups; use of "adaptive research farmer" created new social tensions

NOTE: SFR, soil fertility replenishment.

per hectare in 1965 to 3.48 tons per hectare by 2000–01. Not all of this increase can be attributed to MVs: during this time, TVs increased their yields to 2.14 tons per hectare (0.9 percent per year) due to improved fertilizer and water control. However, even controlling for such increases for TVs, the higher yields of MVs produced an additional 13.1 million tons in 2000 compared to the output of TVs. This difference was important due to area constraints on rice cultivation in the context of rapid population growth. In Chapter 3, Hossain et al. calculate that each year the incremental yields of MVs can feed about 59 million people, or 46 percent of the 2000 population. At the same time, the annual income gains directly from rice cultivation are not large (US$237 per year for the average farm size of 0.67 hectares, equivalent to 21 percent of total annual household income). The modest gains are mainly because of small farm sizes and the falling real price of rice—attributable, in part, to the increases in rice production. However, these same falling rice prices increase the real purchasing power of poor consumers.

The polyculture fishponds also increase fish yields to levels substantially above those realized by traditional fishpond practices and by rice cultivation (IFPRI-BIDS-INFS 1998). Adopting households sold three times as much fish per pond area as those using traditional practices. Cash profits for private fishponds averaged US$223 per hectare compared to US$147 per hectare for traditional fishponds.[3] The effect on household income is much more modest because of the long growing cycle (16 months) and because fishponds form a very small portion of the households' livelihood strategies and hence of their income

3. Comparisons for fish are based on mean productivity and profitability of fishponds for matched groups of adopters and nonadopters, controlling for pond size and NGO membership (see IFPRI-BIDS-INFS 1998; Chapter 4). Figures were reported in takas per acre and have been converted to U.S. dollars per hectare at the prevailing exchange rate of US$1 = Tk 46.5.

portfolio. Monthly income from fishponds for adopting households averaged US$1.36, compared to US$0.79 for nonadopting households—a difference that was dwarfed by income from rice and especially off-farm income. Indeed, monthly household income was slightly larger for nonadopters (US$14.61 compared to US$13.68 for adopters). For group (as opposed to private) fishponds, only five of nine operated as planned. Where they were functioning, profits averaged US$168 per hectare, but the large group size (10–27 members) meant that each member received only US$0.38 per month on average. Thus although the poor can sometimes substitute social capital for natural capital, the returns are not necessarily the same for both.

Yields and profits for improved vegetable varieties were not consistently higher than for local varieties of vegetables, but by introducing women to homestead cultivation, the programs did increase vegetable cultivation among landless and land-poor households, with cash profits averaging US$72 per hectare for each crop, or US$36 per hectare per month (US$33 profit after family labor inputs) for program households. These profits were even higher than those from high-yield varieties of rice (US$21 per hectare per crop, or US$5.25 per hectare per month) in the same location, and the vegetable growing season is only half as long as that for rice (IFPRI-BIDS-INFS 1998). But the limited size of homestead plots means that land-poor women cannot expand the area under vegetables, and these crops can supply only a small portion of household income.

The SFR techniques in the Kenya case are associated with mean increases in maize productivity over a no-nutrient control of about 55 percent per each of two or three cropping seasons following a fallow, resulting in more total production over the whole rotation at less cost. The seasonal net gains were between US$36 and US$61 per hectare, depending on species and length of rotation; returns to labor were above US$2.00 per day, which were more than 33 percent higher than the returns to the more common no-input system undertaken by poor farmers. However, the returns per household were low because of the small sizes of plots on which they were applied (averaging less than 0.05 hectares), and measurement error on the small plots led to problems in econometrically estimating the returns per hectare. The returns of soil fertility replenishment used for maize production are likely to underestimate total productivity increases because they do not include the value of the firewood produced or increased production of higher value crops, which is practiced by a subset of the sampled farmers. When the soil fertility techniques were applied to vegetable plots, the average net returns ranged between US$600 and US$1,000 per hectare.

The Zimbabwe case study focused mainly on the second-generation maize hybrids. First-generation maize hybrids were seen as very successful, even for smallholders, and were associated with the doubling of maize production during 1979–85. The second-generation hybrids, developed primarily to increase resistance to drought and diseases of concern to commercial farmers, did not provide such apparent productivity increases to smallholders. But even the new hybrids increased production and productivity per hectare by about 20 percent, with less variability in production as well. Comparing adopters and nonadopters over the period 1995–2000 within a regression framework that controls for such confounders as higher education, skills, and other assets, the mean household-level maize production increased by nearly 4 tons, yield increased by 1.5 tons, and income gained by more than US$600.[4]

Assessing the productivity impact of creolized maize is less straightforward. On the one hand, farmers reported that the improved germplasm was associated with higher yields. Yields of creolized varieties were higher than those of landraces but lower than for improved varieties that had not been creolized. The differences between varieties were small relative to the differences in yields between good and bad years and between favorable and unfavorable locations. On the other hand, yields tell only part of the story, because much of the benefit of creolized varieties derives from their lower level of yield variability and other traits. In both regions, creolized varieties present useful combinations of traits that reduce some of the trade-offs between landraces and improved germplasm.

In Oaxaca most advantages were associated with landraces (resistance to ear rot, ease of shelling, making nixtamal, and useful for pasture); however, both improved and creolized varieties were superior with respect to resistance to lodging—a key vulnerability factor in the area. Creolized varieties were superior for yield by weight to both improved varieties and landraces. In Chiapas hybrids were seen as superior in most characteristics, though creolized varieties were seen as superior in resistance to lodging compared to landraces and resistance to insects in storage with respect to hybrids—two significant sources of vulnerability. These results pertained to men. In Chiapas women's attitudes were overwhelmingly negative toward hybrids, whereas women in Oaxaca were more positive or neutral. The reasons for these results are not clear. Overall, maize production was not seen as a major route out of poverty, but it did contribute to livelihood security—primarily as essential to food security in Oaxaca and providing cash income and food security in Chiapas: "We need it to live; without it we don't eat" (Chapter 7, p. 277).

Across the case studies, small farm sizes were not a constraint to the adoption of the technologies, except for the case of private fishponds in one of the Bangladesh sites. However, those with more land and other assets tend to receive larger direct benefits from improved technologies because they are adopting them over larger areas.

The extent to which benefits accrued to women or to men depended on gender roles in agriculture and efforts to target the technologies to women. Spe-

---

4. The period studied excludes 1994–95, a drought year and the first year of adoption, when only 9 percent of farmers used the new hybrids. Income is calculated in 1992 U.S. dollars converted using the exchange rate of Z$5.1 = US$1.

cial efforts to disseminate improved vegetable and group fishpond technologies through women's groups reached landless women in Bangladesh, despite strong cultural preferences given to men and landed households. In Kenya designing dissemination materials to be simple and understandable even by those with low literacy levels allowed women with very little education to adopt the technology on a par with others.

Although smallholders were not excluded from the technologies, neither were productivity increases from the technologies themselves a major route out of poverty. This is partly because only the Bangladesh rice and first-generation maize in Zimbabwe were technologies that generated large productivity increases for staple crops, and in a context of unmet demand, the price declines were not as large as they would later become. Furthermore, small holding sizes and low returns to agriculture in general meant that the technologies themselves did not contribute greatly to household incomes. This reduced impact is particularly true of more recent staple cereal crops, for which low output prices have been due, in part, to productivity increases induced by agricultural research. Thus the opportunities for direct impacts of staple grains research on the poverty of farm households seem to be diminishing. Diversification out of agriculture is associated with larger income gains in most of the cases studied. However, income gains from agricultural technology can also facilitate diversification, as in the case of Zimbabwe, where higher maize yields of better-off farmers enabled acquisition of livestock, which reduced the vulnerability of adults and children to the effects of drought. Diversification into nonagricultural activities can also reduce vulnerability to fluctuations in income due to weather and pests.

The broader livelihoods analysis indicates that yields and incomes tell only part of the story of the impacts of agricultural research on the welfare of farm households. For many, increases in the stability of production were also very important (for example, creolized maize in Mexico). Even for households diversifying out of agriculture, the continued household production for home consumption and the generation of cash income provided needed stability and a launching pad for some members to branch out into other activities.

OTHER DIRECT IMPACTS. Some of the case studies identified other direct impacts of agricultural research that are less tangible than effects on yield and income. Adoption of knowledge-intensive practices is associated with increases in human capital skills or generalized knowledge. In Kenya, for example, increased knowledge of SFR practices carried over into better understanding of soil fertility on the whole farm. Those successfully adopting improved fishponds in Bangladesh reported seeing themselves as "scientists." Women cultivating improved vegetables in Bangladesh also reported an improved ability to deal with traders and their husbands, and survey data found statistically significant empowerment effects in terms of freedom of movement, freedom from physical violence, and political knowledge and awareness. Sharing the increases in vegetables and fish with friends and family was also reported in the qualitative results in Bangladesh. Sharing may not contribute measurably to quantitative results on household income or nutrition of the adopting households (compared to nonadopters), but it helps build social capital by strengthening ties among households—a vital asset, especially for the poor.

As discussed earlier, agricultural technologies can also affect social capital formation, particularly where technologies are disseminated through groups. Such was the case for agroforestry approaches in Kenya and vegetable and fishpond technologies in Bangladesh. Where the technology is successful and groups function effectively, we observed increases in social capital, which can have other benefits for households and communities, such as mutual insurance. However, if things did not go well with the technology dissemination, it caused strains in the community and loss of social capital. In particular, problems arose in Bangladesh when fishpond groups broke up, or where the NGO or other organization delivering the technology had technical problems or lost the trust of the community.

Indirect Impacts

Looking at the effects of new agricultural technologies only on those who adopt the technologies gives a partial picture of their impact. There may also be substantial impacts on farm and nonfarm households through direct sharing of output (for example, to neighbors and family members) and through labor, output, and food market effects.

Of the micro-oriented case studies, the study of the effects of modern rice research in Bangladesh provides the clearest evidence for the indirect impacts of agricultural research, at least in part because yield increases were large and rice was a widely grown crop. As a result there were large spillover effects on other households. The higher labor intensity of modern rice varieties increased demand for agricultural labor, thereby increasing employment of agricultural wage laborers.[5] It also contributed to an increase in the leasing out of land, so that more households could become cultivators. A shift from daily wage rates to piecework contracts for laborers and from sharecropping to fixed-rate tenancies also allowed laborers and tenants to earn more from rice. The new varieties also shifted production into the dry season, thereby reducing vulnerability by increasing employment and food availability in the "hungry season." The higher employment, combined with lower rice prices that resulted from increased production meant that the rice equivalent wage grew by 4.8 percent per year during 1987–2000.

5. In 1987, farmers used 206 person-days per hectare for MVs, compared to 142 for traditional varieties. By 2000, that difference had shrunk to 133 and 110 person-days per hectare, because of mechanization in response to growing labor shortages (Chapter 3).

In both Bangladesh cases, there were substantial increases in nonagricultural employment. Some of the growth is directly attributable to the new technologies, such as increases in transporting rice or selling fish fry (small stock). Other increases are more generally attributable to rising prosperity, to which improved agricultural productivity has contributed.

Another type of indirect effect is seen when farmers adapt and then diffuse technologies. The clearest example is found in Mexico, where farmers crossed improved germplasm with their own varieties, which spread to many farmers who would not have bought improved varieties. Similarly in Bangladesh, a few years after the dissemination of new vegetable varieties, it was difficult to distinguish adopting from nonadopting households, because the original package had been adapted in many different ways, with seeds and the knowledge about how to use them disseminated to neighbors.

But not all indirect effects have been positive. In the Bangladesh rice case study, focus groups repeatedly identified declining soil fertility as a problem caused by the intensification of rice production. Poor men and women also expressed concern about the decline in availability of wild green leafy vegetables that had grown on common land or fallows but were squeezed out by the intensification of rice culture, and about declining wild fish availability due to pesticide use. Lower output prices could also hurt nonadopting farmers (unless there is a price premium for traditional varieties).

At the national scale, the impacts of agricultural research on the poor have also been measured for India and China over the past few decades. At this scale of analysis, qualitative measures of poverty are less useful because they cannot be meaningfully aggregated across households and communities to the national level or compared over long periods of time. The India and China case studies thus relied on econometric analysis of official income-based poverty data. Strengths of the approach include an ability to track the different channels through which agricultural research and development (R&D) impacts on the poor in rural and urban areas, control for other factors that influence the outcome, analyze the sources of change over long periods of time, and compare investments in agricultural R&D to other governmental investments. Weaknesses include an inability to capture other important dimensions of poverty (for example, vulnerability, power relations) or to triangulate findings against more in-depth, micro-based evidence. One result, for example, is that the macro measures do not deal with gender dimensions of agricultural research, and hence do not direct attention to the differential needs of women and men.

The results for both countries show that agricultural research played a key role in the dramatic decline in rural poverty in the early decades of the introduction of high-yield varieties in India and China. In India, the rural poverty rate fell from about two-thirds of the rural population in the early 1960s to about one-third by the late 1980s; in China rural poverty fell from about one-third of the rural population in 1970 to about 10 percent by 1984 (Fan, Hazell, and

Thorat 1999; Chapter 8). These reductions were particularly striking given continuing rural population growth. After controlling for different factors, including a wide array of public policies and investments, agricultural research investments are shown to be one of the most important drivers of agricultural productivity growth and rural poverty reduction (Fan, Hazell, and Thorat 1999).

The importance of different contributing factors has changed over time (for example, irrigation investments are much less productive today than they were in the 1970s), but agricultural research investments continue to give high returns and favorable poverty impacts. In fact, in both countries today, additional agricultural research investments give higher productivity returns than any other public investment in rural areas, and they have very favorable poverty impacts. The size of the poverty impact of these public investments is second only to rural infrastructure and education in China and to rural roads in India. Moreover, the contributions of agricultural research are not limited to reducing rural poverty: it has also made a major contribution to reducing urban poverty. For every 1 percent increase in agricultural research investment, urban poverty declined by 0.064 percent in China and 0.021 percent in India. Such benefits are an increasingly important consideration in rapidly urbanizing developing countries.

In terms of the pathways through which agricultural research affects the poor, increases in agricultural productivity proved to be the most important in both countries. Improved productivity led to direct on-farm benefits, but also contributed to higher wages and greater employment in rural labor markets (farm and nonfarm) and lower food prices. The latter impact also reduces urban poverty (the urban poor spend approximately half of their income on food). In India there was some tendency for higher agricultural productivity to increase landlessness when, in the initial stages of the green revolution, larger farmers with better access to capital, technology, and credit bought up land, but this effect was subsequently offset by programs to increase smallholders' access to credit and technology (Hazell and Ramasamy 1991).

Additional analysis was undertaken to trace some of the benefits of the CGIAR's own research for China and India (Chapter 8). The analysis has been completed for rice, for which the parentage of rice varieties has been traced in both countries and, combined with available yield trials data and the econometric analysis reported above, has been used to calculate the share of the productivity growth and poverty impacts attributable to improved genetic material received from the International Rice Research Institute (IRRI). The results indicate that rice improvement research has contributed significantly to rice production in both countries. The annual benefits from total rice research (national plus IRRI) were about 20 percent of the annual value of national rice production in both countries during the 1980s and 1990s and exceeded total rice research investment in these countries by a factor of 10. IRRI's research made important contributions to these gains. Even using a conservative attribution rule (the geometric rule) for crediting plant variety ancestors, IRRI's research can be attributed with 1.7–6.8 percent of the annual rice research benefits in China during 1991–2000, and with 18.1–56.4 percent in India (Chapter 8). These benefits are sufficient to have paid the full costs of IRRI's global rice program more than 20 times over during the past decade.

The India and China studies (Chapter 8) indicate that rice research in India and China has helped large numbers of rural people move out of poverty. In India, about 1.5 million poor escaped poverty each year between 1991 and 1999 as a result of rice variety research, and about one-third of that improvement was due to IRRI's research. In China the number of poor who came out of poverty as a result of rice research declined over the years, from 5 million in 1991 to 1.4 million in 1999, of which only about 5 percent was attributable to IRRI's research. For every US$1 million invested by IRRI in 1999 in its global rice research program, some 800 poor people in China and 15,000 in India rose above the poverty line (poverty benefits will have been generated in other countries as well). Most of these benefits are the results of research conducted prior to 1990. IRRI's rice research investment has declined since then, and so has the corresponding growth in experimental farm yields.

Conclusions and Implications

The evidence of the seven case studies presented in this volume points to a diverse set of impacts of agricultural research on poverty. We now turn to implications of our studies for methodologies for impact assessment, and how we can learn from such impact assessments so that agricultural research can make a stronger contribution to poverty reduction in the future. What should researchers, policymakers, and disseminators learn, and what should they change, as a result of the insights and findings from this research? The conclusions and recommendations that follow speak to methods and approaches for studying poverty impact and to conducting agricultural research and dissemination.

Implications for Impact Assessment Research

Capturing the range of effects on poverty requires a research design that operates at different scales of analysis (intrahousehold, household, village, regional, and national), with different research methods more and less appropriate to these various scales. For example, measurement of indirect benefits arising from intersectoral growth linkages and less costly food requires economy-wide models and analysis, whereas measurement of household benefits within adopting regions must take into account the diversity and complexity of people's lives. The case studies in this book comprise a serious attempt to generate a body of new knowledge and understanding about the impact of agricultural research on the poor that is comprehensive in its coverage of scale, measures poverty in several economic and social dimensions, provides comparability across case studies of different types of technologies in different country situations, and rigorously controls for conditioning and confounding factors. Although not all impact studies can achieve these goals, they are standards that such studies should do their best to meet. The studies provided many insights and lessons with respect to research methods, as discussed below.

THE CHALLENGE OF ASSESSING CAUSALITY. Lack of significant attention to the design features of an evaluation can severely undermine the ability of the researchers to draw inferences about causation. Ultimately, the ability to do so rests on controlling for unobservable factors between groups that have adopted the technology and those that have not (another interesting comparison is among different dissemination pathways). Also required are samples of sufficient size, given the variability in outcomes, to detect statistically significant differences. Finally, one must be able to control for the endogeneity of variables that may affect the outcomes of interest. Ideally, from a scientific perspective, one would want to randomly allocate technology to a control and treatment group, prevent diffusion of technology between groups, and have sufficient sample size to detect a difference in key indicators. These conditions are rarely —if ever—present in practice when evaluating agricultural research, and there are ethical problems concerning supplying a useful technology to some and purposely withholding it from others. However, random allocation of new technologies can be given serious consideration, where diffusion, for logistical or financial reasons, has to be done in phases.[6]

If these "ideal" conditions cannot be attained, attention must be given to constructing a control or comparison group that is matched closely on observable and unobservable characteristics before the adoption of the technology. The effort is fairly straightforward for observable characteristics: the Bangladesh fishpond and vegetable researchers constructed a control most thoroughly by the use of a village census with such questions as "would you adopt if this technology were available?" The other four micro-level case studies in Mexico, Zimbabwe, Bangladesh, and Kenya had to explicitly model the decision to adopt and then incorporate the results in the equations used to assess the impact on poverty. This methodology was most robust when panel data were available, such as in the Zimbabwe and both Bangladesh case studies, as it allows controlling for household and community unobserved effects that remain relatively fixed over time.

The qualitative research used in-depth interviews that enabled the researcher to draw inferences about causality based on the insights and reflections people offer about their lives before and after the technology was available (or adopted). Although people themselves may not be aware of all cause-and-effect relationships or control for confounding factors, in many cases they are

6. See Skoufias and McClafferty (2003) for this approach in the context of a social program.

able to identify linkages based on their own experiences and those of people they know. Disaggregating data collection by different social groups, such as gender, class, and ethnicity, also enables inferences about the different effects of these differences. Inferences about causality can also be drawn from qualitative research comparing adopters with non- and disadopters. Because of their different strengths and weaknesses, qualitative and quantitative methods should be seen as complements, and not substitutes.

CONSIDERING DIRECT AND INDIRECT IMPACTS. The analysis of a portfolio of impacts should always be undertaken, as illustrated in Table 10.3. It is important to avoid restricting analysis to variables and impacts that can be easily quantified. If researchers had focused on direct impacts only, the studies presented in this volume would have missed food price impacts and the wage rate and employment effects that were observed for rice in Bangladesh and in the India and China case studies more generally, and effects on community-wide social capital, both positive and negative.

If the studies had focused solely on quantitative measures, researchers would have been puzzled as to why vegetable and fishpond adoption in Bangladesh was not more widespread (concern about vulnerability) and would have caught the wage changes but missed the changes in the employment arrangements in the Bangladesh rice case study. Also missed would have been the effects on women's power through vegetable growing in Bangladesh and the farmers' feelings of being scientists. The spillover from the greater attention paid by Kenyan farmers to soil fertility on non-SFR crops would not have been captured. The Mexico study of maize in particular discovered that evaluating the impact of varieties must go beyond yield, to see what other traits are valued and supplied by different varieties, the trade-offs among them, and the impact of decreasing these trade-offs. The study also reveals the importance of viewing yield in a more complex manner, differentiating yield by weight, volume, and varied usage.

But for many of the critical impacts, it is not just a question of qualitative or quantitative data, but rather better overall data quality, particularly that which captures changes over time. It is no accident that evidence of indirect effects

TABLE 10.3 Matrix of possible agricultural technology impacts (selected examples)

Impact type	Quantifiable	Qualitative
Direct	Productivity, reduced trade-offs among traits	Perceptions of vulnerability
Indirect	Price changes, wage rate changes, employment changes, agriculture-related investment opportunities	Community-wide changes in women's empowerment

was only available for those cases (notably Bangladesh rice and the China and India studies) that had data for longer times, with widespread adoption of the technologies. Panel or other comparable time series data are also important to examining the riskiness of technologies and their effects on vulnerability. Qualitative time series data (such as when anthropologists repeat studies in the same villages) can also document and explain technology-linked changes in households and communities over time.

USING LIVELIHOODS APPROACHES. Livelihoods analysis enabled researchers to think about the multiple and interactive influences on livelihoods and was a means of communicating across disciplines. The sustainable livelihoods framework draws on many concepts from other frameworks, paradigms, and disciplines with which the interdisciplinary teams of researchers were familiar to varying extents (indeed, some researchers felt that the same issues could have been covered without this framework). Nevertheless, it guaranteed that all research teams considered a very wide range of issues that are normally excluded in conventional impact assessments—helping to ensure that important explanatory factors were not overlooked and enabling comparability of issues and results across the case studies. This approach implies a willingness to acknowledge that livelihoods—and the processes that make interventions effective or not—are complex. Nevertheless, the framework still does not include some concepts important to understanding adoption and outcomes (for example, culture, power, experience) and could not accommodate the nuances of some situations. Other concepts from sociology, anthropology, and economics were thus integrated as needed.

MIXING DISCIPLINES AND METHODS. The case studies confirmed that mixing disciplines from the social sciences—economics, sociology, and anthropology—and using mixed methods from within these disciplines—panel surveys, qualitative interviews, focus groups, and ethnographic methods—are essential to conducting reliable impact assessments. Economic methods provided the basis for measurement of adoption and many types of impacts, especially broad-scale effects on poverty reduction, whereas sociology and anthropology contributed to an explanation of these findings, including an understanding of the perceptions of vulnerability and of social relationships based on gender, class, power, culture, and normative frameworks. Rigor is not simply a matter of establishing proper counterfactuals and controls. The integration and triangulation of qualitative research methods provides a different type of rigor that is needed for a study of how technologies interact with livelihoods and affect poverty. What mediates the translation of yield and productivity gains into poverty reduction is very complex but absolutely critical to address, or rather begin to address, as no single research program will achieve closure on this complicated and context-specific question. Although the studies described in this volume went far toward capturing the benefits and synergies of using mixed research methods, time and funding were found to be insufficient for researchers from the different disciplines to fully integrate their analyses, and the studies thus missed some of the potential value of mixed-method research.

We found that donors can be a positive force for institutional change, if they are committed to such innovation. The set of poverty impact assessment studies presented in this book were originally conceived as economic studies, measuring poverty in terms of income, expenditure, and nutrition. One of the study's main donors, the United Kingdom's Department for International Development, recognized the need for a multidisciplinary, mixed-method approach for assessing poverty impact and supported the total redesign of the study with substantial additional funding for social analysis and early-stage technical support for the integration. It also made this redesign a condition of funding. Although conditionality of funding is often resisted in research, it can be a shock that is sometimes required to provoke a change in institutional culture.

Notable contributions of (noneconomic) social scientists in agricultural research include cases in which social research has increased the adoption or adaptation of technologies and improved the outcomes of natural resource practices, especially through co-management of resources. In both types of cases, increased involvement of farmers or communities was often an important factor. Social research has also played an important role in developing new partnerships and forms of collaboration, such as increasing farmer participation in technology development (Cernea and Kassam 2005). These partnerships are essential if the agricultural research results are to be effectively disseminated, particularly if the international and national agricultural research centers are to move beyond "pipeline" research to become active partners in increasingly complex innovation systems (Dunn et al. 1996; Hall et al. 2001).

Implications for Agricultural Research to Reduce Poverty

Agricultural research has made a significant contribution to poverty reduction, through direct and indirect channels. The largest contributions have been through improvements in staple food crops, but the returns to staple crops have been declining. Diversification into other crops and foods is increasingly sought, but often poses greater challenges, both for agricultural research and for adoption by the poor. The studies in this volume identified many factors that researchers should know in advance (or at least in the very early stages of the evolution of research) to address these challenges, related to the vulnerabilities, priorities, opportunities, and preferences of poor people; social dynamics; the structure of production; assets; and other factors.

It is essential to know the extent and sources of poor people's vulnerability, the priority they place on managing risk, and their capacity to do so. Anything that increases their vulnerability—assuming debt to purchase inputs, dependence on government or NGOs that are undependable, trying a new variety without first seeing it perform—in the absence of insurance or recourse mechanisms will seem less attractive, even if it is productivity-enhancing. Although all farmers, including the poor, do take many risks, agricultural research needs to be aware of the trade-offs between increases in productivity and various sources of vulnerability.

Social differentiation is another factor that deserves attention. It significantly affects the adoption of different technologies, and the impacts on those who adopt, others in their households (for example, women), and those who do not adopt. Technologies with low input requirements and low cash inputs are likely to be especially attractive to poor people. Women and men may be differently affected, so paying particular attention to cultural norms with regard to women and the assets available to them—and building upon those assets (for example, homestead production)—can be particularly beneficial for poverty reduction. Although agricultural research alone cannot overcome inequality of gender, class, or ethnicity, at a minimum it should not exacerbate these inequalities by favoring the already advantaged. New agricultural technologies can be empowering; hence it is important to know who is being empowered. Furthermore, it is vital to link agricultural technology with broader poverty reduction strategies involving many other development agents and instruments (for example, infrastructure, poverty reduction strategy plans, community-driven development, financial services available to the poor, education, and health services and programs). If the goal of agricultural research is solely to increase food production, then NARS and ministries of agriculture may be sufficient partners, but addressing poverty requires building partnerships across sectors.

Researchers should learn what traits farmers value beyond yield, including such factors as stability in yield; taste and texture; and resistance to weather, pests, and disease. Different varieties involve trade-offs, and new varieties can reduce these trade-offs, but local priorities must be understood. Learning about these priorities can be key to producing impacts that actually help the intended beneficiaries.

The value and availability of labor is another significant variable. The assumption that developing country farmers, especially poor farmers, have a relative surplus of labor available is not necessarily valid. Our studies found that many poor agricultural producers face severe time constraints, particularly in environments where AIDS has killed or disabled much of the working-age population and increased the demands on time for caring for the afflicted. Labor-saving technologies may reduce employment opportunities in some contexts; in other settings they may allow households to diversify into other income-earning activities or devote more time to child care or caring for ill family members.

The structure of production and income across and within households, and across gender and age groups, should also be understood. Particularly important is an understanding of the role of agriculture and nonfarm income in livelihood strategies, so that technology can be tailored to fit those strategies rather than assuming that families farm full-time and be applied where agriculture still plays a significant role in the lives of poor people. Also important is the value of homestead production for women in contexts of curtailed spatial mobility. Production close to home is more tractable, given other production activities, such as child care, and also reduces women's vulnerability to harassment.

Such insights are, to varying degrees, context-specific. Some, such as the importance of vulnerability and multiple livelihood strategies for the poor, are widespread, whereas others, such as the extent of farmer-to-farmer dissemination, varied more among our cases. However, the payoffs from an improved understanding of how agricultural research affects the livelihoods of the poor can be high. In this study, we have attempted to go beyond quantifying the impact of agricultural research on poverty, to understanding the processes underlying impact, positive or negative. The costs of the methods used for the present studies are not trivial, but in our experience neither are they prohibitive, being on the order of US$200,000 per case study. Although this amount is a small fraction of the investment that goes into the development of the technology itself, the challenge now is to evolve and adopt cost- and time-effective approaches and methods to enable scientists and other decisionmakers to learn about and appreciate poor people's conditions and priorities, anticipate impacts, and tailor their research accordingly. Ex ante assessments to identify critical factors affecting adoption or impact do not need to meet the same standards of rigor as expected for ex post impact assessments. Secondary sources on the target regions, the use of students engaged in thesis research or employed to carry out interviews and observations, and a range of rapid appraisal techniques (see Ashby 1995) in carefully sampled sites (disaggregated by social groups) can all offer cost-effective ways to gather the information to inform this process. Importantly, the cost of *not* obtaining this information—if technologies are not adopted, not adopted by the target group, or adopted but have more negative impacts than benefits—could be much higher than making investments in understanding the target groups up front.

One clear finding across most of the case studies is how much the dissemination process matters for adoption and for generation of direct and indirect impacts. More thought needs to be given to the dissemination strategies at an early stage in the research design. Technology would be likelier to reach intended groups and unintended effects would be diminished by such foresight. For example, different social environments will be more or less amenable to individual, group-based, or informal dissemination activities. Although farmers' groups can be an effective means of diffusion, they often reproduce power relationships, and it is important to try to understand these power dynamics before relying on groups and when considering the roles and compositions of groups. Similarly, adaptive research farmers are important for testing and adapting technologies, but the social dynamics unleashed through this system must be understood. More time should be invested in facilitating community participation in the selection process and ongoing contacts. Social barriers to inclusion can be taken into consideration and sometimes circumvented.

The formation of new partnerships is essential if agricultural research organizations are to be informed by and have an impact on the livelihoods of poor people. Even though international research centers cannot tailor all technologies for the enormous range of context-specific livelihood strategies found throughout the poor regions of the world, they must facilitate such tailoring by creating meaningful partnerships with institutions that better understand local livelihood strategies and can tap into the knowledge from local men and women. Such facilitation may include partnerships with NGOs and producer organizations, as well as national research and extension systems, as illustrated in the Kenya and Bangladesh fish and vegetables cases. The participatory research strategies that have been developing over the past two decades are a promising development in this regard, because farmers are not only recipients of the technology but also participate in setting the priorities and even in the technology development itself.[7] These partnerships help develop research processes that lead to rapidly adopted technologies with the potential for supporting and strengthening livelihood strategies and reducing vulnerability (Johnson, Lilja, and Ashby 2000). The partnerships will also increase that potential by helping poor people tailor the technology to increase their power both within and outside the agricultural realm.

Forming these partnerships is not easy. National governments remain essential partners, but where they maintain traditional top-down, male-dominated approaches, they will not be helpful, though capacity can be strengthened and joint learning can take place through these partnerships. Some government departments and NARS have moved to more farmer-centered, gender-sensitive, and participatory approaches (such as in western Kenya), and these departments should be sought as partners. Efforts to reach out beyond the agricultural sector is also important to address poverty and devise new approaches. For example, health, nutrition, or education departments or programs dealing with microfinance and women's self-help groups may have more effective ways to reach women farmers. National NGOs are an important resource for dissemination, because they are often closer to the ground and have different perspectives on local environments than do governmental agents. But like government, their performance is also variable and their approaches and performance should be evaluated as well. The strategic choice of partners is a crucial decision in targeting agricultural technology to the poor. Other decisions, such as which areas to work in or which crops to target, will be important, but the choice of partners for the dissemination and development of the technology can be even more

7. For literature on this development, see www.prgaprogram.org.

important. Engaging the right partners should be considered an integral part of the research process.

Farmers themselves are important sources of dissemination and innovation. Reforms of agricultural extension programs (for example, in Uganda) that treat farmers as clients rather than as beneficiaries are an important step toward empowering them, but involving farmers in priority setting and even in conducting the research experiments can be even more useful for tapping into farmers' expertise. The innovations systems paradigm, which includes multiple partners and sources of innovation, including international and national public and private sectors, provides a useful approach to such partnerships (Dunn et al. 1996; Berdegué and Escobar 2002).

Institutional Learning and Change

Finally—and perhaps most important to increasing the impact of this work—we have to be willing to use research and impact assessment for institutional learning and change.[8] Change is promoted through critical self-awareness and processes for reflection. Researchers need to welcome opportunities to learn through mistakes, rather than looking only for positive impacts. Instead of following the conventional pipeline approach, research needs to be participatory with many stakeholders, those more and less obvious, and be iterative, interactive, reflective, and adaptive. Such flexibility is consistent with the changing CGIAR mandate to be increasingly focused on poverty. As Robert Chambers said at an IFPRI workshop in 2003,

> We are talking about change which is paradigmatic in the sense of linked concepts, words, values, methods, behaviors, and relationships. . . . [For these poverty impact studies] the overarching goal is reduction of poverty, to be achieved by learning about how research and dissemination can have bigger and better effects, and what personal, professional, and institutional changes are needed for this to occur. . . . the prestige of the CGIAR gives it a comparative advantage (responsibility?) in the development and spread of high-yielding methodologies (HYMs). This includes methodologies for ILAC. (Chambers 2003, 19–21)

As part of impact assessment, research organizations should ask how technology development and dissemination could have been done differently, given what we now know about its impact, and what aspects of the institution constrain a different approach. If poverty reduction is the goal, then impact assessment should less often ask "how much poverty have we reduced?" and more often ask "what have we been missing and how can we do better?" Organiza-

8. For more on institutional learning and change, see Horton and MacKay (2003) and Watts et al. (2004).

tions should then act upon the findings. In addition to studying the factors underlying success, this requires a willingness for research managers to be self-critical, recognize changing clientele and their needs, acknowledge mistakes and failures, and benefit from them by seeking lessons. It also involves a critical examination of institutional rules and norms that may or may not facilitate organizational learning.

One of the lessons gained from the studies presented here was that if research scientists and managers had been involved early on and through all stages of these impact assessments, and if reflection and adaptation processes were better incorporated, these studies would have been positioned to make more of an impact on the institutional practices of the respective centers. The overall poverty impact project coordinated by IFPRI met its objectives related to developing, testing, and refining methods for integrated economic and social analysis, and to better understanding how agricultural research affects livelihoods and poverty. However—in the spirit of acknowledging weaknesses and learning from them—the project only partially met its third objective of "strengthen[ing] the capacity of CG[IAR] centers and NARS to undertake integrated economic and social poverty impact assessments and to internalize a poverty impact assessment culture for the future" (IFPRI 2000, 3). Appreciation of and capacity for such analysis was built within the centers that led the studies, and other centers subsequently expressed an interest in applying a similar approach.[9] But reaching centers outside the study and influencing institutional culture within them were beyond what this project could achieve. We hope that the interdisciplinary, mixed-method impact studies, such as those presented here, can help agricultural researchers increase their ability to reach poor farmers with technologies that make a positive difference in their lives.

References

Adato, M., and R. Meinzen-Dick. 2003. Assessing the impact of agricultural research on poverty and livelihoods. *Quarterly Journal of International Agriculture* 42 (2): 149–166.

Agarwal, B. 2001. Participatory exclusions, community forestry, and gender: An analysis of south Asia and a conceptual framework. *World Development* 29 (10): 1623–1648.

Ashby, J. A. 1995. Institutionalizing a rapid appraisal approach to participatory, client-driven research and technology development in agriculture. *Development and Change* 26 (4): 753–770.

9. However, one point made by Adato and Meinzen-Dick (2003) is that long-term capacity would have been strengthened far more if sociologists and anthropologists were brought on staff in the centers, rather than relying on external collaborators (that is, outsourcing of the social analysis).

Baland, J. M., and J.-P. Platteau. 1996. *Halting degradation of natural resources: Is there a role for rural communities?* Oxford: Food and Agriculture Organization of the United Nations and Clarendon Press.

Bebbington, A. J., D. Merrill-Sands, and F. J. Farrington. 1994. *Farmer and community organizations in agricultural research and extension: Functions, impacts and questions.* Agricultural Administration (Research and Extension) Network Paper 47. London: Overseas Development Institute.

Berdegué, J. A., and G. Escobar. 2002. *Rural diversity, agricultural innovation policies and poverty reduction.* AGREN Working Paper 122. London: Overseas Development Institute.

Bourdillon, M., P. Hebinck, and J. Hoddinott. 2002. *Assessing the impact of HYV maize in resettlement areas of Zimbabwe.* Final report. Washington, D.C.: International Food Policy Research Institute.

Cernea, M., and A. Kassam, eds. 2005. *Researching the culture in agri-culture.* Wallingford, U.K.: CAB International.

Chambers, R. 2003. Reflections on ILAC. In *Institutional learning and change in the CGIAR: Summary record of the workshop held at IFPRI, Washington, D.C., February 4–6, 2003,* D. Horton, and R. McKay, eds. Impact Assessment Discussion Paper 18. Washington, D.C.: International Food Policy Research Institute.

Dunn, T., L. Humphreys, W. Muirhead, M. Plunkett, N. Croker, and M. Nickl. 1996. Changing paradigms for farmer-researcher-extensionist relationships: Exploring methods and theories of farmer participation in research. *European Journal of Agricultural Education and Extension* 3 (3): 167–181.

Fan, S., P. Hazell, and S. Thorat. 1999. *Linkages between government spending, growth and poverty in rural India.* IFPRI Research Report 110. Washington, D.C.: International Food Policy Research Institute.

Guijt, I., and K. M. Shah, eds. 1998. *The myth of community: Gender issues in participatory development.* London: Intermediate Technology Publications.

Hall, A., G. Bockett, S. Taylor, M.V. K. Sivamohan, and N. Clark. 2001. Why research partnerships really matter: Innovation theory, institutional arrangements and implications for developing new technology for the poor. *World Development* 29 (5): 783–797.

Hoddinott, J., and A. Quisumbing. 2003a. Methods for microeconometric risk and vulnerability assessments. Washington, D.C.: International Food Policy Research Institute. Photocopy.

———. 2003b. Data sources for microeconometric risk and vulnerability assessments. Washington, D.C.: International Food Policy Research Institute. Photocopy.

Horton, D., and R. McKay, eds. 2003. *Institutional learning and change in the CGIAR: Summary record of the workshop held at IFPRI, Washington, D.C., February 4–6, 2003.* Impact Assessment Discussion Paper 18. Washington, D.C.: International Food Policy Research Institute.

IFPRI (International Food Policy Research Institute). 2000. Impact of agricultural research on poverty reduction: An integrated economic and social analysis. Phase II. Proposal prepared on behalf of the ISAEG/SPIA by the International Food Policy Research Institute, June 8, 2000.

IFPRI-BIDS-INFS (International Food Policy Research Institute–Bangladesh Institute of Development Studies–Institute of Nutrition and Food Science). 1998. Commercial vegetable and polyculture fish production in Bangladesh: Their impacts on income, household resource allocation, and nutrition. Donor report. Washington, D.C.: International Food Policy Research Institute.

Johnson, N., N. Lilja, and J. A. Ashby. 2000. Characterizing and measuring the effects of incorporating stakeholder participation in natural resource management research: Analysis of research benefits and costs in three case studies. Working Document 17. Cali, Colombia: Participatory Research and Gender Analysis Program.

Meinzen-Dick, R., and M. Zwarteveen. 2001. Gender dimensions of community resource management: The case of water users' associations in South Asia. In *Natural communities? Ethnicity, ender, and the state in community-based conservation,* C. C. Gibson and A. Agrawal, eds. New Brunswick, N.J.: Rutgers University Press.

Ostrom, E. 1990. *Governing the commons: The evolution of institutions for collective action.* Cambridge: Cambridge University Press.

Place, F., G. Kariuki, J. Wangila, P. Kristjanson, A. Makauki, and J. Ndubi. 2002. *Assessing the factors underlying differences in group performance: Methodological issues and empirical findings from the highlands of central Kenya.* CAPRi Working Paper 25. Washington, D.C.: International Food Policy Research Institute.

Place, F., M. Adato, P. Hebinck, and M. Omosa. 2005. *The impact of agroforestry-based soil fertility replenishment practices on the poor in western Kenya.* IFPRI Research Report 141. Washington, D.C.: International Food Policy Research Institute.

Röling, N. 1988. *Extension science.* Cambridge: Cambridge University Press.

Skoufias, E. 2003. Economic crises and natural disasters: Coping strategies and policy implications. *World Development* 31 (7): 1087–1102.

Skoufias, E., and B. McClafferty. 2003. Is PROGRESA working? Summary of the results of an evaluation by IFPRI. In *Household decisions, gender, and development: A synthesis of recent research,* A. R. Quisumbing, ed. Washington, D.C.: International Food Policy Research Institute.

Watts, J., R. Mackay, D. Horton, A. Hall, B. Douthwaite, R. Chambers, and A. Acosta. 2004. Institutional learning and change: An introduction. Paper prepared for the International Food Policy Research Institute. Photocopy.

Contributors

Michelle Adato (m.adato@cgiar.org) is a senior research fellow in the Food Consumption and Nutrition Division of the International Food Policy Research Institute, Washington, D.C.

Javier Becerril (jbecerr@food-econ.uni-kiel.de) is a doctoral candidate with the Institute of Food Economics and Consumption Studies at the University of Kiel, Germany. At the time he contributed to this work, he was a research assistant at the International Maize and Wheat Improvement Center, based in Texcoco, Mexico.

Suraiya Begum (suraiya@bttb.net.bd) is a research associate with Research Initiatives Bangladesh, Dhaka.

Mauricio R. Bellon (m.bellon@cgiar.org) is director of the Diversity for Livelihoods Programme at Bioversity International, Rome. At the time he contributed to this work, he was a senior scientist at the International Maize and Wheat Improvement Center, based in Texcoco, Mexico.

Manik L. Bose (m.bose@cgiar.org) is a research fellow at WorldFish Center, Penang, Malaysia. At the time he contributed to this work, he was a postdoctoral fellow at the International Rice Research Institute, Los Banos, the Philippines.

Michael F. C. Bourdillon (mbourdillon@mango.zw) is professor emeritus in the Department of Sociology at the University of Zimbabwe, Harare.

Connie Chan-Kang (chan0576@umn.edu) is a research associate with the International Science and Technology Practice and Policy Center of the Department of Applied Economics at the University of Minnesota, St. Paul. At the time she contributed to this work, she was a research analyst in the Environment and Production Technology Division of the International Food Policy Research Institute.

Alamgir Chowdhury (abirch@bangla.net) is a director and participatory research specialist at Socioconsult Limited, Dhaka, Bangladesh.

Shenggen Fan (s.fan@cgiar.org) is the director of the Development Strategy and Governance Division at the International Food Policy Research Institute. At the time he contributed to this work, he was a senior research fellow with the Institute's Environment and Production Technology Division.

Lawrence Haddad (l.haddad@ids.ac.uk) is the director of the Institute of Development Studies at the University of Sussex, Brighton, U.K. At the time he contributed to this work, he was the director of the Food Consumption and Nutrition Division of the International Food Policy Research Institute.

Kelly Hallman (khallman@popcouncil.org) is a research associate in the Policy Research Division of the Population Council, New York. At the time she contributed to this work, she was a research fellow in the Food Consumption and Nutrition Division of the International Food Policy Research Institute.

Peter Hazell (p.hazell@imperial.ac.uk) is a visiting professor at the Imperial College, London. At the time he contributed to this work, he was the director of the Environment and Production Technology Division of the International Food Policy Research Institute.

Paul Hebinck (paul.hebinck@wur.nl) is a senior lecturer at Wageningen University, The Netherlands.

John Hoddinott (j.hoddinott@cgiar.org) is a senior research fellow in the Food Consumption and Nutrition Division of the International Food Policy Research Institute.

Mahabub Hossain (m.hossain@cgiar.org) is an economist and head of the Social Sciences Division at the International Rice Research Institute, Los Banos, the Philippines.

Bill Kinsey (bkinsey@feweb.vu.nl) is a senior research fellow at Vrije Universitiet, Amsterdam, The Netherlands, and the Institute of Development Studies at the University of Zimbabwe, Harare.

K. Krishnaiah is a former director of the Directorate of Rice Research, Hyderabad, India.

David Lewis (d.lewis@lse.ac.uk) is a reader in social policy at the London School of Economics.

John Marondo (marondoj@yahoo.co.uk) is a research specialist with the National Poverty Assessment Study Survey, jointly funded by the United Nations Development Programme and the Government of Zimbabwe. At the time he contributed to this work, he was an independent consultant.

Ruth Meinzen-Dick (r.meinzen-dick@cgiar.org) is a senior research fellow in the Environment and Production Technology Division of the International Food Policy Research Institute.

Dubravka Mindek (dmindek@axtel.net) is an assistant director with the Division of Domestic Violence Prevention at the Women and Health Program of the Ministry of Health, Mexico. At the time she contributed to this work, she was an independent consultant.

Netsayi Mudege (netsayi@mweb.co.zw) is a lecturer at the University of Zimbabwe, Harare. At the time she contributed to this work, she was a masters student at the University of Zimbabwe.

Mary Omosa (momosa@uonbi.ac.ke) is a senior research fellow with the Institute for Development Studies at the University of Nairobi, Kenya.

Trudy Owens (Trudy.Owens@Nottingham.ac.uk) is a lecturer at the University of Nottingham, U.K. At the time she contributed to this work, she was a research officer with the Centre for the Study of African Economies at the University of Oxford, U.K.

Frank Place (f.place@cgiar.org) is an economist at the World Agroforestry Centre, Nairobi, Kenya.

Keming Qian (keming@agri.gov.cn) is the director of the Agricultural Trade Promotion Center of the Ministry of Agriculture, Beijing. At the time he contributed to this work, he was director general of the Institute of Agricultural Economics at the Chinese Academy of Agricultural Sciences, Beijing.

Index

Page numbers for entries occurring in boxes are followed by a *b;* those for entries in figures, by an *f;* those for entries in notes, by an *n;* and those for entries in tables, by a *t.*